Proteins and Peptides

DRUGS AND THE PHARMACEUTICAL SCIENCES
A Series of Textbooks and Monographs

Executive Editor

James Swarbrick
PharmaceuTech, Inc.
Pinehurst, North Carolina

Advisory Board

For information on volumes 1–152 in the *Drugs and Pharmaceutical Science*
Series, please visit www.informahealthcare.com

Proteins and Peptides
Pharmacokinetic, Pharmacodynamic, and Metabolic Outcomes

edited by
Randall J. Mrsny
University of Bath
Bath, UK

Ann Daugherty
Genentech
South San Francisco, California, USA

CRC Press
Taylor & Francis Group
Boca Raton London New York

CRC Press is an imprint of the
Taylor & Francis Group, an **informa** business

CRC Press
Taylor & Francis Group
6000 Broken Sound Parkway NW, Suite 300
Boca Raton, FL 33487-2742

First issued in paperback 2019

ISBN-13: 978-1-4200-7806-0 (hbk)
ISBN-13: 978-0-367-38505-7 (pbk)

This book contains information obtained from authentic and highly regarded sources. While all reasonable efforts have been made to publish reliable data and information, neither the author[s] nor the publisher can accept any legal responsibility or liability for any errors or omissions that may be made. The publishers wish to make clear that any views or opinions expressed in this book by individual editors, authors or contributors are personal to them and do not necessarily reflect the views/opinions of the publishers. The information or guidance contained in this book is intended for use by medical, scientific or healthcare professionals and is provided strictly as a supplement to the medical or other professional's own judgement, their knowledge of the patient's medical history, relevant manufacturer's instructions and the appropriate best practice guidelines. Because of the rapid advances in medical science, any information or advice on dosages, procedures or diagnoses should be independently verified. The reader is strongly urged to consult the relevant national drug formulary and the drug companies' and device or material manufacturers' printed instructions, and their websites, before administering or utilizing any of the drugs, devices or materials mentioned in this book. This book does not indicate whether a particular treatment is appropriate or suitable for a particular individual. Ultimately it is the sole responsibility of the medical professional to make his or her own professional judgements, so as to advise and treat patients appropriately. The authors and publishers have also attempted to trace the copyright holders of all material reproduced in this publication and apologize to copyright holders if permission to publish in this form has not been obtained. If any copyright material has not been acknowledged please write and let us know so we may rectify in any future reprint.

Library of Congress Cataloging-in-Publication Data

Proteins and peptides : pharmacokinetic, pharmacodynamic, and metabolic outcomes / edited by Randall J. Mrsny, Ann Daugherty.
    p. ; cm. — (Drugs and the pharmaceutical sciences ; 202)
  Includes bibliographical references and index.
  ISBN-13: 978-1-4200-7806-0 (hardcover : alk. paper)
  ISBN-10: 1-4200-7806-2 (hardcover : alk. paper) 1. Proteins—Therapeutic use. 2. Proteins—Metabolism. 3. Peptide drugs. I. Mrsny, Randall J., 1955- II. Daugherty, Ann L., 1954- III. Series: Drugs and the pharmaceutical sciences, 202.
  [DNLM: 1. Peptides—pharmacokinetics. 2. Proteins—pharmacokinetics. 3. Drug Delivery Systems—methods. 4. Drug Discovery—methods. 5. Peptides—therapeutic use. 6. Proteins—therapeutic use. W1 DR893B v.202 2009 / QU 55 P9645 2009]
  RM666.P87P76 2009
  615'.7—dc22

                                      2009028689

Visit the Taylor & Francis Web site at
http://www.taylorandfrancis.com

and the CRC Press Web site at
http://www.crcpress.com

Preface

The promise of biotechnology and an increased understanding of the human genome have resulted in an explosion of protein and peptide therapeutics entering research and development pipelines in biopharmaceutical as well as traditional pharmaceutical companies. Increasingly, these protein and peptide candidates are being designed to address previously untreatable diseases and conditions. To advance these molecules into clinical trials, however, an understanding of their pharmacokinetic, pharmacodynamic, and metabolic fate is required. The study of these events represents emerging disciplines with issues and challenges distinct from those of small molecules, on which many of the principles of these fields were initially developed.

That many of the protein and peptide therapeutics being evaluated are endogenous or emulate an endogenous material potentially defines preestablished pharmacokinetic, pharmacodynamic, and metabolic parameters for these molecules in the human model. Unfortunately, or fortunately, initial preclinical testing of potential protein and peptide therapeutics requires obtaining information on safety and preclinical efficacy in (typically) several nonhuman animal models. Such studies are complicated, or rather compromised, by the fact that nonhuman models may not express critical elements such as receptors, binding proteins, and enzyme activities that function to define pharmacokinetic, pharmacodynamic, and metabolic parameters in humans. Additionally, human disease states being emulated, but never fully recapitulated, in nonhuman animal models may or may not faithfully describe conditions that will be confronted in the clinic. All of these issues are further complicated by the fact that these peptide and protein therapeutic candidates must be formulated for long-term storage stability and delivered at concentrations and in locations that are likely very different from endogenous events.

Most protein and peptide therapeutics are administered by injection, usually being formulated with a strategy to minimize the frequency of these injections. In this regard, recent studies have identified several alternative routes of administration for proteins and peptides that were previously not considered a viable option for delivery. Although the size and labile nature of protein and peptide therapeutic candidates typically impede their passage across most biological barriers, intranasal and pulmonary delivery for therapeutic proteins and peptides is now a commercial reality. Additionally, tremendous progress has been made for the delivery of proteins and peptides via transdermal and oral routes as well as delivery to the eye and brain. All of these routes pose unique pharmacokinetic, pharmacodynamic, and metabolic challenges for the protein or peptide being delivered, each of which might be altered in the human model of disease relative to the nonhuman models initially examined.

The chapter topics that follow were selected to provide an overall road-map for understanding our current understanding of parameters that define the pharmacokinetic, pharmacodynamic, and metabolic challenges for the delivery of protein or peptide therapeutics. In general, these chapters recite lessons learned for the major areas of proteins and peptide therapeutics that have been successfully taken to market, for example, antibodies, interleukins, interferons, growth factors, and peptide hormones. Additionally, chapters have been included that explore innovations for protein and peptide delivery that include needle-less delivery and strategies to deliver these molecules to locations such as the eye and brain. Although all of the chapters were written with a forward-looking perspective, with the goal of identifying issues that are likely to become increasingly important in the future. It is hoped that the information shared in these chapters will provide the reader with an increased understanding of issues critical for successfully guiding a protein or peptide therapeutic candidate through the maze of issues that define the pharmacokinetics, pharmacodynamics, and metabolism of these molecules.

We would like to take this opportunity to again thank our authors, who represent key contributors in these areas, for the expertise that they have shared. It is our sincere wish that the knowledge put forth in the following chapters will have a positive impact on the development of new drugs that will improve health and alleviate suffering.

Randall J. Mrsny
Ann Daugherty

Contents

Contributors

James M. Anderson Institute of Pathology, Case Western Reserve University, Cleveland, Ohio, U.S.A.

C. Andrew Boswell Research and Early Development, Genentech, Inc., South San Francisco, California, U.S.A.

Gordon C. Brandt MDRNA, Inc., Bothell, Washington, U.S.A.

William G. Brodbeck Institute of Pathology, Case Western Reserve University, Cleveland, Ohio, U.S.A.

Henry R. Costantino MDRNA, Inc., Bothell, Washington, U.S.A.

David Cregut Sequence/Structure Bioinformatics, Drug Discovery Informatics, Merck Serono International, Geneva, Switzerland

Rong Deng Research and Early Development, Genentech, Inc., South San Francisco, California, U.S.A.

Mark S. Dennis Department of Antibody Engineering, Genentech, Inc., South San Francisco, California, U.S.A.

Steven Dinh Noven Pharmaceuticals, Inc., Miami, Florida, U.S.A.

Paul J. Fielder Research and Early Development, Genentech, Inc., South San Francisco, California, U.S.A.

Diane Frank MDRNA, Inc., Bothell, Washington, U.S.A.

Amita Joshi Research and Early Development, Genentech, Inc., South San Francisco, California, U.S.A.

Robert F. Kelley Departments of Antibody Engineering and Protein Engineering, Genentech, Inc., South San Francisco, California, U.S.A.

Leslie A. Khawli Research and Early Development, Genentech, Inc., South San Francisco, California, U.S.A.

Cecile M. Krejsa Applied Pharmacology LLC, Seattle, Washington, U.S.A.

Cherry Lei   Research and Early Development, Genentech, Inc., South San Francisco, California, U.S.A.

Chingyuan Li   MDRNA, Inc., Bothell, Washington, U.S.A.

Kedan Lin   Research and Early Development, Genentech, Inc., South San Francisco, California, U.S.A.

Puchun Liu   Noven Pharmaceuticals, Inc., Miami, Florida, U.S.A.

Henry B. Lowman   Department of Antibody Engineering, Genentech, Inc., South San Francisco, California, U.S.A.

Iftekhar Mahmood   Office of Blood Review and Research, Center for Biological Evaluation and Research, Food and Drug Administration, Bethesda, Maryland, U.S.A.

Maria Mitsi   Departments of Biochemistry and Ophthalmology, Boston University School of Medicine, Boston, Massachusetts, U.S.A.

Graham Molineux   Hematology-Oncology Research, Amgen, Inc., Thousand Oaks, California, U.S.A.

Emmanuel Monnet   Neurology and Autoimmune and Inflammatory Diseases, Stratified Medicine–Exploratory Medicine, Medical Sciences and Innovation, Merck Serono International, Geneva, Switzerland

Randall J. Mrsny   Department of Pharmacy and Pharmacology, University of Bath, Bath, U.K.

Matthew A. Nugent   Departments of Biochemistry and Ophthalmology, Boston University School of Medicine, and Department of Biomedical Engineering, Boston University, Boston, Massachusetts, U.S.A.

Wendy S. Putnam   Research and Early Development, Genentech, Inc., South San Francisco, California, U.S.A.

Steven C. Quay   MDRNA, Inc., Bothell, Washington, U.S.A.

Anthony P. Sileno   MDRNA, Inc., Bothell, Washington, U.S.A.

Jean L. Spencer   Departments of Biochemistry and Ophthalmology, Boston University School of Medicine, Boston, Massachusetts, U.S.A.

Stephen J. Szilvassy   Hematology-Oncology Research, Amgen, Inc., Thousand Oaks, California, U.S.A.

Mike Templin   MDRNA, Inc., Bothell, Washington, U.S.A.

Frank-Peter Theil   Research and Early Development, Genentech, Inc., South San Francisco, California, U.S.A.

Arto Urtti   Centre for Drug Research, University of Helsinki, Helsinki, Finland

Yik A. Yeung   Department of Antibody Engineering, Genentech, Inc., South San Francisco, California, U.S.A.

1

In Vitro/In Vivo Correlations of Pharmacokinetics, Pharmacodynamics, and Metabolism for Hematologic Growth Factors and Cytokines

Graham Molineux and Stephen J. Szilvassy
Hematology-Oncology Research, Amgen, Inc., Thousand Oaks, California, U.S.A.

HEMATOPOIETIC LINEAGES AND THE CONTROL OF CELL PRODUCTION

Blood comprises approximately 55% liquid and 45% cellular material and fulfills many recognized functions in mammalian physiology. The most important of these is oxygenation of bodily tissues followed by an important secondary function in combating disease, particularly infectious disease, via both cell-based and humoral mechanisms.

Blood has been the subject of scientific inquiry from prehistory and because of its ready accessibility and liquid nature has lent itself to early dissection of both organization and function. In the early part of the 20th century, Carnot pioneered the idea that blood composition was controlled by a humoral factor (1), which was ultimately identified as erythropoietin (EPO). This work was founded on the observations made by Viault (2), who followed changes in red blood cell count as he and his traveling companions (human and animal) ascended to altitude. From early in the last century it was thus suspected that blood composition may be subject to change in response to environmental variation and that humoral factors may be the mediators of this effect.

The cellular constituents of blood had, of course, been observed by Anthony van Leeuwenhoek in the 17th century in one of the first applications of his newly invented microscope. Hence, the idea that there are various types of blood cells and that their production is under humoral control is not really new, nor is it confined to the era of recombinant proteins, which began in the 1970s. However, that epoch did provoke unprecedented advances in understanding cytokines in general and hematopoietic cytokines in particular, culminating in the cloning of the first hematopoietic cytokine [interleukin-3 (IL-3)] in 1984 (3).

Understanding the basis of cellular diversity in blood had meanwhile undergone equally important advances with the description of the first quantitative assays for murine hematopoietic "stem" cells in 1961 (4). Although spleen colony-forming units (CFU-S) first described by Till and McCulloch were ultimately demonstrated not to exhibit all of the hallmark properties that characterize the most primitive hematopoietic stem cells (i.e., most CFU-S lacked lymphoid differentiation potential and exhibited only a limited capacity for self-renewal), this assay and the cell type it detected is viewed by many to have ushered in the modern era of stem cell biology. The first in vitro colony formation assays for hematopoietic progenitor cells were described in 1965 and 1966 (5,6). In these assays, bone marrow cells that were otherwise unrecognizable were cultured in semisolid medium in the presence of crude preparations of body fluids, tissue extracts, or medium "conditioned" by various cells. Since these extracts (and later their components) stimulated the formation of blood cell colonies, they acquired the descriptive name of "colony-stimulating factors" (CSFs) and their cell targets, the equally unsurprising epithet "colony-forming cells."

Although spectacular progress had been made in the three previous decades, work in the early 1990s provided a remarkable leap in our insight into the organization and control of hematopoiesis; an understanding that to date has still to be equaled for any other tissue in the body. The hematopoietic cell hierarchy, as it was defined at that time and as it is still understood today, is represented by, at its root, a self-sustaining stem cell pool. Maintenance and selected expansion of this pool occurs through processes of asymmetric cell division, and some would say deterministic, others would say stochastic, cell fate decisions that yield a heterogeneous pool of differentially committed progenitor cells. At one extreme, these precursor cells may have the potential to develop into any of the six blood cell lineages, and at the other extreme, they may be capable of responding in one of only two ways—either by dying (a process referred to as apoptosis) or by developing into a single type of mature blood cell. Stem cell self-renewal is largely regulated by intracellular transcription factors that control the expression of an array of "stemness" genes. Oppositely, later processes of hematopoietic development are under the control of extracellular humoral regulators—variously called the CSFs, growth factors, interleukins, or cytokines. These cytokines act either alone or in concert to control the number and type of blood cells that are produced. Some of them act on relatively primitive cells with multilineage differentiation potential [e.g., IL-3 or stem cell factor (SCF)], while others act only on more committed cells in the later stages of blood cell production (e.g., EPO).

Many of these cytokines have been purified and cloned and are available in pharmaceutically useful quantities in recombinant form. Since they are large molecules that cannot be absorbed intact through the gut or skin, recombinant cytokines must be administered via intravenous or subcutaneous injection. While some of these cytokines have been deployed as therapeutics used in millions of patients, others have found little application in medicine and have thus far remained useful only as laboratory reagents or research tools. Of those that have found clinical utility, several have been reengineered to enhance their drug-like attributes, while others remain essentially identical to the native proteins purified from tissue sources.

RECOMBINANT HEMATOPOIETIC CYTOKINES OF THERAPEUTIC IMPORTANCE

The discovery of hematopoietic cytokines, predominantly in the 1970s and 1980s, followed the development of assays to detect their activity like the in vitro colony-forming cell assays introduced above. However, the larger challenge at that time was purifying proteins with separate activities from the complex biological fluids used as the starting material. Macrophage (M)-CSF (also known as CSF-1) was the first hematopoietic growth factor to be purified, initially from human urine and later from medium conditioned by a murine fibroblast cell line (7). This was followed in the same year by the discovery of granulocyte-macrophage (GM)-CSF in medium conditioned by tissues from the lungs of mice previously treated with bacterial lipopolysaccharide (8). A few years later, a third myeloid growth factor was identified: granulocyte (G)-CSF (9). It was after some years that the genes that encoded these proteins were cloned—cloning was a relatively nascent technology at that time; thus, 1985 saw the cloning of human M-CSF (10), EPO (11,12), and GM-CSF (13,14), and 1986 saw the cloning of G-CSF (15,16), IL-3 (17), and IL-5 (18).

The natural versions of most hematopoietic cytokines are glycosylated, for example, IL-3 (17), IL-5 (19), IL-6 (20), IL-7 (21), GM-CSF (13), G-CSF (22), M-CSF (23), SCF (24), and EPO (25). In several cases, however, the carbohydrate has been shown not to be required to maintain activity, for example, the O-linked carbohydrate at threonine 133 on natural G-CSF. In one celebrated case however, that of EPO, the carbohydrate component was found to be not only obligatory for in vivo action but also amenable to manipulation to therapeutic advantage (26). Endogenous cytokines are frequently heterogeneous at some level, often because of posttranslational modifications such as glycosylation, sulfation, proteolytic cleavage, etc. Recombinant forms may not therefore be identical to the natural prototype and will vary markedly depending on the host cell in which they are produced, method of purification, and a number of other factors. Overall, the precise biochemical nature and activity of endogenous cytokines remain largely unknown as does their comparability with recombinant preparations. Comparisons can be made to define relative potency, but other aspects of product performance, for example, pharmacokinetics, safety, etc., must be studied carefully in animals or humans and often in large numbers of subjects and over extended periods before their safety and efficacy can be definitively established.

With respect to the clinical development and subsequent consideration of therapeutic proteins by regulatory agencies, it has been suggested that the protein product is in essence the process used to manufacture it. This perspective presents a considerable hurdle in comparing related products like, for example, follow-on biologics (FOBs), subsequent entry biologics, or biosimilars intended to offer alternative products after innovator patent expiry. Thus, the term "generic" is difficult to apply given the likely nonidentity of proteins produced in different host cell systems that are purified and formulated using different methods—presenting an interesting challenge for regulatory authorities for which differing solutions are being developed in different countries.

From a drug development perspective, the general observation that has emerged from the medical exploitation of hematopoietic cytokines is that pleiotropy is an undesirable property for such agents. More lineage-restricted cytokines have, in general, proven more useful (27), as exemplified by the clinical utility of EPO (28), G-CSF (29), and GM-CSF (30) and the promise of a thrombopoietin (TPO) mimetic. In the following sections, the discovery and development of these hematopoietic growth factors with demonstrated clinical utility, and their pharmacokinetic (PK) and pharmacodynamic (PD) properties, will be discussed.

STEM CELL FACTOR (STEMGEN®)

Also known as mast cell growth factor (MGF), kit ligand (KL), and steel factor, SCF is the ligand for the cognate tyrosine kinase receptor c-kit. It is approved for clinical use in limited countries as a coadministration with G-CSF for hematopoietic stem and progenitor cell mobilization based on phase 3 clinical trial data in breast cancer patients (31). Despite its use in stem cell mobilization, all patients require prophylactic administration of H1 and H2 antihistamines and a bronchodilator to ameliorate the collateral effects of SCF in stimulating mast cells.

The PK parameters of SCF in humans have not been extensively studied but appear relatively unremarkable. A phase 1 trial in cancer patients indicated a predose serum SCF level of around 1 μg/mL, with a C_{max} 12 to 17 hours after first administration, reducing with subsequent injections (32). Clearance was linear, with a half-life of approximately 35 hours. More intriguing were the data obtained for recombinant SCF administered to mice. Following intravenous administration, radiolabeled material distributed very quickly to the lungs of treated mice and was then eliminated via the kidney and liver with a half-life of around two hours. Sl/Sl^d mice, which lack mast cells because of a genetic lesion in the SCF gene, also accumulated SCF in the lungs but did not suffer the effects of mast cell degranulation seen in their wild-type littermates (33).

The link between the PK and PD of SCF is not particularly clear. The major PD endpoint measured in phase 3 trials was the mobilization of CD34$^+$ cells. However, mobilization is an indirect result of neutrophil-derived proteases cleaving adhesion molecules that tether stem and progenitor cells to the bone marrow stroma (34). Thus, mobilization is mechanistically related to the granulocyte *response* rather than a direct effect of SCF. Since SCF has been shown to interact with intracellular G-CSF signaling (35), the phenomenon observed and exploited in patients is understandable. This outcome may not be directly linked to SCF, and so it may be causally distinct from the PK. In contrast, the side effects (or at least unintended effects) on mast cells are better understood and more satisfactorily linked to drug exposure.

GRANULOCYTE-MACROPHAGE COLONY-STIMULATING FACTOR (LEUKINE®)

GM-CSF is one of two myeloid cytokines approved for clinical use in cancer patients in the European Union and the United States, the other being G-CSF. GM-CSF does not have the breadth of application that G-CSF has, with its approved clinical uses being confined to acute leukemia and in transplant settings. As the name implies, GM-CSF is more pleiotropic than G-CSF. Among the documented effects of GM-CSF are stimulation of progenitor cell proliferation (36), neutrophil function (37,38), monocyte activation (39), and dendritic cell function (40), especially, as a vaccine adjuvant.

In a recent study (41), GM-CSF was administered daily for 10 days to cancer patients; PK analysis showed a dose-dependent increase in drug level several hours after the first administration when none had been detectable beforehand. By the time of the next daily dose, about half the patients still had low but detectable GM-CSF in their blood. In common with many cytokines, SC administration prolonged the half-life of GM-CSF, possibly via delayed absorption, with nonlinear clearance for escalating doses (42). With repeated administration, clearance of GM-CSF gradually increases (41,43,44). Though the mechanism for this effect is not well defined, it may include target cell–mediated clearance as will be discussed later for M-CSF, G-CSF, and TPO. Intravenous administration of GM-CSF illustrates two distinct phases of disposition: the first, presumably representing initial distribution, is quick ($T_{1/2}$ less than five minutes); the second phase is slower, with $T_{1/2}$ of two to three hours (45) representing clearance.

Hematological (PD) responses to administration of GM-CSF include increases in circulating lymphocytes, monocytes, neutrophils, and eosinophils, with small or no changes in erythrocytes and platelets (41). Though these effects, especially on neutrophils, may be used to define the PK/PD relationship in, for example, neutropenia after bone marrow transplantation, the desired PD in other settings may not be so clear. For instance, in the deployment of GM-CSF for immunotherapy applications, the increased leukocyte count, which relates to both dose and duration of GM-CSF treatment, correlated positively with the absolute number of putative immune effector (GM-CSFRα^+/CD14$^+$, GM-CSFRα^+/CD66b$^+$) cells. In contrast, high doses of GM-CSF impaired antibody-dependent cellular cytotoxicity (ADCC) in in vitro assays of harvested cells. This suggests that dose and schedule need to be optimized for this application, but the predictable PK of GM-CSF should make this relatively straightforward as long as the nature of the desired biological effect is well defined. In practice, the cell types required to elicit optimal immune function are not fully understood and will require further study to define the desired PD of GM-CSF in what would appear to be its most useful application.

MACROPHAGE COLONY-STIMULATING FACTOR

M-CSF is approved for clinical use in some countries under the name Leuko-prol® (mirimostim). It was originally cloned in 1985 but was one of the first cytokines studied in the 1960s and had been purified from urine by 1975(46). As the name would suggest, M-CSF was first shown to stimulate the growth of bone marrow–derived monocyte/macrophage cells in vitro (7) but was subsequently found to play a role in inflammation (47), bone remodeling (48), reproduction (49), the central nervous system (50–52), and cancer (53–56).

PK studies using M-CSF created a new paradigm for understanding the relationship between the PK and PD of hematopoietic cytokines. This new understanding centers on the ability of these cytokines to stimulate the production of their appropriate target cells, in this case monocytes/macrophages, only then to have those very cells consume and ultimately clear the stimulator from the serum as their numbers increase. This model has been extended to TPO (57,58), EPO (59,60), G-CSF (61), and perhaps even GM-CSF (41,43,44), but rests on insight gained from the study of M-CSF (62).

Mice normally have detectable levels of M-CSF in their serum, and studies performed using radiolabeled M-CSF demonstrated the serum half-life of this cytokine to be about 10 minutes. Approximately 96% of the cleared M-CSF could be accounted for by splenic or hepatic macrophages, the remainder was eliminated in the urine. Upon analysis of a number of parameters, including the effect of lysosomal protease inhibitors, it was apparent that internalization and degradation in macrophages via the cell surface M-CSF receptor, c-fms, was the predominant mechanism of M-CSF clearance.

The implications of this mechanism are clear. First, the clearance of physiological amounts of cytokines can be quite rapid, being mediated by the normal population of receptor positive cells. Second, pharmacological levels of exogenous cytokine can quickly saturate this clearance mechanism, leading to prolonged exposure and increasing the relative contribution of nonspecific clearance mechanisms, for example, renal filtration. Third, as the PD response to the cytokine accumulates over time, the capacity of the selective clearance mechanism will increase, reducing the relative role of nonspecific pathways. Fourth, in the absence of a target cell response, the clearance of a cytokine might be rather slow, increasing as the response mounts. This model is very attractive to explain homeostatic regulation of cytokine levels and target cell populations, and has ramifications for therapeutic administration of recombinant cytokines that share much of their biology with their endogenous prototypes. Indeed, this exact mechanism was used to develop therapeutically enhanced versions of G-CSF, as is outlined later.

GRANULOCYTE COLONY-STIMULATING FACTOR (NEUPOGEN®, FILGRASTIM)

G-CSF was one of the earliest cytokines to be biologically and biochemically characterized by the Australian CSF pioneers at the Walter and Eliza Hall Institute of Medical Research in Melbourne under the guidance of such giants in the field as Don Metcalf and Richard Stanley. It is due only to the insight of these pioneers that human G-CSF could be purified (22) and cloned (15,16) elsewhere and subsequently developed into a major therapeutic drug that has been administered to several million cancer patients since its launch in 1991.

Some of the early studies were confounded by incomplete separation of GM-CSF and G-CSF, and the seminal paper describing the activity of purified human G-CSF referred to it as a pluripotent factor (22), possibly in error because of assaying it on impure cell preparations. Nevertheless, from the early days, experiments where G-CSF was used as a single activity showed that although it was a modest CSF, it was highly selective in its actions on neutrophilic progenitor cells (9,63). As it turned out, the modesty of its in vitro actions was misleading, but its selectivity was probably not (for review see Ref. 29). The dominant clinical effect of G-CSF action is neutrophilia, though minor or sporadic effects on other blood cells have been reported. Most notably, G-CSF is well documented to increase monocyte proliferation (64,65), which may be linked also to reports of increased osteoclast-mediated bone turnover (66,67). These data illustrate that increased bone turnover, at least in rodents, results from expanded osteoclast activity after treatment (68). Whether this is related to the profound effects of G-CSF on monocyte production kinetics awaits definition of the relationship between these monocytes and osteoclast development.

Humans injected with G-CSF can expect a neutrophil response within one to two days (69–71). However, this is not the case after cancer chemotherapy where G-CSF is normally used to treat neutropenia, because the marrow is often not capable of responding on that timescale (69). This PD response is driven by a rapid absorption of typically SC administered G-CSF, wherein peak concentrations are noted within two to eight hours. The elimination half-life after either SC or IV administration is two to four hours depending on dose and neutrophil count (61,72). As G-CSF is administered daily, the neutrophil count increases, and in parallel, the clearance time of G-CSF is shortened; a relationship that was correlated even in early studies with receptor number on neutrophils (73). As noted above, this appears to be a very similar mechanism to that suggested for the M-CSF PK/PD relationship, that is, the cellular response to a cytokine in turn selectively clears that very cytokine, while in parallel a less saturable pathway (renal clearance) accounts for the balance of the elimination.

In an extension of this very satisfying model, a novel form of G-CSF was engineered specifically to evade the nonselective clearance pathway, yielding a new drug tailored to effect a neutrophil response that could only be cleared by those very neutrophils once they accumulate to a sufficient level (74–78). This form (pegfilgrastim) was designed for use in patients undergoing cancer chemotherapy and in whom support for neutrophil production was required. The underlying hypothesis in designing a form of G-CSF that would not be cleared by the kidney yet would remain sensitive to neutrophil-mediated clearance was that a degree of self-regulation would be an intrinsic feature of the molecule. This was proven to be correct first in animal and then in clinical studies. During neutropenia, the drug has an extended half-life; upon neutrophil recovery clearance is reactivated (75). Thus, for the first time, a drug that offered "automated" control of neutrophil counts was developed. This exciting mechanism of action has led to the broad uptake of pegfilgrastim in medical practice, but has yet to be applied to other therapeutics.

ERYTHROPOIETIN (EPOGEN®, EPOETIN ALFA)

EPO is widely used in the treatment of anemia since it is the central regulator of erythropoiesis. The major quantitative site of EPO production is the kidney, so patients with declining renal function were the first and are still the most obvious candidates for EPO therapy (79). Use in anemia associated with cancer treatment is also common. Although controversial, a number of other experimental uses have emerged since EPO was approved for use in 1989 (80), including stroke, nerve crush injury, heart failure, myocardial infarction, immunomodulation, and for improving cognitive function. It remains unclear how these latter effects work in the absence of EPO receptor on many of the target tissues (see Ref. 81 for a critique of methods used to claim otherwise).

Confining our discussion to the effects of EPO on erythropoiesis, it must be borne in mind how highly dynamic is the process of red blood cell production. A normal 70-kg human produces on the order of 2.5×10^{11} erythrocytes per day, and this rate of production is maintained by a basal EPO level of around 10 to 20 mU/mL (82,83). Pharmacological administration of EPO at a dose intended to sustain a three times per week dosing cycle (150 U/kg) or a weekly treatment cycle (40,000 U/kg) leads to a C_{max} of 150 or 850 mU/mL, respectively (84). Reticulocytes are released earlier than normal from the bone marrow and reside for a disproportionately longer fraction of their life span in the blood following EPO therapy. Despite this being the first PD readout of EPO administration, the more important result is a change in hemoglobin concentration. In the same study (84), the reticulocyte shift could be clearly seen in the blood by five days and a readily discernable change in hemoglobin by day 8—the two dosing regimens being approximately the same despite the 30% dose increment with the weekly regimen. This inefficiency is suggested to be driven by the nonlinearity of EPO PK, which seems to lean toward reduced clearance at higher doses. In this case, it is likely that the similar PD response was driven by the accumulated time above the concentration threshold required for pharmacological action, which was similar between the two regimens.

A model was expounded in the early 1990s (85,86) that still yields a satisfactory explanation of the relationship between EPO exposure and response. Furthermore, this model has to date proven satisfactory to explain the PD response to all erythropoiesis-stimulating agents (ESAs). The model states, in essence, that the time between administrations during which the ESA serum level exceeds the threshold for response is the sole driver of efficacy. Of course, the details of the model parameters change with intrinsic potency of the ESA, dose, and clearance parameters, but the model remains the same across all ESAs. The implication is that all ESAs perform similarly when matched for the time above this threshold level. Inefficiency does become a factor as the interval between injections gets longer—explaining the 30% dose penalty with EPO administered once versus three times per week, as shown in the above study. Longer-acting analogs of EPO specifically engineered to improve half-life [darbepoetin alfa (87)] and pegylated EPO [e.g., PEG-EPOβ (88)] are not hampered by this inefficiency until after a longer interval and are, therefore, able to sustain a desired clinical outcome for up to three or four weeks between injections. It remains to be seen how dosing of a non-EPO-based ESA may be approved by regulatory authorities (89), but initial observations suggest adherence to the same PK/PD model.

THROMBOPOIETIN

Despite being named in 1958 (90), TPO was not isolated until 1994 when this was achieved simultaneously by five groups (91–95). TPO is the seminal regulator of platelet production, which, like M-CSF and G-CSF, is consumed by its target cells (megakaryocytes and platelets) that express the c-mpl receptor (57,58). Two forms of recombinant human TPO were initially examined in clinical studies: a full-length and glycosylated molecule that is equivalent to the native growth factor (rHuTPO) and a truncated and pegylated version known as megakaryocyte growth and development factor (MGDF). None of these "first-generation" agents attained regulatory approval mainly because of the production of antibodies by the human immune system that were directed against the administered therapeutic (96,97). These antibodies were also capable of neutralizing endogenous TPO causing extended-term refractory thrombocytopenia. This spurred the creation of novel mpl ligands, seven of which have been recently discussed (98) and all of which have the feature of no overlap in amino acid sequence with endogenous TPO.

The PK of MGDF is reflected in a predictable absorption and elimination profile (99), with C_{max} being observed three to four days after a single SC administration. Elimination is, as mentioned above, affected by the PD response to the drug (57,58). In monkeys, the C_{max} is attained in about 3 hours and MGDF is eliminated with a half-life of around 8 to 13 hours (100). The PK and PD characteristics of full-length recombinant human TPO and MGDF are similar (101). Elimination half-lives are 24 to 40 hours for rHuTPO and 31 hours for MGDF in humans.

The platelet response to administered MGDF is not immediate (102), taking three to four days before even reticulated platelets (a controversial though acceptable measure of early platelet increases) are detected in the circulation, with platelet counts peaking only after around 13 to 15 days. This probably reflects the indirect nature of mpl agonism on thrombocytopoiesis, the main action being confined to an increase in megakaryocyte ploidy and maturation rather than platelet formation (99). Similar kinetics are also exhibited by AMG 531 (Nplate®, romiplostim), one of the third-generation synthetic peptide mpl agonists (103). The medical exploitation of mpl ligands is not yet complete, with several third-generation molecules being developed for the treatment of immune thrombocytopenic purpura (ITP). As with many biopharmaceuticals, it is still unclear for which diseases they will finally be used and how the disease setting will affect their PK/PD.

SUMMARY

Emerging from the confusion of the early days of hematopoietic cytokine discovery was a simple view that for each type of blood cell there would be a single lineage-specific regulator and for each cytokine there would be a specific and defined function. This has not turned out to be the case—blood cell lineages are affected by many different cytokines throughout their development. In addition, all cytokines have been found to have a diverse array of actions, some direct, others indirect, even for the most selective of agents, EPO and G-CSF. Other cytokines have very complex actions, especially as part of overlapping cytokine networks with hereto unforeseen interactions and interdependencies.

In general, most hematopoietic cytokines are short lived in the blood and require repeated frequent injections to clearly see their actions. To improve their utility as therapeutics, the exposure profile of some have been modified by relatively simple pegylation, for example, G-CSF [pegfilgrastim (76)] and EPO [PEG-EPOβ (88)] and that of others by more complex glycoengineering, for example, EPO [darbepoetin alfa (104)]. Some have been mimicked by peptides, for example, EPO [hematide (89)], TPO (AMG 531), or even small molecules, for example, TPO [eltrombopag (98)], while others have been conjugated into chimeric molecules, for example, G-CSF and Flt-3 ligand [progenipoietin (105)].

The field of hematopoietic cytokine biology continues to develop as complex pathways are deconvoluted, and surprises continue to emerge (27). For a number of these factors, end-cell regulation has emerged as a common method of homeostatic control of cellular pathways, with cytokines serving as the central humoral mediators. It remains to be seen how this will be exploited further for the development of cytokine therapeutics with utility in human medicine.

REFERENCES

1. Carnot P, Deflandre C. Sur l'activité cytopoietique du sang et des organs regeneres au cours du regeneration du sang. C R Acad Sci (Paris) 1906; 143:432–435.
2. Viault F. Sur la quantité d'oxygen contenue dans le sang des animaux des hauts pleateaux de L'Amerique du Sud. C R Acad Sci (Paris) 1891; 112:295–298.
3. Fung MC, Hapel AJ, Ymer S, et al. Molecular cloning of cDNA for murine interleukin-3. Nature 1984; 307(5948):233–237.
4. Till JE, McCulloch EA. A direct measurement of the radiation sensitivity of normal mouse bone marrow cells. Radiat Res 1961; 14:213–222.
5. Pluznik DH, Sachs L. The cloning of normal "mast" cells in tissue culture. J Cell Physiol 1965; 66(3):319–324.
6. Bradley TR, Metcalf D. The growth of mouse bone marrow cells in vitro. Aust J Exp Biol Med Sci 1966; 44(3):287–299.
7. Stanley ER, Heard PM. Factors regulating macrophage production and growth purification and some properties of the colony stimulating factor from medium conditioned by mouse L neoplastic fibroblast cells. J Biol Chem 1977; 252(12):4305–4312.
8. Burgess AW, Camakaris J, Metcalf D. Purification and properties of colony stimulating factor from mouse lung conditioned medium. J Biol Chem 1977; 252(6): 1998–2003.

9. Nicola NA, Metcalf D, Matsumoto M, et al. Purification of a factor inducing differentiation in murine myelomonocytic leukemia cells identification as granulocyte colony stimulating factor. J Biol Chem 1983; 258(14):9017–9023.

10. Kawasaki ES, Ladner MB, Wang AM, et al. Molecular cloning of a complementary DNA encoding human macrophage-specific colony-stimulating factor Csf-1. Science 1985; 230(4723):291–296.

11. Jacobs K, Shoemaker C, Rudersdorf R, et al. Isolation and characterization of genomic and complementary DNA clones of human erythropoietin. Nature 1985; 313(6005):806–810.

12. Lin FK, Suggs S, Lin CH, et al. Cloning and expression of the human erythropoietin gene. Proc Natl Acad Sci U S A 1985; 82(22):7580–7584.

13. Wong GG, Witek JS, Temple PA, et al. Human granulocyte-macrophage colony-stimulating factor molecular cloning of the complementary DNA and purification of the natural and recombinant proteins. Science 1985; 228(4701):810–815.

14. Lee F, Yokota T, Otsuka T, et al. Isolation of complementary DNA for a human granulocyte-macrophage colony-stimulating factor by functional expression in mammalian cells. Proc Natl Acad Sci U S A 1985; 82(13):4360–7364.

15. Souza LM, Boone TC, Gabrilove J, et al. Recombinant human granulocyte colony-stimulating factor: effects on normal and leukemic myeloid cells. Science 1986; 232 (4746):61–65.

16. Nagata S, Tsuchiya M, Asano S, et al. Molecular cloning and expression of cDNA for human granulocyte colony-stimulating factor. Nature 1986; 319(6052):415–418.

17. Yang YC, Ciarletta AB, Temple PA, et al. Human interleukin 3 multi-colony-stimulating factor identification by expression cloning of a novel hematopoietic growth factor related to murine interleukin 3. Cell 1986; 47(1):3–10.

18. Kinashi T, Harada N, Severinson E, et al. Cloning of complementary DNA encoding T cell replacing factor and identity with B cell growth factor II. Nature 1986; 324 (6092):70–73.

19. Tominaga A, Takahashi T, Kikuchi Y, et al. Role of carbohydrate moiety of IL-5 effect of tunicamycin on the glycosylation of IL-5 and the biologic activity of deglycosylated IL-5. J Immunol 1990; 144(4):1345–1352.

20. Santhanam U, Ghrayer J, Sehgal PB, et al. Post-translational modifications of human interleukin-6. Arch Biochem Biophys 1989; 274(1):161–170.

21. Namen AE, Schmierer AE, March CJ, et al. B cell precursor growth-promoting activity purification and characterization of a growth factor active on lymphocyte precursors. J Exp Med 1988; 167(3):988–1002.

22. Welte K, Platzer E, Lu L, et al. Purification and biochemical characterization of human pluripotent hematopoietic colony-stimulating factor. Proc Natl Acad Sci U S A 1985; 82(5):1526–1530.

23. Das SK, Stanley ER. Structure function studies of a colony stimulating factor CSF-1. J Biol Chem 1982; 257(22):13679–13684.

24. Zsebo KM, Wypych J, McNiece IK, et al. Identification, purification, and biological characterization of hematopoietic stem cell factor from buffalo rat liver—conditioned medium. Cell 1990; 63(1):195–201.

25. Dordal MS, Wang FF, Goldwasser E. The role of carbohydrate in erythropoietin action. Endocrinology 1985; 116(6):2293–2299.

26. Egrie JC, Browne JK. Development and characterization of novel erythropoiesis stimulating protein (NESP). Nephrol Dial Transplant 2001; 16(suppl 3):3–13.

27. Metcalf D. Hematopoietic cytokines. Blood 2008; 111(2):485–491.

28. Molineux G, Foote MA, Elliott SG. Erythropoietins and erythropoiesis: molecular, cellular, preclinical, and clinical biology. In: Molineux G, Foote MA, Elliott SG, eds. Erythropoietins and Erythropoiesis: Molecular, Cellular, Preclinical, and Clinical Biology. Cambridge: Birkhaeuser Publishing Limited, 2003:i–xi, 1–269.

29. Welte K, Gabrilove J, Bronchud MH, et al. Filgrastim (r-metHuG-CSF): the first 10 years. Blood 1996; 88(6):1907–1929.

30. Hamilton JA, Anderson GP. GM-CSF biology. Growth Factors 2004; 22(4):225–231.

31. Shpall EJ, Wheeler CA, Turner SA, et al. A randomized phase 3 study of peripheral blood progenitor cell mobilization with stem cell factor and filgrastim in high-risk breast cancer patients. Blood 1999; 93(8):2491–2501.

32. Young JD, Crawford J, Gordon M, et al. Pharmacokinetics (pk) of recombinant methionyl human stem cell factor (SCF) in patients (pts) with lung or breast cancer in phase I trials. Proceedings of the American Association for Cancer Research Annual Meeting 1993; 34:217.

33. Lynch DH, Jacobs C, Dupont D, et al. Pharmacokinetic parameters of recombinant mast cell growth factor (rMGF). Lymphokine Cytokine Res 1992; 11(5):233–243.

34. Levesque J-P, Liu F, Simmons PJ, et al. Characterization of hematopoietic progenitor mobilization in protease-deficient mice. Blood 2004; 104(1):65–72.

35. Duarte RF, Franf DA. The synergy between stem cell factor (SCF) and granulocyte colony-stimulating factor (G-CSF): molecular basis and clinical relevance. Leuk Lymphoma 2002; 43(6):1179–1187.

36. Metcalf D, Begley CG, Johnson GR, et al. Biological properties in-vitro of a recombinant human granulocyte-macrophage colony-stimulating factor. Blood 1986; 67(1):37–45.

37. Kenny PA, McDonald PJ, Finlay-Jones JJ. The effect of cytokines on bactericidal activity of murine neutrophils. FEMS Immunol Med Microbiol 1993; 7(3):271–279.

38. Kapp A, Zeck-Kapp G, Danner M, et al. Human granulocyte-macrophage colony stimulating factor an effective direct activator of human polymorphonuclear neutrophilic granulocytes. J Invest Dermatol 1988; 91(1):49–55.

39. Jones TC. The effect of granulocyte-macrophage colony stimulating factor (rGM-CSF) on macrophage function in microbial disease. Med Oncol 1996; 13(3):141–147.

40. Markowicz S, Engleman EG. Granulocyte-macrophage colony-stimulating factor promotes differentiation and survival of human peripheral blood dendritic cells in-vitro. J Clin Invest 1990; 85(3):955–961.

41. Liljefors M, Nilsson B, Mellstedt H, et al. Influence of varying doses of granulocyte-macrophage colony-stimulating factor on pharmacokinetics and antibody-dependent cellular cytotoxicity. Cancer Immunol Immunother 2008; 57(3):379–388.

42. Cebon JS, Bury RW, Lieschke GJ, et al. The effects of dose and route of administration on the pharmacokinetics of granulocyte-macrophage colony-stimulating factor. Eur J Cancer 1990; 26(10):1064–1069.

43. Mueller CE, Mukodzi S, Reddemann H. Relationships of cytokine (GM-CSF) serum concentration to blood cell count and the inflammatory parameters in children with malignant diseases. Pediatr Hematol Oncol 1999; 16(6):509–518.

44. Stute N, Furman WL, Schell M, et al. Pharmacokinetics of recombinant human granulocyte-macrophage colony-stimulating factor in children after intravenous and subcutaneous administration. J Pharm Sci 1995; 84(7):824–828.

45. Cebon J, Dempsey P, Fox R, et al. Pharmacokinetics of human granulocyte-macrophage colony-stimulating factor using a sensitive immunoassay. Blood 1988; 72(4):1340–1347.

46. Stanley ER, Hansen G, Woodcock J, et al. Colony stimulating factor and the regulation of granulopoiesis and macrophage production. Fed Proc 1975; 34(13):2272–2278.

47. Fixe P, Praloran V. M-CSF: haematopoietic growth factor or inflammatory cytokine? Cytokine 1998; 10(1):32–37.

48. Ai-Aql ZS, Alagl AS, Graves DT, et al. Molecular mechanisms controlling bone formation during fracture healing and distraction osteogenesis. J Dent Res 2008; 87(2):107–118.

49. Shimada-Hiratsuka M, Naito M, Kaizu C, et al. Defective macrophage recruitment and clearance of apoptotic cells in the uterus of osteopetrotic mutant mice lacking macrophage colony-stimulating factor (M-CSF). J Submicrosc Cytol Pathol 2000; 32(2):297–307.

50. Yagihashi A, Sekiya T, Suzuki S. Macrophage colony stimulating factor (M-CSF) protects spiral ganglion neurons following auditory nerve injury: morphological and functional evidence. Exp Neurol 2005; 192(1):167–177.

51. Murase S-I, Hayashi Y. Expression pattern and neurotrophic role of the c-fms proto-oncogene M-CSF receptor in rodent Purkinje cells. J Neurosci 1998; 15:10481–10492.

52. Tkachuk M, Gisler RH. The promoter of macrophage colony-stimulating factor receptor is active in astrocytes. Neurosci Lett 1997; 225(2):121–125.

53. Uemura Y, Kobayashi M, Nakata H, et al. Effects of GM-CSF and M-CSF on tumor progression of lung cancer: roles of MEK1/ERK and AKT/PKB pathways. Int J Mol Med 2006; 18(2):365–373.

54. Pederson L, Winding B, Foged NT, et al. Identification of breast cancer cell line-derived paracrine factors that stimulate osteoclast activity. Cancer Res 1999; 59(22):5849–5855.

55. Takagi A, Takeda S, Matsuoka K, et al. Macrophage colony-stimulating factor (M-CSF) production in vivo and in vitro in gynecologic malignancies. Int J Clin Oncol 1999; 4(3):142–147.

56. Ramakrishnan S, Xu FJ, Brandt SJ, et al. Constitutive production of macrophage colony-stimulating factor by human ovarian and breast cancer cell lines. J Clin Invest 1989; 83(3):921–926.

57. Tanaka H, Takama H, Arai Y, et al. Pharmacokinetics of pegylated recombinant human megakaryocyte growth and development factor in healthy volunteers and patients with hematological disorders. Eur J Haematol 2004; 73(4):269–279.

58. Li J, Xia Y, Kuter DJ. Interaction of thrombopoietin with the platelet c-mpl receptor in plasma: binding, internalization, stability and pharmacokinetics. Br J Haematol 1999; 106(2):345–356.

59. Chapel S, Veng-Pedersen P, Hohl RJ, et al. Changes in erythropoietin pharmaco-kinetics following busulfan-induced bone marrow ablation in sheep: evidence for bone marrow as a major erythropoietin elimination pathway. J Pharmacol Exp Ther 2001; 298(2):820–824.

60. Agoram B, Molineux G, Jang G, et al. Effects of altered receptor binding activity on the clearance of erythropoiesis-stimulating proteins: a minor role of erythropoietin receptor-mediated pathways? Nephrol Dial Transplant 2006(4):303–304.

61. Layton JE, Hockman H, Sheridan WP, et al. Evidence for a novel in vivo control mechanism of granulopoiesis: mature cell-related control of a regulatory growth factor. Blood 1989; 74(4):1303–1307.

62. Bartocci A, Mastrogiannis DS, Migliorati G, et al. Macrophages specifically regulate the concentration of their own growth factor in the circulation. Proc Natl Acad Sci U S A 1987; 84(17):6179–6183.

63. Metcalf D, Nicola NA. Proliferative effects of purified granulocyte colony stimulating factor on normal mouse hemopoietic cells. J Cell Physiol 1983; 116(2):198–206.

64. Lord BI. Myeloid cell kinetics in response to haemopoietic growth factors. Baillieres Clin Haematol 1992; 5(3):533–550.

65. Lord BI, Molineux G, Pojda Z, et al. Myeloid cell kinetics in mice treated with recombinant interleukin-3, granulocyte colony-stimulating factor (CSF), or gran-ulocyte-macrophage CSF in vivo. Blood 1991; 77(10):2154–2159.

66. Takamatsu Y, Simmons PJ, Moore RJ, et al. Osteoclast-mediated bone resorption is stimulated during short-term administration of granulocyte colony-stimulating factor but it not responsible for hematopoietic progenitor cell mobilization. Blood 1998; 92(9):3465–373.

67. Purton LE, Lee MY, Torok-Storb B. Normal human peripheral blood mononuclear cells mobilized with granulocyte colony-stimulating factor have increased osteo-clastogenic potential compared to nonmobilized blood. Blood 1996; 87(5):1802–1808.

68. Lee MY, Fukunaga R, Lee TJ, et al. Bone modulation in sustained hematopoietic stimulation in mice. Blood 1991; 77(10):2135–2141.

69. Bronchud MH, Scarffe JH, Thatcher N, et al. Phase I/II study of recombinant human granulocyte colony-stimulating factor in patients receiving intensive chemotherapy for small cell lung cancer. Br J Cancer 1987; 56(6):809–813.

70. Morstyn G, Campbell L, Souza LM, et al. Effect of granulocyte colony stimulating factor on neutropenia induced by cytotoxic chemotherapy. Lancet 1988; 1(8587):667–672.

71. Morstyn G, Campbell L, Duhrsen U, et al. Clinical studies with granulocyte colony stimulating factor G-Csf in patients receiving cytotoxic chemotherapy. Behring Inst Mitt 1988; 83:234–239.

72. Roskos L, Cheung E, Vincent M, et al. Pharmacology of Filgrastim (r-metHuG-CSF). In: Morstyn G, Dexter TM, Foote M, eds. Filgrastim (r-metHuG-CSF) in Clinical Practice. 2nd ed. New York: Marcel Dekker, 1998:51–72.
73. Terashi K, Oka M, Ohdo S, et al. Close association between clearance of recombinant human granulocyte colony-stimulating factor (G-CSF) and G-CSF receptor on neutrophils in cancer patients. Antimicrob Agents Chemother 1999; 43(1):21–24.
74. Holmes FA, Jones SE, O'Shaughnessy J, et al. Once-per-cycle pegylated filgrastim (SD/01) is as effective and safe as daily filgrastim in reducing chemotherapy-induced neutropenia over multiple cycles of therapy. Breast Cancer Res Treat 2000; 64(1):89.
75. Johnston E, Crawford J, Blackwell S, et al. Randomized, dose-escalation study of SD/01 compared with daily filgrastim in patients receiving chemotherapy. J Clin Oncol 2000; 18(13):2522–2528.
76. Molineux G, Kinstler O, Briddell B, et al. A new form of Filgrastim with sustained duration in vivo and enhanced ability to mobilize PBPC in both mice and humans. Exp Hematol 1999; 27(12):1724–1734.
77. Roskos LK, Yang B, Schwab G, et al. A cytokinetic model describes the granulopoietic effects of r-metHuG-CSF-SD/01 (SD/01) and the homeostatic regulation of SD/01 clearance in normal volunteers. Clin Pharmacol Ther 1999; 65(2):196.
78. Green M. A single, fixed-dose of Pegfilgrastim given once-per-chemotherapy cycle is as effective as daily Filgrastim in the management of neutropenia in high-risk breast cancer. Eur J Cancer 2001; 37(suppl 6):S146–S147.
79. Eschbach JW, Egrie JC, Downing MR, et al. Correction of the anemia of end-stage renal disease with recombinant human erythropoietin results of a combined phase I and II clinical trial. N Engl J Med 1987; 316(2):73–78.
80. Arcasoy MO. The non-haematopoietic biological effects of erythropoietin. Br J Haematol 2008; 141(1):14–31.
81. Elliott S, Busse L, Bass MB, et al. Anti-Epo receptor antibodies do not predict Epo receptor expression. Blood 2006; 107(5):1892–1895.
82. Sun CH, Ward HJ, Paul WL, et al. Serum erythropoietin levels after renal transplantation. N Engl J Med 1989; 321(3):151–157.
83. Wide L, Bengtsson C, Birgegard G. Circadian rhythm of erythropoietin in human serum. Br J Haematol 1989; 72(1):85–90.
84. Cheung W, Minton N, Gunawardena K. Pharmacokinetics and pharmacodynamics of epoetin alfa once weekly and three times weekly. Eur J Clin Pharmacol 2001; 57(5):411–418.
85. Besarab A, Flaharty KK, Erslev AJ, et al. Clinical pharmacology and economics of recombinant human erythropoietin in end-stage renal disease the case for subcutaneous administration. J Am Soc Nephrol 1992; 2(9):1405–1416.
86. Jumbe NL, Rossi G, Heatherington AC. The science of erythropoiesis: quantification of factors influencing response by pharmacokinetic (PK) and pharmacodynamic (PD) modeling. Blood 2002; 100:9b.
87. Macdougall IC. Optimizing the use of erythropoietic agents—pharmacokinetic and pharmacodynamic considerations. Nephrol Dial Transplant 2002; 17(suppl 5):66–70.
88. Topf J. CERA: third-generation erythropoiesis-stimulating agent. Expert Opin Pharmacother 2008; 9(5):839–849.
89. Wiecek A, Macdougall IC, Mikhail A, et al. Long-term safety, tolerability, and pharmacodynamics of hematide (TM) a synthetic peptide-based erythropoiesis stimulating agent in a phase II, multi-dose study in patients with chronic kidney disease. Nephrol Dial Transplant 2006; 21(4):155.
90. Keleman E, Cserhati I, Tanos B. Demonstration of some properties of human thrombopoietin in thrombocythaemic sera. Acta Haematol 1958; 20(6):350–355.
91. Kato T, Ogami K, Shimada Y, et al. Purification and characterization of thrombopoietin. J Biochem 1995; 118(1):229–236.
92. Bartley TD, Bogenberger J, Hunt P, et al. Identification and cloning of a megakaryocyte growth and development factor that is a ligand for the cytokine receptor Mpl. Cell 1994; 77(7):1117–1124.

93. De Sauvage FJ, Hass PE, Spencer SD, et al. Stimulation of megakaryocytopoiesis and thrombopoiesis by the c-Mpl ligand. Nature 1994; 369(6481):533–538.

94. Lok S, Kaushansky K, Holly RD, et al. Cloning and expression of murine thrombopoietin cDNA and stimulation of platelet production in vivo. Nature 1994; 369 (6481):565–568.

95. Kuter DJ, Beeler DL, Rosenberg RD. The purification of megapoietin: a physiological regulator of megakaryocyte growth and platelet production. Proc Natl Acad Sci U S A 1994; 91(23):11104–11108.

96. Basser RL, O'Flaherty E, Green M, et al. Development of pancytopenia with neutralizing antibodies to thrombopoietin after multicycle chemotherapy supported by megakaryocyte growth and development factor. Blood 2002; 99(7):2599–2602.

97. Li J, Yang C, Xia Y, et al. Thrombocytopenia caused by the development of antibodies to thrombopoietin. Blood 2001; 98(12):3241–3248.

98. Kuter DJ. New thrombopoietic growth factors. Blood 2007; 109(11):4607–4616.

99. Harker LA, Roskos LK, Marzec UM, et al. Effects of megakaryocyte growth and development factor on platelet production, platelet life span, and platelet function in healthy human volunteers. Blood 2000; 95(8):2514–2522.

100. Sola MC, Christensen RD, Hutson AD, et al. Pharmacokinetics, pharmacodynamics, and safety of administering pegylated recombinant megakaryocyte growth and development factor to newborn rhesus monkeys. Pediatr Res 2000; 47(2):208–214.

101. Kuter D. Thrombopoietin factors. In: Morstyn G, Foote M, Lieschke G, eds. Cancer Drug Discovery and Development Hematopoietic Growth Factors in Oncology: Basic Science and Clinical Therapeutics. Totawa, NJ: Humana Press, 2004:125–152.

102. Begley CG. Clinical studies with megakaryocyte growth and development factor (Mpl-ligand). Thromb Haemost 1997; 78(1):42–46.

103. Kuter D, Bussel JB, Aledort LM, et al. A phase 2 placebo controlled study evaluating the platelet response and safety of weekly dosing with a novel thrombopoietic protein (AMG531) in thrombocytopenic adult patients (pts) with immune thrombocytopenic purpura (ITP). Blood 2004; 104(11 pt 1):148a

104. Egrie JC, Browne JK. Development and characterization of novel erythropoiesis stimulating protein (NESP). Br J Cancer 2001; 84(suppl 1):3–10.

105. Streeter PR, Kahn LE, Joy WD, et al. Progenipoietin-G, a multifunctional agonist of human flt-3 and G-CSF receptors. Blood 1997; 90(suppl 1, part 1):57a.

2 | In Vitro–In Vivo Correlations of Pharmacokinetics, Pharmacodynamics, and Metabolism for Antibody Therapeutics

C. Andrew Boswell, Rong Deng, Kedan Lin, Wendy S. Putnam, Cherry Lei, Frank-Peter Theil, Amita Joshi, Paul J. Fielder, and Leslie A. Khawli
Research and Early Development, Genentech, Inc., South San Francisco, California, U.S.A.

INTRODUCTION

In contrast to small molecular weight drugs (SMD), the development of in vitro–in vivo correlations (IVIVCs) for monoclonal antibody (mAb) therapeutics is still in its infancy. High-throughput in vitro assays with established IVIVC that currently exist for SMD have not been so strongly pursued for antibody (Ab) therapeutics, reflecting to a certain degree the substantially lower attrition rate and the associated lower numbers of potential clinical mAb candidates. This chapter represents one of the first attempts at summarizing the available data correlating preclinical in vitro and in vivo data of mAbs.

In recent years, the correlation between in vitro and in vivo properties of chemically derived SMD has become an established methodology to predict single or multiple components of in vivo pharmacokinetics (PK), pharmacodynamics (PD), safety/toxicity, and efficacy (1–3). Accessibility to physiologically relevant in vitro assays has fostered the development of IVIVC involving critical PK, PD, safety, and efficacy data, allowing prediction of in vivo behavior prior to in vivo studies. In fact, because of the often high-throughput capabilities of such assays, thousands of new compounds can be screened for their potential to become viable clinical drug candidates in a highly efficient manner. Given the relatively high attrition rate of SMD, the pressure to combine in vitro screening tools with established IVIVC has increased the demand for tools that facilitate a more successful and efficient selection process of drug-able clinical candidate molecules. Besides the impact on the overall drug discovery and development productivity, in vitro assays provide mechanistic insight into subprocesses that are often not easily assessable under in vivo conditions. Complex mechanistic processes can be isolated into mechanistic subprocesses at organ, cellular, and subcellular levels. One example of a mechanistic dissection of complex in vivo processes is the development of in vitro methodologies to characterize the essential components of PK, namely absorption, distribution, metabolism, and excretion (ADME). Predictive in vitro and even in silico tools have been recently developed to obtain estimates for individual ADME parameters that describe the manner in which drug candidates are being absorbed, distributed, and eliminated under in vivo conditions. In a second step, the predicted ADME parameters can be combined to estimate the overall in vivo PK as a composite of the individual ADME processes. Physiologically based pharmacokinetic (PBPK) modeling tools have been developed to predict the in vivo PK solely based on in vitro and in silico ADME data together with established physiological information required to describe the mammalian body (4,5). This concept of predicting human PK on the basis of in vitro and in silico data using mechanistic PBPK models has been successfully implemented in the realm of drug discovery and development, thereby partly replacing empirical prediction methods. The establishment of IVIVC derived from mechanistic models has resulted in a somewhat paradigm shift in that the human PK of SMD is now predicted primarily on the basis of in vitro and in silico data, whereas in vivo data from preclinical animal species are rather used to verify prediction success as surrogate for human prior to the actual prediction of human PK. In addition to IVIVC established for PK parameters, very successful applications of similar types of methods are known in the area of PD, safety, and efficacy of SMD (see section "Modeling of Antibody In Vitro–In Vivo Correlations").

In contrast, IVIVC for mAb therapeutics are not as well established as that for SMD. Still, the concept of developing and using IVIVC involving mAbs is not a new one. Seminal work reported by Day et al. in 1974 probed the relationship between in vitro binding and in vivo localization of ^{125}I-labeled Abs; the results of this work helped elucidate the blood-brain barrier (BBB) concept for these molecules (6). In present day, the selection of appropriate in vitro and in vivo biological assays in evaluation of biotechnological products remains a critical challenge (7–9). In theory, relationships between in vitro and in vivo data may encompass a broad spectrum of situations ranging from qualitative explanations to true statistical evaluations. In practice, however, the vast majority of cases are best described as qualitative correlations or trends, as demonstrated by the diverse examples discussed herein.

There are more than 20 therapeutic Abs, fragments, and immunoconjugates approved by the Food and Drug Administration (FDA) (Table 1), and hundreds more are in development (10–14). For a mAb to enter clinical trials, it must be first tested preclinically using a battery of methods; several of these are listed in Table 2 (15,16). Because of the interdependencies that exist between these and other in vitro and in vivo phenomena, it is possible to establish correlations between various combinations of these measurements. For example, the in vitro information on binding characteristics is of particular interest considering that the Ab PK and PD can be influenced by crystallizable fragment (Fc) receptor binding and several other factors (11,17). In addition, the binding affinity to cognate target antigens and the level of target expression are known to influence PK/PD behavior.

Within this chapter, we will initially present a brief overview of mAb structures and functions, followed by an examination of the specific approaches used to correlate in vitro–in vivo data related to PK, PD, and metabolism of mAb therapeutics. Because of the growing number of mAb derivatives that are being generated, the quantity of correlative data is very large, making it difficult to discuss all study cases in this work. As such, representative examples involving mostly preclinical studies throughout the literature will be described in this chapter, but an emphasis on more recent work will be apparent in most cases. In this context, we will examine three major in vitro properties: (*i*) binding, (*ii*) potency, and (*iii*) cellular metabolism. In parallel, we will consider four major in vivo phenomena: (*i*) binding and/or targeting, (*ii*) PD and/or potency, (*iii*) PK, and (*iv*) in vivo metabolism. Because IVIVC can be assessed using

TABLE 1 Current Approved Antibodies for the Parenteral Use in Treatment of Disease

Generic name	Trade name	Binding target/antibody class and isotype	Indication	Approval
Naked Antibodies				
Rituximab	Rituxan®	Anti-CD20/chimeric IgG_1	B-cell lymphoma	1997
Trastuzumab	Herceptin®	Anti-HER2/humanized IgG_1	Breast cancer	1998
Alemtuzumab	CamPath®	Anti-CD52/humanized IgG_1	Chronic lymphocytic leukemia	2001
Cetuximab	Erbitux®	Anti-EGFR/chimeric IgG_1	Colorectal cancer	2004
			Head/neck cancer	2006
Bevacizumab	Avastin®	Anti-VEGF/chimeric IgG_1	Colorectal cancer	2004
			Lung cancer	2006
Panitumumab	Vectibix®	Anti-EGFR/humanized IgG_2	Colorectal cancer	2006
Abciximab	ReoPro®	Fab anti-CD41 7E3/chimeric IgG_1	Cardiovascular disease	1994
Adalimumab	Humira®	Anti-TNF-α/chimeric IgG_1	Autoimmune disorders	2002
Basiliximab	Simulect®	Anti-CD25/chimeric IgG_1	Transplant rejection	1998
Daclizumab	Xenapax®	Anti-CD25/humanized IgG_1	Transplant rejection	1997
Eculizumab	Soliris®	Anti-complement protein C5/humanized $IgG_{2/4\kappa}$	Inflammatory diseases	2007
Efalizumab	Raptiva®	Anti-CD11a/humanized IgG_1	Psoriasis	2002
Edrecolomab	Panorex®	Anti-Ag17-1a/murine IgG_{2a}	Colorectal cancer	1994[a]
Infliximab	Remicade®	Anti-TNF-α/chimeric IgG_1	Autoimmune disorders	1998
Muromonab-CD3	Orthoclone® OKT3	Anti–T cell receptor–CD3 complex/murine IgG_{2a}	Transplant rejection	1986
Natalizumab	Tysabri®	Anti-$\alpha 4$ integrin/humanized IgG_4	Autoimmune-related multiple sclerosis therapy	2006
Omalizumab	Xolair®	Anti-IgE/humanized $IgG_{1\kappa}$	Allergy-related asthma therapy	2004
Palivizumab	Synagis®	Anti-F protein of RSV/humanized IgG_1	RSV	1998
Radioimmunoconjugates				
Ibritumomab tiuxetan	Zevalin®	Anti-CD20/^{90}Y-murine IgG_1 + rituximab	B-cell lymphoma	2002
Tositumomab "Anti-B1"	Bexxar®	Anti-CD20/^{131}I-murine IgG_{2a} + unlabeled tositumomab	B-cell lymphoma	2003
Tumor necrosis therapy	Cotara®	Anti-nuclear histone H_1/^{131}I-chimeric IgG_{2a}	Lung cancer	2003[b]
Antibody-Drug Conjugates				
Gemtuzumab ozogamicin	Mylotarg®	Anti-CD33/humanized IgG_4 conjugated to calicheamicin	Acute myelogenous leukemia	2000

[a] Approved by Germany.
[b] Approved by China.
Abbreviations: CD, cluster of differentiation; HER2, human epidermal growth factor receptor 2; EGFR, epidermal growth factor receptor; VEGF, vascular endothelial growth factor; Ig, immunoglobulin; Fab, fragment antigen binding; TNF, tumor necrosis factor; RSV, respiratory syncytial virus.
Source: From Refs. 10–14.

TABLE 2 In Vitro and In Vivo Methods Used in Preclinical Studies of Antibody Therapeutics

In Vitro Methods	In Vivo Methods
Antigen identification and quantification on target cells	Pharmacokinetics studies
Normal tissue screen	Target/tissue distribution studies
Evaluation of target specificity	Dosimetry studies
Immunoreactivity/avidity binding constant	Toxicology studies
Potency assay (e.g., cytotoxicity, etc.)	Assessment of toxicological model validity (target homology, expression, distribution)
Fc receptor binding affinity (e.g., FcRn, FcγR)	Metabolism studies
Target antigen–binding affinity	Imaging studies in animal disease model
Cellular metabolism studies	Determination of circulating antigen content
	Therapy studies in animal disease model

Abbreviations: FcRn, neonatal fragment crystallizable receptor; FcγR, fragment crystallizable γ receptors.

numerous combinations of these parameters, we will simplify the presentation by organizing topics primarily on the basis of the in vitro data. Specific and recent examples of cited correlative studies involving both naked Abs and radioimmunoconjugates may be found throughout these various sections. In addition, we will explore in detail correlative studies involving selected Abs conjugated to potent cytotoxic drugs termed antibody-drug conjugates (ADCs). These armed immunoconjugates possess unique properties and are designed to achieve excellent localization of a small-molecule drug at the desired site to optimize the therapeutic effect of the agent.

This chapter appears to be one of the first attempts at summarizing the available PK, PD, and metabolism information in the exciting area of IVIVC of mAbs. Despite the apparent slow progress in this area, the corresponding success for SMD illustrates the potential benefit of using these IVIVC tools to drive toward a better mechanistic understanding of the PK, PD, and metabolism of mAbs as well as to accelerate the drug discovery and development processes, in particular the selection of successful drug candidates.

OVERVIEW OF ANTIBODY THERAPEUTICS

Abs are multifunctional glycoproteins that are produced by B lymphocytes and are capable of binding to foreign substances known as antigens. Abs, also known as immunoglobulins (Ig), are primarily classified by structural differences in their heavy-chain constant domains. In humans, five distinct classes of Abs exist and have been designated IgG (γ), IgA (α), IgD (δ), IgE (ε), and IgM (μ), including four subclasses of IgG and two subclasses of IgA (17). All five classes are associated with two light-chain isotypes, C_κ and C_λ (17). These classes differ from each other in size, charge, amino acid composition, and carbohydrate content. The IgG class, whose molecular weight is approximately 150 kDa, comprises most of the Ig in normal human serum (85%) as well as most Ab therapeutics currently marketed (Table 1) or in development (17,18).

An IgG Ab consists of two identical fragment antigen bindings (Fabs) and an Fc domain (Fig. 1). Structural features of the Fc, which comprises the carboxy-terminus of the heavy chain, ultimately influence the Fc-mediated downstream consequences (e.g., effector functions) of antigen binding (17). Effector functions induced by binding of an Ab to its antigen can lead to Fc-mediated destruction of

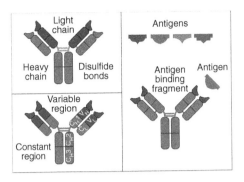
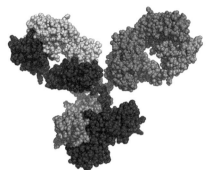

FIGURE 1 Structural representations of antibody structure. [The structure cartoon (*left*) was derived by manipulation of images retrieved from Genentech's Digital Media Services Asset Bank.] The structural model (*right*) of IgG1 (PDB code, 1 igy) was generated using The PyMOL Molecular Graphics System by Kiran Mukhyala of the Bioinformatics Department at Genentech, Inc. *Abbreviation*: PDB, Protein Data Bank. *Source*: From Ref. 19.

the antigen: (*i*) through recruitment of Fc-binding molecules of the complement cascade, that is, complement-dependent cytotoxicity (CDC) and (*ii*) through phagocytosis triggered by engagement of fragment crystallizable γ receptors (FcγR) on macrophages or other immune effector cells, that is, Ab-dependent cell-mediated cytotoxicity (ADCC) (18). In addition, an Ab may simply block signaling by binding receptors, thus preventing receptor-ligand interactions. The Fc domain of IgG also contains binding sites for the neonatal fragment crystallizable receptor (FcRn), which provides protection from lysosomal catabolism and also assists in the transport of Abs across some cellular membranes (20).

Two identical polypeptide heavy chains (having four domains) and two identical light chains (having two domains) define each IgG molecule (17). The heavy-chain domains include V_H, C_H1, C_H2, and C_H3, where V denotes "variable" and C denotes "constant" regions (17,21). The light-chain domains are designated as V_L and C_L. Intertwining of the V_H and V_L domains typically form the antigen-binding pocket, which includes three short stretches of peptide in each V domain known as complementarity-determining regions (CDR) that are located at the amino-terminus. Interchain disulfide bonds between C_H1 and C_L stabilize the heavy- and light-chain association (17). Multiple inter-heavy-chain disulfide bonds are located at the hinge region, which resides between C_H1 and C_H2. The hinge region determines the flexibility of the Fabs relative to the Fc and affects the extent to which an Ab can bind antigens divalently (17).

Methods of Ab engineering and production can greatly influence the degree of success of Ab therapeutics in treatment of disease. Murine monoclonal Abs were initially derived from mouse hybridomas, immortal Ab-producing cells formed by fusion of a B cell with a myeloma cell (22). Such hybridoma technology enables these B-cell/myeloma hybrids to produce Abs of a desired specificity, which are then cloned and expanded for large-scale mAb manufacture. This process has resulted in the identification of a considerable number of new target-associated antigens, as well as Abs against different epitopes of the same antigen. These murine-derived Abs have proven useful in the in vitro diagnostics market; however, injection into humans for therapeutic purposes has often induced human anti-mouse antibody (HAMA) responses (18,21,23). A HAMA response typically occurs two to three weeks postinjection, is usually unrelated to the dose and rate of administration, may lead to infusion reactions that significantly limit the possibility of retreatment, and can alter the PK of a mAb, potentially increasing clearance rate and impairing targeting and/or therapy (18,21,23). To avoid HAMA responses, Abs were developed through genetic engineering by fusing the variable region of mouse Ab with the constant region of a human Ab (24). Further reductions in immunogenicity were realized by developing humanized Abs, grafting portions of the murine Ab into human frameworks, while more recent methods have generated fully human mAbs (24).

Because of the domain structure of Abs, molecular biology has provided the means to produce several Ab fragments including F(ab')$_2$, Fab, and single-chain variable fragment (scFv); these can be generated by removing parts or all of the Fc constant region (25). For example, scFv (30 kDa), one of smallest forms of Ab derivatives, possesses a single binding site consisting of variable light and heavy domains connected by a flexible small peptide linker. The ability of the scFv to bind to tumor receptors lies in a fine balance between its ability to penetrate tumor tissues due to its small size and its fast renal clearance (26). To further optimize tumor binding, another variation of Ab fragment termed "minibody" (e.g., diabody or triabody molecule) has been devised. The diabody (60 kDa) consists of two single-chain molecules joined by a very short linker, whereas the triabody is prepared without the linker, thereby preventing dimer formation and forcing trimerization (90 kDa) (27,28). The disadvantage of all of the above fragments is that, while they retain their antigen-binding properties, they exhibit no induction of Fc effector functions including ADCC, which is a relevant antitumoral mechanism of therapeutic Abs (25,29). Furthermore, some intact human IgG$_1$ Abs have a half-life of three to four weeks, while Ab fragments undergo blood clearance in less than 24 hours because of relatively fast renal excretion (30). These features make Ab fragments more useful in diagnostic applications and receptor target imaging where high target to background ratios are desired (31).

Another class of Ab variants, immunoconjugates, may be derived through chemical conjugation of Abs to several different types of natural or synthetic compounds. Both radioimmunoconjugates and ADCs fall into this broad class of Ab-based agents. Arming Abs with drugs, toxins, or therapeutic radionuclides can significantly improve their potential therapeutic benefits. Several chemotherapeutics drugs including doxorubicin, calicheamicin, maytansine, methotrexate, chlorambucil, and many others have been linked to Abs using various chemical conjugation methods (31,32). The successful use of an ADC depends on a number of factors such as its in vivo stability, its ability to localize and persist at the tumor site with little localization in normal tissues, and its intratumoral and intracellular penetration and metabolism. The sensitivity of the tumor cells themselves to the conjugated drug also influences the overall therapeutic efficacy of this approach (15). In addition, radionuclides have been used to enhance the cytotoxic effects of mAbs while simultaneously providing a means for detection. Radioiodination may be achieved directly through reaction with tyrosine residues or, alternatively, by indirect methods such as the use of the prosthetic group, N-succinimidyl-3-[^{125}I]iodo-benzoate ([^{125}I]SIB) (33). For the attachment of radiometals via bifunctional chelating agents, chemical linkage methodologies comprising thiourea, thioether, amide, ester, and disulfide bonds are available (12). For example, radioimmunoscintigraphy using γ cameras or single-photon emission computed tomography (SPECT) requires coupling of γ-emitting isotopes (e.g., ^{99m}Tc, ^{123}I, ^{111}In) to mAbs (34); positron emission tomography (PET) relies on attachment of positron emitters (e.g., ^{18}F, ^{64}Cu, ^{68}Ga, ^{86}Y, or ^{124}I) (35). For radioimmunotherapy, however, β emitters (e.g., ^{131}I, ^{90}Y, ^{177}Lu, ^{67}Cu) or α-emitting radionuclides (e.g., ^{213}Bi and ^{211}At) are generally considered to be the radionuclides of choice (13). It is important, however, that the tumor receives sufficient doses of radiation for radiolabeled Abs to have clinical efficacy. Boswell and Brechbiel have reviewed the field of radio-immunotherapeutic and diagnostic Abs (12); in addition, Keller et al. have highlighted the field of targeted radioimmunotherapy (36).

The establishment of methods to produce humanized and fully human mAbs has greatly aided the ongoing pursuit toward the development of immunotherapeutic agents capable of targeting and treating many diseases (25). As mentioned earlier, there are already more than 20 FDA-approved therapeutic mAbs and fragments (Table 1). As biotherapeutic agents, mAbs possess several ideal properties as they are homogeneous, they recognize specific antigenic determinants, and they can be mass produced; however, the development of Abs into clinically useful biotherapeutic drugs is a daunting challenge because of their molecular size, complex structure, and composition. To aid in this process, radioactive and fluorescent probes, enzyme-linked immunosorbent assays (ELISA), and several other analytical methods have been utilized in the evaluation of mAbs for clinical application. In addition, various methods for the large-scale production and purification of mAbs continue to improve, enabling clinicians to use reagents with lower immunogenicity and higher purity (25). In general, these results have fostered new hope that mAbs will be able to treat a wide range of clinical disorders and diseases. However, to realize the full potential of these reagents, a great deal of preclinical work is required. Therefore, the present chapter will provide recent and representative examples to illustrate the current preclinical strategies being employed for the correlation of in vitro and in vivo properties of several Abs, paying particular attention to radioimmunoconjugates and ADCs, both of which have entered the mainstream of clinical medicine for the diagnosis and treatment of many diseases in human.

IN VITRO BINDING CORRELATED TO IN VIVO PROPERTIES

In vitro cell-binding experiments employ highly simplified model systems lacking the physiological complexities of living organisms. This distinction underscores a critical uncertainty that trends observed in binding to cultured cells expressing the target antigen alone may or may not necessarily translate into the same trend with respect to targeting, efficacy, or other in vivo behaviors. In some regards, in vitro models are best-case scenarios in which the chances are stacked in favor of successful binding because of an inherent lack of physical barriers between the Ab and its receptor. In certain cases, however, specific in vivo factors or conditions that are absent in vitro may also cause discrepancies between the two measurements. To demonstrate these concepts, we will cite examples of Ab studies that relate in vitro binding to in vivo PK, in vivo targeting, and in vivo PD.

In Vitro Binding Correlated to In Vivo Pharmacokinetics

Correlations between in vitro antigen binding and in vivo PK must be carefully interpreted because binding events do not exclusively determine Ab PK. In addition to the effects of antigen binding, the valency, shape, size, isoelectric pH (pI), and blood pool concentration are all factors that govern the speed, magnitude, and depth (penetration) of tissue uptake and clearance of a given Ab (37). In some cases, however, differences in in vitro binding affinities can be correlated to differences in PK profiles in vivo assuming that Ab valency, shape, size, and pI are similar and that antigen density remains constant. Both the binding affinity of a given Ab for its antigen and the expression levels of the target antigen can heavily influence PK; however, nontarget specific binding interactions between Fc-binding domains of a given Ab and endogenous Fc receptors can also play an important role.

Target Antigen Binding Correlated to Antibody Pharmacokinetics
The exploration of cluster of differentiation (CD)11a as a target for psoriasis treatment illustrates how target antigen expression can affect Ab PK. Efalizumab (anti-CD11a, Raptiva®) has been approved for treatment of moderate to severe psoriasis, a chronic skin disease involving T cells, and its PK and PD have been reviewed by Joshi et al. (38). This Ab binds the subunit CD11a of the human integrin lymphocyte function–associated antigen-1, and interferes with the T-cell infiltration, activation, migration to the skin, and reactivation process, all hallmarks of the disease. Coffey et al. used an in vitro human T-cell model to study cellular uptake and clearance of the anti-CD11a Ab (39). Flow cytometric analysis was used to examine the cell surface and intracellular expression of CD11a following in vitro incubation of anti-CD11a with human T cells, demonstrating downmodulation of both surface and intracellular CD11a expression over time. Addition of a secondary cross-linking Ab revealed CD11a internalization in vitro, and it was rationalized that a similar phenomenon helps mediate the in vivo clearance of anti-CD11a (39). In vivo experiments were conducted by intravenously administering mice with M17, a rat anti-mouse CD11a mAb, and muM17 (40). Similar flow cytometric analysis of muM17 binding to CD11a on $CD3^+$, $CD4^+$, and $CD8^+$ T cells and cellular clearance in vivo was performed. The results were consistent with the in vitro results, suggesting that anti-CD11a clearance in vivo was driven by receptor-mediated internalization and lysosomal degradation by cells expressing CD11a (40).

Fcγ Receptor Binding Correlated to Antibody Pharmacokinetics
Nearly every cell type of the immune system expresses FcγRs, including B lymphocytes, dendritic cells, macrophages, natural killer (NK) cells, neutrophils, mast cells, platelets, Langerhans cells, eosinophils, mesangial cells, and endothelial cells (41). Three different FcγRs exist: FcγRI (CD64), which demonstrates high affinity for monomeric IgG ($K_d \approx 10^{-9}$ M), and two low-affinity receptors, FcγRII (CD32) and FcγRIII (CD16), which bind to monomeric IgG with a dissociation constant (K_d) of approximately 10^{-6} M but have high avidity for IgG-opsonized particles and IgG immune complexes (42). FcγRs are known to play a critical role in linking IgG Ab-mediated immune response with cellular effector functions. In addition, Ab PK may also be impacted through FcγR-mediated elimination.

The correlation of FcγR binding affinity to mAb PK is complicated and dependent on many factors. Target expression is a major determinant of Ab PK; however, when comparing the PK between a wild-type (WT) and an Fcγ mutant Ab (having equal antigen expression levels), Fcγ affinity becomes an important parameter. Some Abs bind to *soluble targets* to form soluble immune complexes that can promote binding to the low-affinity FcγRII or FcγRIII. Once the soluble immune complex has been engaged by FcγRs, the complexes are internalized by hepatocytes including Kupffer cells and sinusoidal endothelial cells. A change in binding affinity to the FcγR will noticeably impact PK if the target antigen concentration is high relative to the mAb of interest when the contribution of the target-mediated clearance to the total clearance is significant. The binding of Ab to antigen is a dynamic process, and Fcγ influences not only the PK of the mAb but also the Ab-antigen complex. As such, an Fcγ-driven decrease in clearance of the Ab-antigen complex will result in a concomitant decrease in Ab clearance. For example, anti-IgE mAb (IgE, a soluble target with high endogenous concentration) with reduced FcγR binding affinity demonstrated slower clearance compared with its WT counterpart in monkeys. However, another undisclosed mAb against a different target (soluble target with low endogenous concentration) with reduced FcγR binding affinity had similar clearance compared with its WT counterpart in mice (Internal data, Genentech, Inc., unpublished data).

In contrast, Ab binding to *cell membrane–bound targets* can result in ADCC-driven cellular depletion through binding to the FcγRs on NK cells or other immune cells; the cell membrane–bound Ab is also destroyed in this process. Again, changes in binding affinity to FcγRs will significantly impact Ab PK only if the target expression level is high. For example, murine mAbs against OX40L (OX40L, cell-bound targets with low expression level) with reduced FcγR binding affinity have similar clearance when compared with their WT counterparts in mice (Internal data, Genentech, Inc., unpublished data). Gillies et al. reported an inverse correlation between the affinity of the interleukin-2 (IL-2) fusion protein for FcγRI and its half-life in Balb/C mice (43). The target for the IL-2 fusion protein is the IL-2 receptor expressed on cancer cells. The amount of IgG1-IL-2 mutant, which has reduced affinity to mouse FcγRI, remaining in the circulation at all time points, was significantly higher than that in the original IgG1-IL2. Campath-1H is a humanized mAb that reacts with CD52 antigen present on human lymphoid and myeloid cells. The blood concentration for the Fc mutants with a greatly reduced capacity to interact with FcγR was approximately two- to threefold higher than that of Campath-1H WT; this effect is due to both decreased clearance and a decrease in normal tissue uptake. In particular, the mutants showed significantly less spleen, liver, and bone uptake, with an overall trend toward lower normal tissue to blood ratios but with higher tumor uptake (44).

In the case of membrane-bound target antigens that undergo rapid internalization upon mAb binding, differences in binding affinity to FcγR have only a minor impact on mAb PK. For example, keliximab and clenoliximab are monkey/human chimeric antihuman CD4 Abs with similar binding affinities to human CD4. Keliximab shows strong binding to FcγRs in vitro and can deplete CD4$^+$ T cells. Clenoliximab has an approximately 10- to 100-fold lower affinity to FcγRs compared with keliximab and cannot deplete CD4$^+$ T cells (45). Once an anti-CD4 mAb binds to CD4$^+$ T cells, it will either be internalized or the anti-CD4-coated T cell will be cleared through an FcγR-mediated clearance pathway (e.g., phagocytosis). Since the internalization pathway appears much faster than FcγR-mediated clearance pathway and because the internalization pathway is not dependent on the Fcγ affinity, the PK profiles of keliximab and clenoliximab should be similar. Accordingly, these two mAbs showed similar PK profiles in transgenic mice bearing human CD4; however, the PD profiles (CD4$^+$ T cells) were different (45). Binding of the anti-CD4 mAb to cell surface CD4 resulted in rapid internalization of the complex, and this internalization clearance pathway dominated the total clearance of the mAb. As such, although Fcγ does play a substantial role for anti-CD4 mAb PD, the binding affinity to FcγRs has only a minor impact on anti-CD4 mAb PK due to rapid receptor-mediated internalization. It should be noted, however, that receptor saturation may be an additional contributing factor to similarity in PK between keliximab and clenoliximab. In addition, internalization induced by Ab binding also plays a critical role in the PK of another anti-CD4 Ab, TRX1 (see section "In Vitro Binding Correlated to In Vivo Pharmacodynamics").

In addition, the relative contribution of the FcγR-mediated clearance pathway to the total clearance of mAb compared with other factors also has impact on the correlation of FcγR binding affinity to mAb PK. For instance, fucosylation may impact FcγR binding affinity, as the removal of the core fucose from the biantennary complex-type oligosaccharides attached to Fc regions results in dramatically enhanced ADCC of Abs via improved Fcγ binding. Core-fucosylated and nonfucosylated antihuman CD20 IgG1 from Chinese hamster ovary (CHO) cells with similar binding affinity to FcRn demonstrated the same PK in mice, although the nonfucosylated molecule has a threefold increased binding affinity to FcγRs compared with the core-fucosylated one (46). Other relevant factors include the size and, more importantly, the complexity (i.e., number of antigen and Ab molecules in a complex) of the immune complex (47). Multivalent immune complexes tend to be cleared more rapidly than simpler immune complexes because of increased Fc receptor–driven elimination resulting from multivalent Fc interactions (47). In summary, a number of factors must be considered for the effects of FcγR binding affinity on the mAb PK, including antigen location (i.e., membrane vs. soluble), antigen size, antigen expression level, internalization rate, and the relative contribution of the FcγR-mediated clearance pathway to the total clearance of mAb compared with other factors.

FcRn Binding Correlated to Antibody Pharmacokinetics

Recent reports have demonstrated that the FcRn receptor is a prime determinant of the disposition of IgG Abs (48–50). FcRn, which protects IgG from catabolism and contributes to the long plasma half-life of IgG, was first postulated by Brambell in 1964 (51) and cloned in the late 1980s (52,53). FcRn is a heterodimer consisting of a β2m light chain and a major histocompatibility (MHC) class I–like heavy chain. The receptor is widely expressed in cells and tissues. Several studies have shown that IgG clearance in β2m knockout mice (48,49) and FcRn-heavy chain knockout mice (54) is increased 10- to 15-fold, with no changes in the elimination of other Ig. The FcRn receptor binds to IgG in a pH-dependent manner, binding to IgG within acidic endosomes (pH ~ 6.0) and releasing IgG in the plasma (pH ~ 7.4) or interstitial (pH ~ 7.0) compartments. Any unbound IgG proceeds to the lysosome and undergoes proteolysis, therefore, IgG clearance is dependent on its affinity to FcRn receptors. The shorter half-life of the IgG3 isotype was attributed to its lower binding affinity to the FcRn receptor (50,55). Murine mAbs generally have much shorter serum half-lives (few days) than human mAbs (few weeks) due to their low binding affinity to the human FcRn receptor. It is also reported that human FcRn binds human, rabbit, and guinea pig IgG but not rat, mouse, sheep, and bovine IgG; however, mouse FcRn binds IgG from all of these species (56).

Interestingly, human IgG1 has been shown to have an eightfold higher affinity to murine FcRn than that for human FcRn (57–59), indicating a potential limitation in using mice as preclinical models for human IgG1 PK evaluation. In this context, Vaccaro et al. confirmed that an engineered human IgG1 had disparate properties in murine and human systems (60). Engineered IgG with higher affinity to human FcRn receptor had two- to threefold longer half-lives in some instances compared with WT in human FcRn-expressing mice and monkeys (57,61). Hinton et al. found that the half-life of an IgG1 FcRn mutant with increased binding affinity to human FcRn at pH 6.0 was about 2.5-fold longer than that of the WT Ab in monkey (Table 3) (61). However, Dall'Acqua et al. reported that an increase in binding affinity for FcRn at both pH 6.0 and pH 7.4 in mice was paralleled by a decrease in serum concentration of such variants (65). They proposed that a higher affinity to FcRn at pH 7.4 adversely affects release into the serum and offsets the benefit of the enhanced binding at pH 6.0. Furthermore, unlike previously reported PK studies with Abs having the same or similar FcRn mutations, Datta-Mannan et al. did not observe a direct relationship between increased binding affinity to FcRn and improved PK properties for their Abs in the limited number of mice and monkeys tested (62,63). In Table 3, the published PK properties of humanized IgG1 variants with differential binding properties to FcRn are summarized. In general, a number of factors may impact the influence of FcRn binding affinity on mAb PK: differences in the absolute IgG-FcRn affinity at pH 6.0 and pH 7.4, the kinetics of IgG/FcRn interaction at pH 6.0 versus pH 7.4, and the relative contribution of the FcRn-mediated clearance pathway to the total clearance of mAb compared with other factors. Another excellent example of FcRn binding correlated to PK in primates was recently published (64).

Albumin Binding Correlated to Antibody Pharmacokinetics

Albumin is the most abundant (50 mg/mL) plasma protein and has a half-life of about 19 days in humans, similar to a typical IgG1 (67). Albumin binding (AB) can be an effective strategy to improve the PK properties of otherwise short-lived molecules such as small molecules (68), peptides (69), and fragments of IgG (e.g., Fab) (70–72). Dennis et al. demonstrated that the clearance of an Fab fragment can be dramatically decreased through association with albumin (71). They developed one high-affinity AB peptide and fused it to an anti-tissue factor Fab D3H44. The clearance of AB-Fab (AB-Fab D3H44-L) decreased approximately 58-fold, and the half-life increased approximately 40-fold to 32.4 hours compared with 0.8 hours for Fab D3H44 in rabbits (71). Building on Dennis's work, Nguyen et al. found that the PK of an anti–human epidermal growth factor receptor 2 (HER2) AB-Fab 4D5 could be modulated as a function of affinity for albumin (72). There was an inverse correlation between affinity for albumin (ranging from 0.04 to 2.5 μM) and clearance (ranging over ~50-fold in rats and ~20-fold in rabbits) for AB-Fab variants (72). Later, Dennis et al. also showed that AB-Fab 4D5 rapidly targeted tumors, achieved tumor concentrations comparable to that of IgG, and quickly achieved higher tumor to normal tissue ratios compared with IgG (70). They also found that AB-Fab 4D5 did not accumulate in the kidney, suggesting that association with albumin leads to an altered route of clearance and metabolism (70). Overall, the increased half-life of Fab through association with albumin is consistent with the longer half-life of

TABLE 3 Humanized IgG$_1$ Variants with Differential Binding Properties to the Neonatal Crystallizable Fragment Receptor: Relationship to Pharmacokinetics

Molecule	K_d[a] or percentage bound pH 6.0	pH 7.4	PK parameters Half-life	CL	Note
Hinton et al. (61) Target: hepatitis B surface antigen					
WT	1×	NA	336 ± 34 hr	0.190 ± 0.022 mL/hr/kg	PK study in rhesus monkeys
FcRn mutant T250Q/M428L	29×[b]	NA	838 ± 187 hr	0.0811 ± 0.0191 mL/hr/kg	
Petkova et al. (57) Target: HER2					
WT	1×	1×	1.72 ± 0.075 days	NA	PK study in human FcRn transgenic mice
FcRn mutant N434A	1.6×[b]	1×[c]	4.35 ± 0.53 days	NA	
FcRn mutant T307A/ E380A/N434A	2.3×[b]	2.4×[c]	3.85 ± 0.55 days	NA	
Datta-Mannan et al. (62,63) Target: TNFα					
WT	209 ± 11 nM	0.5 ± 3[d]	121 ± 16 hr	0.8 ± 0.2 mL/hr/kg	PK study in cynomolgus monkeys
FcRn mutant P257I/Q311I	2.6 ± 0.5 nM	7 ± 2[d]	114 ± 9 hr	0.9 ± 0.1 mL/hr/kg	
FcRn mutant T250Q/M428L	5.2 ± 0.1 nM	0.7 ± 0.1[d]	112 ± 11 hr	0.9 ± 0.4 mL/hr/kg	
Datta-Mannan et al. (62,63) Target: TNFα					
WT	118 ± 7 nM	6 ± 3[d]	15.3 days	0.7 mL/hr/kg	PK study in CD-1 mice
FcRn mutant T250Q/ M428L	0.23 ± 0.05 nM	14 ± 2[d]	16.7 days	0.3 mL/hr/kg	
Dall'Acqua et al. (65) Target: respiratory syncytial virus					
WT	269 ± 1 nM	No binding	3.2 ± 0.5 days	Serum concentration at 6 hr: WT > G235D/ Q386P/N389S > M252Y/ S254T/T256E	PK study in Balb/ C mice
FcRn mutant G235D/Q386P/N389S	187 ± 10 nM	Binding	2.8 ± 0.4 days		
FcRn mutant M252Y/S254T/T256E	27 ± 6 nM	Binding	3.0 ± 0.5 days		
Dall'Acqua et al. (66) Target: respiratory syncytial virus					
WT	1196 ± 170 nM	No binding	5.7 ± 1.4 days	11.8 mL/day/kg	PK study in cynomolgus monkeys
FcRn mutant	134 ± 5 nM	No binding	21.2 ± 9.1 days	2.45 mL/day/kg	

[a]The assay method can greatly influence measured affinity values; therefore, affinity comparisons should only be made within a given study unless identical assay methods are used.
[b]Compared with WT at pH 6.0.
[c]Compared with WT at pH 7.4.
[d]Percentage bound to FcRn at pH 7.4.
Abbreviations: K_d, dissociation constant; PK, pharmacokinetics; CD, cluster of differentiation; WT, wild type; FcRn, neonatal fragment crystallizable receptor.

albumin. It was proposed that the MHC-related Fc receptor for IgG (FcRn) protects albumin from intracellular catabolic degradation, as it does for IgG, accounting for the uniquely long half-lives of both molecules and explaining their similar concentration-catabolism relationships (73). Both albumin and IgG do indeed bind to FcRn; however, it should be noted that they possess different binding sites so that binding to one does not influence that to the other.

Impact of Charge (pI), Size, and Valency on Binding/Pharmacokinetic Correlations

Valency, size, pI, and affinity for target antigen are all important parameters of a given mAb, affecting the extent and penetration/permeation kinetics of target uptake. Originally, the concept that changes in the pI value of an Ab, realized through charge modification, could improve target distribution was investigated in 1991 by Khawli et al. who used both in vitro and in vivo assays to confirm a preservation of antigen binding following chemical modification of chimeric tumor necrosis treatment or therapy (chTNT)-1 mAb (74–76). This chimeric antibody (chAb; mouse variable, human IgG$_1$/kappa constant regions, Cotara®) targets solid tumors by binding to common histone antigens found in the central necrotic core. Using several experimental approaches including in vitro immunoreactivities, binding affinities, and serum stabilities, the investigators related the effects of chemical and charge modification of chTNT, through conjugation to a small organic molecule (i.e., biotin), on the PK and targeting characteristics with the intent of determining their clinical potentials (76–78). Avidity binding studies demonstrated that biotinylation did not interfere with in vitro binding to fixed Raji cells, and immunoreactivities of both biotin-modified and unmodified chTNT Abs were 65% to 69%, respectively (76). PK studies also revealed that the charge-modified Abs exhibited a faster blood clearance, while in vivo biodistribution studies indicated a preservation of tumor uptake with lower normal organ uptake (75,76). These phenomena were attributed to a reduction in the net positive charge of the Ab, which lessens the influence of electrostatic interactions with negatively charged mammalian cell membranes (79).

Several investigators have similarly demonstrated that modification of the net charge of an Ab also correlates with an altered PK without affecting the in vitro antigen binding. For instance, Lee and Pardridge have related the immunoreactivities of charge-modified anti–epidermal growth factor receptor (EGFR) Abs to their PK profiles (80). The 528 murine mAb recognizing the human EGFR was sequentially cationized with hexamethylenediamine and conjugated with diethylenetriaminepentaacetic acid (DTPA) as a potential radioimmunoconjugate for imaging EGFR-expressing cancer (80). An immunoradiometric assay showed comparable affinities for the modified and native anti-hEGFR 528 mAb; however, significant differences were observed in the PK behaviors of the two formulations (80). In this example, the discrepancy between in vitro and in vivo data is rationalized by the fact that the charge modification influenced the in vivo PK profiles without affecting the in vitro binding affinities. More recently, a similar in vitro–in vivo approach was applied to a series of genetically engineered mAb chTNT-3 derivatives of varying size and valency. In this regard, several variants were developed and evaluated including the scFv, diabody, triabody, Fab, and F(ab')$_2$ (81,82). In general, results from these studies demonstrated a preservation of immunoreactivity and better tumor penetration for these variants; however, lower overall tumor uptakes were observed in vivo because of faster elimination kinetics (81).

In Vitro Binding Correlated to In Vivo Binding/Targeting

Preclinical evaluation of mAbs as candidates for diagnosis and therapy often includes the correlation of in vitro antigen binding to in vivo tissue distribution. In this context, a commonly utilized strategy involves an in vitro cell- or antigen-binding assay followed by assessment of in vivo targeting by molecular imaging. Radioactive (12) and/or fluorescent (83) probes are often employed because of high sensitivity for both in vitro and in vivo measurements. Although high in vitro binding affinity is often a strong early indicator of in vivo targeting ability, data interpretation in such scenarios must be approached with caution because of the influence of stroma or other connective tissues, the presence/absence of endogenous factors, and the influence of antigen shedding. In addition, receptor expression levels tend to fluctuate by varying degrees between in vitro and in vivo models and can affect serum clearance and half-life of mAbs in clinical situations (18,84). A further complication limiting the utility of in vitro models is the absence of several physiological (in vivo) parameters including blood flow, vascular permeability, and interstitial pressure (85–87). In addition, the quantitative correlation of in vitro and in vivo targeting using fluorescent probes is quite challenging because of fluorescence attenuation by tissues, quantum yield modulation by local conditions (e.g., low pH in lysosomes), and other complicating factors. Radioactive probes lack such complications, typically require less overall structural modification than fluorescent probes, and are therefore the preferred method of detection for many applications requiring high sensitivity with maximally preserved immunoreactivity.

Binding/Targeting Correlations

Several studies have demonstrated that in vitro antigen-binding affinities can have an important bearing on in vivo mAb targeting. For example, d-related human leukocyte antigen (HLA-Dr), an antigen expressed on all human B cells, monocytes, and activated T cells, has been explored as a target of the mAb Lym-1. Khawli et al. related the in vitro binding (immunoreactivities via a live cell–binding assay) of this murine monoclonal IgG_{2a} Lym-1 labeled through different methods of radioiodination to their in vivo targeting abilities (via biodistribution and imaging studies in tumor-bearing mice) (74). Furthermore, the same group also related the in vitro immunoreactivities of a chimeric Lym-1 (IgG_1, κ) and other Abs to their in vivo clearance rates (75). A similar approach was applied by Lewis et al. to 1A3, an IgG1 of the κ isotype that binds to an unspecified colorectal adenocarcinoma antigen. They related the immunoreactivities and serum stabilities to the in vivo distributions of 1A3 conjugated to ^{64}Cu-TETA (1,4,8,11-tetraazacyclotetradecane-1,4,8,11-tetraacetic acid) using two different linkers (88). Overall, the in vitro serum stabilities and immunoreactivities of the two different preparations were in good agreement, and the in vivo biodistribution profiles of both radioimmunoconjugates were also nearly identical (88). In general, the results from both Lym-1 and 1A3 demonstrated favorable binding both in vitro and in vivo and supported the concept that the in vitro binding is a valuable predictor of specific Ab targeting.

Binding/Targeting Noncorrelations

Some examples of targets with poor IVIVC have also been revealed, including the failure of in vitro antigen-binding assays to predict in vivo localization of the anti-phosphocholine murine IgM, HPCM2 (hybridoma phosphocholine monoclonal 2) (89). In these studies, Rodwell et al. compared in vitro binding and in vivo targeting of HPCM2 labeled by different methods of attachment (through oligosaccharides, tyrosines, or lysines) using ^{125}I or ^{111}In-DTPA (89). For instance, results from this work showed greater localization efficiency for both indirect ^{125}I-labeled and ^{111}In-DTPA-labeled radioimmunoconjugates (having ^{125}I or ^{111}In attached through residualizing probes) relative to the corresponding conjugate labeled with ^{125}I nonselectively on tyrosines (89). These findings did not agree well with in vitro cell-binding data that did not distinguish between the two methods of radioiodination, demonstrating that in vivo localization to small subcutaneous xenografts is a more stringent test of Ab binding than in vitro cell-binding studies (89). In this case, the chemical method of radionuclide attachment is ultimately responsible for this discrepancy; direct radioiodination of tyrosine residues results in aryl-halogen bonds that are susceptible to in vivo enzymatic cleavage by dehalogenases with subsequent clearance from target sites, whereas radioiodinated oligosaccharides and radiometal-chelate complexes like ^{111}In-DTPA are residualizing, meaning that the residualizing label persists indefinitely within target sites even upon Ab metabolism, typically as an adduct with the amino acid to which it was conjugated (e.g., lysine or cysteine).

In recent work, a lack of correlation between in vitro and in vivo binding was attributed to the high receptor binding affinity of 14C5, a murine IgG1 mAb directed against a yet undefined molecule involved in cell substrate adhesion originally discovered on the surface of cells within malignant breast cancer tissue (90). Burvenich and colleagues studied the in vitro (internalization) and in vivo (biodistribution) targeting properties of radioiodinated 14C5 Ab in non-small-cell lung cancer and colon carcinoma models (90). A significant difference in the level of binding and internalization of 14C5 into cultured A545 human lung carcinoma (high antigen-expressing) and LoVo human colon carcinoma (low antigen-expressing) cells was demonstrated by an in vitro radioimmunoassay and by confocal microscopy (90). Nevertheless, both the high- and low-14C5-expressing tumors showed good in vivo tumor uptake, both having approximately 10% injected dose per gram of tumor tissue at 24 hours postinjection, and this result was attributed to the high binding affinity of 14C5 for its antigen (90). It has been suggested that high-affinity Abs are capable to effectively bind both high- and low-density antigen, whereas a low-affinity Ab only binds appreciably to high-density antigen because of a requirement for the avidity conferred by divalent binding for effective attachment (91).

Another lack of IVIVC was observed upon evaluation of three IgG mAbs prepared by standard hydridoma methods using osteosarcoma cells resected from an untreated patient (92). Sakahara et al. explored the relationship between the levels of in vitro binding and in vivo tumor accumulation of these radiolabeled anti-osteosarcoma mAbs, concluding a lack of IVIVC (92). Discrepancies between cell binding and in vivo tumor uptake were observed and were related to very different blood clearance rates (92). These results suggest that, while binding studies may be used to exclude Abs having poor in vitro binding, in vivo serum clearance may be a better test for choosing Abs with similar binding.

In summary, a number of factors can lead to poor correlation between in vitro binding and in vivo targeting. The selected examples above described noncorrelations resulting from (*i*) the chemical nature and metabolic stability of the probe, (*ii*) the interplay between binding affinity and antigen density, and (*iii*) limited exposure due to rapid blood clearance rate. In addition, in all of these cases, the noncorrelation can be attributed to inherent differences between in vitro and in vivo models. For instance, (*i*) dehalogenation may be more pronounced in vivo because of higher enzyme levels and greater opportunity for subsequent clearance, (*ii*) different thresholds for avidity may exist between in vitro and in vivo models because of variations in antigen density, and (*iii*) the availability for antigen binding can be PK limited because of unfavorably rapid clearance.

In Vitro Binding Correlated to In Vivo Pharmacodynamics

In general, efficacy can be defined as the relationship between antigen occupancy and the ability to initiate a pharmacological (i.e., PD) response at the molecular, cellular, tissue, or system level. On the other hand, potency is a measure of the concentration of a drug at which it is effective. Bridging these two concepts is PD, the study of the biochemical and physiological drug effects, the mechanisms of drug action, and the relationship between drug concentration and effect.

Correlation of in vivo PD exclusively to in vitro binding is a somewhat rare scenario, as most studies tend to also examine the in vitro potency. One of the few examples is TRX1, a nondepleting humanized antihuman CD4 monoclonal IgG1 Ab being developed to induce tolerance by blocking CD4-mediated functions (93). Ng et al. related TRX1-induced internalization of CD4 and subcellular localization of T cell internalized TRX1 Ab in human T cells to the in vivo downmodulation of CD4 and clearance of TRX1 (93). TRX1 displayed nonlinear PK behavior, and TRX1 treatment induced saturation and downmodulation of CD4 receptors on T cells (93). Results from in vitro studies using purified human T cells suggested that CD4-mediated internalization may constitute one pathway by which TRX1 is cleared in vivo. The observed in vivo PD effect, downmodulation of CD4, was also explained by the in vitro studies that indicated binding and subsequent internalization of TRX1 into purified human T cells accompanied by TRX1-induced internalization of CD4 (93).

The anti-general control nondepressible (GCN)4 scFv fragment directed against the Gcn4p dimerization domain represents another example in which in vivo PD is correlated to an in vitro phenomenon other than potency (94). Wörn et al. investigated the interplay between in vitro stabilities, binding affinities, and potencies of a series of these cytoplasmically expressed scFv intracellular Abs (i.e., intrabodies) (94). The in vivo performance (i.e., cytoplasmic inhibition) of the Ab fragments, measured as decreasing β-galactosidase reporter gene activity, was related to their in vitro stability, measured by denaturant-induced equilibrium unfolding (94).

IN VITRO POTENCY CORRELATED TO IN VIVO PHARMACODYNAMICS

The correlation of in vitro potency and in vivo pharmacological (i.e., PD) response of Abs is one of the most widely pursued studies of its type for a number of reasons. Among the pharmacological properties we have examined, the observed potency at a suitable dose level is perhaps most closely associated with clinical efficacy, especially when appropriate in vivo models are chosen. However, one must recognize that the adjustment of dose levels can effectively overcome differences in potency to achieve similar efficacy. Many of the limitations previously mentioned for binding/targeting IVIVCs also apply to this situation; these include the influence of stroma or other connective tissues, the presence/absence of endogenous factors, and the influence of antigen shedding. Furthermore, variations in receptor expression levels, vascular permeability, interstitial pressure, and other parameters can also play a significant role. In general, efficacy related to cytotoxic effects or other pharmacological responses is a more complicated phenomenon than binding/targeting, typically requiring the successful interplay of a greater number of complex processes. As such, one must exercise caution in drawing conclusions from what may appear to be a relationship between in vitro and in vivo potency or efficacy.

In the present section, we have chosen to highlight Abs that recognize IgE, CD11a, CD20, CD30, CD40, HER2, fibroblast growth factor (FGF)19, tumor necrosis factor (TNF), and vascular endothelial growth factor (VEGF) to demonstrate the various methods used to develop IVIVC for therapeutic Abs in preclinical development.

IgE as a Target to Treat Allergic Asthma

Anti-IgE mAbs provide immunological examples wherein in vitro data have been correlated to in vivo PK and PD, as a measure of in vivo potency. Omalizumab (Xolair®, rhuMAb-E25) is a humanized IgG1 anti-IgE mAb used in the treatment of moderate to severe allergic asthma. This Ab suppresses serum-free IgE levels by selectively binding to IgE, forming small, biologically inert immune complexes (95). By blocking the binding of IgE to high-affinity FcεRI, the Ab inhibits cross-linking of IgE bound to FcεRI, thereby inhibiting the release of histamine and other proinflammatory mediators of the allergic response (96). Fox et al. related the in vitro formation of immune complexes and the stability of omalizumab to PK and immune complex formation in cynomolgus monkeys (97). More recently, Shulman described a similar strategy using the same Ab for the treatment of allergic respiratory disorders; in vitro data on IgE-mediated inflammatory processes were related to clinical PK and PD, in particular suppression of free IgE (98). A correlation between serum-free IgE levels and clinical outcomes was used to optimize the dosing strategy of omalizumab (95).

High affinity anti-IgE 1 (HAE 1) is a second-generation, high-affinity anti-IgE mAb having the same IgG1 framework as omalizumab but nine amino acid differences in the CDR. Binding affinity measurements using Biacore (based on surface plasmon resonance) indicated that the apparent K_d values of the HAE1-Fab and the omalizumab Fab were 0.66 nM and 15.5 nM, respectively, which reflects an approximately 23-fold higher affinity of the HAE-Fab for IgE (99). Interestingly, this difference was mainly due to a much slower dissociation rate of the HAE1-Fab from the IgE, whereas the association rates of the two molecules with IgE were very similar. An in vitro potency assay demonstrated that the higher affinity of HAE1 for IgE increased its ability to inhibit binding of human IgE to the FcεRI receptor by approximately 50-fold (Fig. 2, upper panel).

The increased potency of HAE1 offered two potential benefits relative to current omalizumab therapy: (*i*) treatment of patients with higher baseline IgE

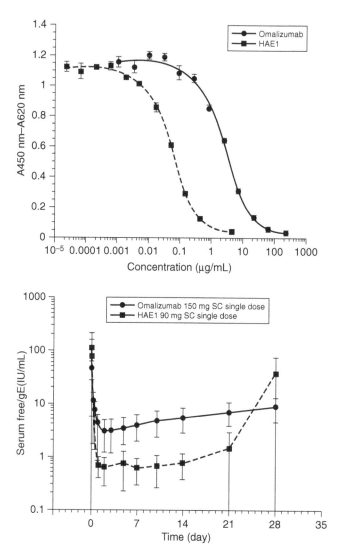

FIGURE 2   Top panel: Comparison of inhibition of binding of human IgE to the high-affinity FcεRI receptor by HAE1 and omalizumab. The absorbance at A450 nm–A620 nm was a measure of biotin IgE binding to FcεRI. The IC50 values for inhibition of biotin IgE binding to FcεRI were 2.69 and 0.053 μg/mL for omalizumab and HAE1, respectively. Bottom panel: Comparison of suppression of free IgE in humans after administration of a single subcutaneous dose of HAE1 at 90 mg or omalizumab at 150 mg. *Abbreviations*: Ig, immunoglobulin; FcεR, fragment crystallizable ε receptor. *Source*: Figure prepared by Wendy Putnam (Department of Pharmacokinetic and Pharmacodynamic Sciences, Genentech, Inc.) and Don Sinclair (Clinical PK/PD-Biotherapeutics Department, Genentech, Inc.).

levels (>700 IU/mL) and (*ii*) improved dosing convenience. Although many asthma patients treated with omalizumab require multiple, biweekly subcutaneous injections, a higher-affinity molecule might require a lower dose and/or less frequent dosing regimen. As Figure 2 lower panel shows, a 90-mg dose of HAE1 achieved greater suppression of free IgE in humans than omalizumab at a 150-mg dose. This data must be interpreted with caution given that the free IgE levels were measured using different assays, and the two anti-IgE molecules were evaluated in different studies. However, the results strongly suggest that the higher affinity and the increased in vitro potency of HAE1 did translate into greater in vivo potency.

CD11a as a Target for Psoriasis Treatment

As discussed earlier in regard to target antigen binding correlated to Ab PK, CD11a is a heavily pursued target for psoriasis treatment. Surrogate Abs are a potential solution to the limited safety testing possible with humanized mAbs with restricted species cross-reactivity. A chimeric mouse/rat anti-mouse CD11a mAb (muM17) was studied in the context of evaluating both in vitro and in vivo as a potential surrogate for efalizumab, a humanized anti-CD11a Ab in development for psoriasis (100). CD11a is a subunit of lymphocyte function–associated antigen-1, an integrin involved in cell-cell interactions important to immune responses and inflammation (100). Clarke et al. related in vitro binding, lymphocyte response, and tissue cross-reactivity of the anti-CD11a Ab muM17 to its in vivo toxicology (100). Results from in vitro and in vivo pharmacology studies showed similar pharmacological and toxicological activities, thereby justifying the use of muM17 as a surrogate for efalizumab (100).

CD20 as a Target for Rheumatoid Arthritis Treatment

The FDA-approved anti-CD20 human-mouse chimeric mAb rituximab (Rituxan®) is effective against a wide variety of disorders from non-Hodgkin lymphoma (NHL) to rheumatoid arthritis. Rituximab recognizes the CD20 antigen on the surface of peripheral mature B cells and is able to induce ADCC, CDC, and apoptosis of the cells (101). Using a combination therapy approach, Daniel et al. correlated in vitro cell viability to in vivo tumor xenograft growth after treatment with rituximab with or without the Apo2 ligand/TNF-related apoptosis-inducing ligand (rhApoL/TRAIL) (102). Proapoptotic activity of rhApo2L/TRAIL was observed for three of the seven NHL cell lines tested in vitro, and rituximab augmented rhApo2L/TRAIL-induced caspase activation in some cases (102). These results correlated well with in vivo studies, demonstrating that rhApo2L/TRAIL and rituximab cooperated to attenuate or reverse growth of tumor xenografts of the corresponding cell lines (102). In another study, Zhang et al. correlated the in vitro induction of apoptosis and caspase-3 activity caused by rituximab, and by its new hyper-cross-linked polymer formulation, to the regression of CD20$^+$ lymphoma xenografts in vivo (103). The results confirmed that the polymer formulation of this therapeutic Ab more effectively suppressed Raji lymphoma growth in vivo relative to the native Ab in accordance with the extent of both surface CD20 clustering and caspase-3 activation, emphasizing the idea that induction of apoptosis by a hyper-cross-linked polymer can be a powerful mechanism for treating lymphomas. In support of this concept, clinical studies by Byrd et al. observed caspase-3 cleavage in lymphoma cells of patients treated with rituximab, suggesting the induction of apoptosis after treatment (104).

CD30 as a Target for Lymphoma Treatment

Another Ab targeting hematological tumor burden is the anti-CD30 mAb 5F11, a promising agent for Ab-based immunotherapy of Hodgkin lymphoma (HL) and anaplastic large-cell lymphoma. This fully humanized mAb showed specific binding to CD30 (cluster A) (105). Borchmann et al. have related in vitro ADCC and growth inhibition by 5F11 to the corresponding in vivo tumor growth inhibition (105). The in vitro ADCC assays indicated dose-dependent lysis of L540 cells only when the Ab was combined with human effector cells, and upon cross-linking, growth inhibition of CD30-expressing cell lines was observed (105). The overall trend of cell growth inhibition correlated well with in vivo studies in which treatment with 5F11 induced a marked growth delay or even a complete regression of established tumor xenografts in mice (105). The success of the in vitro lysis assay was governed by the addition of human effector cells, suggesting that this mechanism of action might also be active in vivo.

CD40 as a Target in the Treatment of B-Cell Malignancies

Cell surface expression of CD40 in B-cell malignancies and multiple solid tumors has raised interest in its potential use as a target for Ab-based cancer therapy. SGN-40, a humanized anti-CD40 mAb, mediates ADCC and inhibits B-cell tumor growth in vitro, and is currently in phase I/II clinical trials for B-cell malignancies. Kelley et al. correlated inhibition of B-cell tumor growth in vitro to in vivo activity and PK properties of SGN-40 in both rodents and cynomolgus monkeys (106). The Ab's ability to mediate ADCC, to induce apoptosis, and to inhibit cell growth in a wide variety of B cell–driven cancer cell lines in vitro suggests that SGN-40's in vivo antitumor mechanism may occur via ADCC and/ or direct cytotoxicity to tumor cells.

HER2 as a Target for Breast Cancer Treatment

The effects of combination therapy can be more carefully examined using a combination of both in vitro and in vivo methods. An example is trastuzumab (Herceptin®), a humanized Ab that targets HER2/neuroblastoma-associated gene (*neu*) and has been approved by the FDA as a treatment for breast cancer (107). Mann et al. related in vitro cell proliferation to in vivo tumor growth in testing the potency of trastuzumab with or without the selective cyclooxygenase 2 (COX-2) inhibitor celecoxib (108). Both in vitro and in vivo assays were used to determine the effects of the selective COX-2 inhibitor, celecoxib, and/or an anti-HER-2/neu mAb (either Herceptin or 2C4) on cell growth (108). In a separate study, Agus et al. demonstrated that the in vitro and in vivo growth of several breast and prostate tumor models is inhibited by treatment with 2C4 (pertuzumab), an Ab that targets ErbB2 in a different manner with respect to trastuzumab (109).

FGF19 as a Target for Colorectal Cancer Treatment

FGF19 normally regulates bile acid homeostasis and gall bladder filling (110); however, therapeutic inactivation of FGF19 could be beneficial for the treatment of colon and liver cancer. FGF19 and its cognate receptor FGFR4 were found to be coexpressed in primary human liver, lung, and colon tumors (111). Desnoyers et al. related in vitro binding, phosphorylation, and cell migration of an anti-FGF19 Ab to inhibition of colon cancer xenografts in vivo (111). This Ab abolished FGF19-mediated activity in vitro, inhibited growth of colon tumor xenografts in vivo, and effectively prevented hepatocellular carcinomas in FGF19 transgenic mice (111).

Tumor Necrosis Factor: a Potency/Efficacy Noncorrelation

Anti-TNF therapy of inflammatory (112), oncological (113), and autoimmune (114) diseases illustrates a complex scenario in which in vitro and in vivo data sometimes seem to contradict one another. For example, in a recent study by Egberts et al., although inhibition of TNF-α with infliximab or etanercept only marginally affected proliferation and invasiveness of pancreatic ductal adenocarcinoma cells in vitro, both reagents exerted strong antitumoral effects in vivo (115). Noncorrelation was also observed by Teng et al., who demonstrated that TNF-producing and TNF-nonproducing tumor cells grow at similar rates in vitro but form different-sized tumors in vivo (116). The authors proposed several explanations for this lack of correlation, including the possibility that tumor cells when grown in vivo, rather than in vitro, were directly susceptible to the secreted TNF or that secondary mediators induced by TNF had an effect on the tumor cells (116). Alternatively, the mechanism of inhibition could be indirect and related to TNF effects on the blood vessels or other nonmalignant stromal components, including inflammatory cells that may be activated by TNF to destroy the tumor (116). Earlier work by Bromberg et al. addressed the discrepancy between the in vivo and in vitro activities of anti-TNF antiserum by proposing the following three possibilities: (*i*) TNF is important for in vivo cellular communication, but different arrays of cells and cytokines juxtaposed in culture are able to activate cells in the relative absence of TNF, (*ii*) low-level TNF–TNF receptor interactions may be of higher affinity than the Ab and override the inhibitory effect, and/or (*iii*) TNF is most important during priming events, therefore anti-TNF will be most effective during the afferent, priming events in immunity (which are strongly inhibited in vivo) that encompasses initial, low-affinity interactions with Ag; conversely, most in vitro assays represent secondary or high-affinity primary responses that seem to produce either relatively low quantities of TNF and/or other cytokines that can activate T cells in the absence of TNF, including IL-1α and IL-6 (117). In contrast to all of the above noncorrelative studies, Waterston et al. employed the B16F10 murine melanoma model to investigate the role of TNF in promoting metastasis (118). TNF autovaccination was able to generate high-titer anti-TNF autoantibodies whose in vitro inhibition of TNF function correlated well with in vivo inhibition of metastasis (118).

VEGF and Other Targets for Antiangiogenic Treatment

An important player in tumor angiogenesis, VEGF-A is targeted by a humanized anti-VEGF-A mAb (bevacizumab, Avastin®) (119), which has been approved by the FDA as a treatment for metastatic colorectal (120), metastatic breast (121), and nonsquamous, non-small-cell lung cancer in combination with chemotherapy (122). Discrepancies between in vitro and in vivo data among various affinity-matured anti-VEGF-A Abs have revealed important factors contributing to this apparent lack of correlation (123). Contributions of both tumor- and stromal cell–derived VEGF-A to vascularization of human tumors grown in immunodeficient mice hinder the direct comparison between the pharmacological effects of anti-VEGF Abs having different binding affinities (123). Even though in vitro studies demonstrated a correlation between Ab binding affinity and potency in assays measuring inhibition of VEGF-stimulated endothelial cell proliferation, various affinity-matured anti-VEGF-A Abs did not show a clearly increased potency and efficacy to block tumor growth in most in vivo models when compared with their lower-affinity counterparts (123). Several factors may help explain the failure of potency to increase with anti-VEGF-A binding affinity. For instance, in contrast to Abs that directly target antigens in the tumor cells, anti-VEGF-A Abs may not need to penetrate the tumor mass to induce pharmacological effects because they interfere with angiogenesis primarily by preventing VEGF-A from binding its receptors within tumor vasculature; this opens the possibility that inhibition of VEGF-A binding to VEGF receptors may be saturated in vivo even when lower-affinity anti-VEGF-A Abs are used (123).

In addition to oncological applications, VEGF-A is also an exciting target in ocular disorders, where IVIVCs play critical roles in the study of its mechanism of action. Neovascular age-related macular degeneration (AMD) is the leading cause of blindness in older adults in the Western world, and ranibizumab (Lucentis®, anti-VEGF-A), a humanized Ab fragment directed against VEGF-A, is FDA approved for the treatment of neovascular AMD. PK studies of ranibizumab (rhuFabV2) revealed a favorable three-day terminal half-life following intravitreal administration to monkeys (124). In subsequent studies, Lowe et al. related the in vitro binding and HUVEC proliferation effects of ranibizumab to its effects on VEGF-A vascular permeability (125).

IN VITRO METABOLISM CORRELATED
TO IN VIVO METABOLISM

The study of Ab metabolism is not a new endeavor, but much has yet to be learned. Because of the complex nature of in vivo models, in vitro systems offer many attractive advantages in the study of Ab metabolism; however, care must be taken to acknowledge certain missing components in these highly artificial systems. In the context of anti–carcinoembryonic antigen (CEA) Abs, Sands and Jones have reviewed the methods for elucidating the mechanisms by which radiolabeled Abs accumulate and undergo metabolism in the tumor and the liver (126). These techniques include the use of isolated perfused rat liver to determine the extent of hepatic protein metabolism in the absence of other organs, although isolated hepatocytes or Kupfer cells may also be employed. Another technique, reticuloendothelial system (RES) blockade using dextran sulfate, demonstrated no effect on hepatic uptake of radiolabeled Ab, suggesting that a true receptor binding event to hepatocytes is involved (126). Dual-labeled Abs can answer important questions regarding metabolism, especially when site-specific labeling of different sites can be achieved. Micropore chambers can be implanted subcutaneously at the site of tumor inoculation to allow sampling of tumor interstitial fluid. Finally, in vitro tissue culture studies offer a simplified, albeit artificial, opportunity to study Ab metabolism (126).

As previously mentioned, radiometals may be chemically attached to Abs via bifunctional chelating agents such as DTPA; this serves not only as a handle to detect Ab uptake in tissues by γ ray counting but also as a means for non-invasive scintigraphic imaging (12). Rogers et al. have identified ^{111}In-DTPA-ε-lysine as the major metabolite of ^{111}In-DTPA-labeled anti-colorectal carcinoma monoclonal IgG 1A3 and a corresponding F(ab')$_2$ in vivo, consistent with previous in vitro studies (127). Intracellular metabolism was studied through the use of ^{111}In-labeled anti-colorectal carcinoma Ab 1A3 that localize to the lysosomes by receptor-mediated endocytosis where they are catabolized, especially in nontarget organs such as the kidneys and liver (127). These studies demonstrated that the main lysosomal metabolite is ^{111}In-chelate-ε-lysine, both in vitro and in vivo (127).

To determine the metabolic profile of another widely used metal chelate, analyses were performed on kidney homogenates from nude mice injected with an ^{111}In-DOTA-Fab (DOTA, 1,4,7,10-tetraazacyclododecane-1,4,7,10-tetraacetic acid) generated enzymatically from the anti-lymphoma intact anti-CD20 Ab Rituxan. Similar to the previously reported in vitro and in vivo data from Rogers et al. (127), Tsai and colleagues reported a lysine metabolite adduct for an ^{111}In-DOTA-conjugated Rituxan Fab (128). The major kidney metabolite was chromatographically identified as ^{111}In-DOTA-amino-lysine by comparison with an authentic synthetic standard, and this end product was also identified in the urine, along with relatively small amounts of ^{111}In-DOTA-Fab (128). Because injection of ^{111}In-DOTA-amino-lysine into nude mice resulted in rapid clearance into the urine without kidney retention, it is likely that the renal retention observed was due to kidney uptake of ^{111}In-DOTA-Fab, followed by lysosomal degradation to ^{111}In-DOTA-amino-lysine, which is only slowly cleared from this compartment (128). This observation is supported by auto-radiographs of the kidney showing rapid localization of radioactivity into the distal regions of the renal cortex (128). To extend this analysis to clinical trials, analysis was performed on urine taken from a patient injected with the intact Ab ^{111}In-DOTA-cT84.66, and the major radioactive species observed was also ^{111}In-DOTA-amino-lysine (128).

SPECIAL TOPIC: ANTIBODY-DRUG CONJUGATES
Overview of Antibody-Drug Conjugates
The ADC concept was envisioned as early as when Ehrlich's magic bullet idea was conceived (129). The target specificity of Abs prompted the idea of using them as vehicles to deliver therapeutic agents at sites of action (130). Because of the toxicity of these therapeutic agents, most ADC development focuses on oncological indications, but this approach could be potentially used for a wide variety of indications including infectious diseases (131). The conceptual composition of an ADC is rather simple: chemically combining a nonselective but effective cytotoxic agent with a highly selective tumor-specific Ab. An ideal ADC would remain inactive in systemic circulation, with its activity being restored upon structural modification at intended delivery sites (132). In vivo, an ADC circulates through the blood in its intact form; upon antigen binding, it internalizes via receptor-mediated endocytosis, and the cytotoxic moiety is released from the lysosome into the cell to restore its activity (133).

There are many potential advantages of ADCs over naked Abs. First of all, the utilization of ADCs greatly expands the list of potential therapeutic targets. Naked Abs work by inducing immune responses, blocking overexpressed growth factor receptors on tumor cells, activating cellular death, binding soluble cytokines/growth factors, or inhibiting angiogenesis; many of these processes require involvement of the targeted antigen in neoplastic process (134). ADC targets, on the other hand, can be expanded into tumor targets with no known function or antitumor activity. Second, the powerful cargo carried by ADCs may potentiate the Ab's activity and give a second chance for well-validated targets. For example, an ADC version of trastuzumab has demonstrated significant activity in a trastuzumab-pretreated population of patients with HER2-positive metastatic breast cancer in a phase I clinical trial (135). Another notable feature of ADCs is that the chemotherapy agents delivered through Abs may bypass drug resistance efflux transporters such as P-glycoprotein (136,137). This may circumvent the tolerance and resistance that often plague the traditional chemotherapy.

Although several ADCs are currently in development, only one cytotoxic drug-linked ADC has been approved for clinical use (Mylotarg®) (Table 1) (138). The simple concept of "Ab-linker-drug" is significantly complicated by the need to optimize three different moieties, often with opposing intentions, and to synchronize them to generate the most desirable biological effects. The Ab itself must be selective to reduce nonspecific toxicity. Upon conjugation, it is crucial that an ADC should preserve its Ab characteristics without impairing its affinity, imparting immunogenicity or altering its PK profile. The possibility of internalization of antigen upon Ab binding is important to bear in mind. The choice of cytotoxic drug and the conjugation chemistry are of great importance. The conjugate should be systemically nontoxic, meaning that the linker must be stable in circulation yet efficiently cleaved inside cancer cells. In vivo efficacy is also largely dependent on the cancer type, target expression, metabolism/catabolism, and PK profiles.

The complexity of chemical and biological variables in ADC development has provided a fertile ground for exploring IVIVCs. The following examples from the literature will serve to highlight specific in vitro/in vivo relationships in the context of ADCs, including not only examples that follow intuitive correlations but also those in which in vivo observations seem to contradict predictions derived from in vitro data.

In Vitro Binding Affinity/Specificity, Internalization, and Expression Levels Related to In Vivo Efficacy

While Ab selection has a definitive impact on efficacy, high affinity does not always translate into superior efficacy. For example, mucin 16 (MUC16) is a cell surface marker for epithelial ovarian adenocarcinoma (139), and ADCs recognizing this antigen have been evaluated for ovarian cancer therapy. Chen and colleagues compared two Abs with monovalent (11D10 binds a unique epitope on the target antigen) or multivalent (3A5 binds to multiple epitopes per target antigen) binding to the extracellular domain (ECD) of MUC16, and the efficacies of their respective conjugates to auristatin were assessed (140). The higher-affinity unique epitope-binding ADC (11D10) actually bound to fewer sites per cell than 3A5 by flow cytometric analysis (Fig. 3A) and was a less potent inhibitor of in vitro ovarian carcinoma (OVCAR)-3 cell growth than its multiple epitope-binding counterpart (3A5) (Fig. 3B). Better in vivo efficacy was also observed for 3A5 in vivo in the OVCAR-3 intraperitoneal xenograft tumor model (Fig. 3C). In this case, increased ADC binding per cell may have resulted in an increased overall delivery of drug into the cells both in vitro and in vivo. Importantly, multivalent Ab-antigen interactions seem to have ultimately overcome any advantage of increased binding affinity.

In Vitro Potency Related to In Vivo Efficacy

In vitro cytotoxicity assays are powerful tools for initial proof of concept and candidate ranking/selection. IC50 values (i.e., concentrations required for 50% growth inhibition) are routinely used as an indicator of potency. However, because of the intrinsic simplicity of the system, in vitro assays do not always predict the magnitude of in vivo responses; in many cases, they lead to underestimation of the in vivo efficacy for ADCs. In studies developing prostate stem cell antigen (PSCA) ADC for prostate cancer therapy, Ross and colleagues compared in vitro cytotoxicity of maytansinoid-conjugated Abs with their in vivo tumor growth inhibition (141). In vitro cytotoxicity assays in cell lines expressing different levels of PSCA showed that the IC50 values increased as the expression level decreased. In some cases, ADCs showed minimal activity in low-expressing lines. However, significant activity was observed in a xenograft model derived from the same cell lines. Both the duration of drug exposure to the tumor and the dose level of the ADC may be responsible for this phenomenon. Similar findings were reported in the colorectal cancer target EphB2, wherein Mao and coworkers related the in vitro binding, internalization, and cytotoxicity to in vivo tumor suppression (142). In the case of EphB2, a 100-fold difference in potency in vitro led to a <5-fold difference in vivo. In these cases, in vitro efficacy is more directly related to expression level, reflecting that the number of binding sites per cell is likely a major determinant in the corresponding cell-killing response. In contrast, in vivo efficacy of ADCs is complicated by multiple other factors including differences in PK, drug exposure, and metabolic pathways. As such, the receptor copy number or expression level is not always the limiting factor for in vivo efficacy.

Evaluation of CD79 as an NHL target provides another interesting example of an IVIVC derived from an ADC. CD79 is a covalent heterodimer consisting of two subunits designated CD79a and CD79b. Both anti-CD79a and anti-CD79b ADCs have been explored using an in vitro cell-killing assay and an in vivo xenograft model (143). The relative efficacies of the ADCs did not correlate with the affinities of the Abs. An anti-CD79a Ab having higher affinity translated to inferior in vivo efficacy relative to CD79b. Anti-CD79b was internalized and specifically targeted to the lysosome-like MHC class II–enriched compartment (i.e., MIIC) where active metabolites were released (143), suggesting that the superiority of anti-CD79b may be related to intracellular drug release mechanisms. The determinant factor for in vivo efficacy in this example is effective ADC metabolism and release of drug inside the cells instead of binding affinity.

Drug-Antibody Ratio and Choice of Linker Related to Efficacy

Two major aspects often lost in translation from in vitro to in vivo are exposure (PK) and metabolism. A compelling example of this scenario involves the impact of drug to antibody ratio (DAR) or drug load on efficacy and toxicity. A higher DAR has been shown to drastically increase in vitro cell-killing capacity. For

(A)

Antibody	Affinity (pmol/L)	Sites per cell
11D10	52	31,700
3A5		
Measurement #1	433	243,400
Measurement #2	288	107,800

(B)

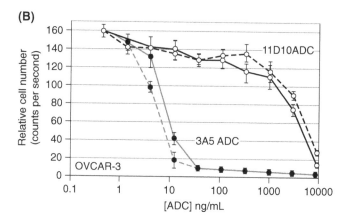

(C)

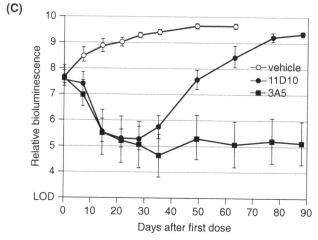

FIGURE 3 The repeat epitope-binding 3A5 ADC is more efficacious than the single epitope-binding 11D10 ADC. (**A**) Scatchard analysis of anti-MUC16 binding to OVCAR-3 cells. (**B**) Inhibition of in vitro cell proliferation by auristatin conjugates of human Fc chimeric 3A5 and 11D10. Conjugates were applied continuously to OVCAR-3 cells for five days. Concentrations are 9 µg/mL, and serial threefold dilutions are down to 1.4 ng/mL. Viable cell numbers were measured using an ATP-dependent luminescent assay and are expressed as counts per second. Data from untreated cells are plotted at 0.45 ng/mL. 3A5-MC-MMAF (*filled diamonds and dashed lines*), 3A5-VC-MMAE (*filled squares and solid lines*), 11D10-MC-MMAF (*open diamonds and dashed lines*), and 11D10-VC-MMAE (*open squares and solid lines*). (**C**) In vivo efficacy of armed anti-MUC16 antibodies (Ab-VC-MMAE conjugates) against OVCAR-3/luc mouse xenografts.

example, CD30 ADCs bearing increasing copies of drug (2, 4, and 8 drugs per Ab) showed a corresponding trend toward lower IC50s. However, the benefit was not observed in a xenograft model, wherein their in vivo antitumor activities were largely comparable. In fact, the therapeutic index actually increased with decreased DAR because of the better tolerability (144). In addition, ADCs with a higher DAR were shown to exhibit more rapid clearance compared with one with a lower drug load, minimizing the exposure at tumor sites (144). One must recognize, however, that the rapid metabolism of an ADC with higher DAR may lead to higher levels of active metabolites in systemic circulation and, consequently, greater toxicity. Given the potency of these cytotoxic agents, once a minimal efficacious threshold is reached, any further increase in drug load may offer very little, if any, benefit. Another possibility is that overmodification of an Ab with lipophilic drugs shifts the tissue distribution toward increased nonspecific uptake (e.g., liver, spleen) and increases normal organ toxicity. The nature of the target must also be considered, including the level of expression, the degree of internalization upon binding, and drug sensitivity.

Optimizing in vitro characteristics in the absence of in vivo context such as PK and metabolism may not yield the benefit one would expect. Another example is the impact of linker stability on efficacy. ADC linker technology has evolved into a vast array of varieties, including labile linkers, such as hydrazone and disulfide linkers, and peptidic linkers that are stable in serum but can be readily degraded in intracellular compartments by specific enzymes including lysosomal proteases (136,145). The mechanism of action for ADCs requires cleavage of the linker; this might possibly occur through enzymatic cleavage or chemical degradation of the linker in plasma, or through catabolism of the Ab within the lysosome or other intracellular organelles (146). In theory, more stable linkers allow for maximal intratumoral drug delivery and more efficient drug release at the target site. However, this must be evaluated in the context of many other factors. In a recent study on peptidic amide linkers for CD70 ADC, a more stable bromoacetamidecaproyl (*bac*) linker was compared with maleimidocaproyl (*mc*) linker. The *bac* linkers showed increased plasma stability and a higher intratumoral drug exposure relative to the *mc* linker, yet they shared similar in vivo activity profiles (147). In this case, further extension of linker half-life rendered limited benefit. Once again, it highlights the importance of identifying rate-limiting steps in translating from in vitro to in vivo.

Animals were dosed once weekly for four total doses after tumors were established and bioluminescence values were stabilized: dosing on days 0, 7, 14, and 21. Both ADCs were dosed at 56 µg/m^2 MMAE equivalent (2.8 mg/kg 3A5 ADC, 2.6 mg/kg 11D10 ADC). Although the 11D10 ADC had an effect, it was transient, while the 3A5 activity was sustained. At day 250, 70% of the 3A5 ADC-treated mice were alive, while all of the 11D10 ADC–treated mice were euthanized before day 200. Note that the vehicle data is only plotted through day 63 because there were too few surviving animals beyond that time. Data are plotted on a log$_{10}$ scale; the error bars are the standard error of the mean (i.e., *SEM*). Tumor burden is plotted as relative bioluminescence. *LOD*, limit of detection [1000 relative luciferase units (*RLU*)]. *Abbreviations*: ADC, antibody-drug conjugate; MUC16, mucin 16; MMAF, monomethyl auristatin F; MMAE, monomethyl auristatin E. *Source*: From Refs. 132 and 140.

In Vitro Plasma Stability Related to In Vivo Pharmacokinetics and Metabolism

Predicting in vivo PK and metabolism is challenging because of a lack of sufficient analytical means to track complex and often heterogeneous ADC molecules; the effort is further hampered by an incomplete understanding of ADC metabolism. Most factors that impact ADC PK are not amenable for in vitro studies; among these are solid tumor burden, tissue distribution, and shed antigen. In contrast, the systemic stability, as one potential clearance pathway, can be assessed easily through the use of in vitro plasma stability studies. High plasma stability encourages the maximum total accumulation at tumor sites and minimizes systemic toxicity due to free drug release. It also provides an initial assessment of linker stability and overall metabolic stability. A good example is the use of plasma stability to aid in the selection of appropriate ADC linkers. In an effort to develop potent mAb-auristatin conjugates for cancer therapy, Doronina et al. compared the in vitro plasma stabilities between a conventional peptide (i.e., amide) linker and a less stable acid-labile hydrazone linker (148). These results were related to in vivo efficacy studies that showed better in vivo efficacy in a tumor model for the more stable linker relative to the acid-labile linker derivatives (148). These results emphasize that in vitro plasma stability may be used as a predictive tool for candidate ranking and selection.

Antibody-Drug Conjugate Metabolism Related to Efficacy

Understanding ADC metabolism is an important requirement in selecting appropriate efficacy and toxicity models and in addressing species comparability. The information gained from in vitro metabolic stability studies will ultimately help guide clinical monitoring. The metabolism of ADCs has been shown to mimic that of radioimmunoconjugates (*vide supra*) in that the major metabolite tends to be an adduct of the drug still linked to the amino acid to which it was originally conjugated. For instance, Doronina et al. demonstrated both in vitro cytotoxicity and in vivo therapeutic efficacy for monomethyl auristatin F (MMAF)-conjugated anti-CD30 ADCs, and the major metabolite was identified by high-performance liquid chromatography (HPLC) and mass spectrometry as the cysteine-linked adduct of MMAF (136). This is consistent with the idea that ADC metabolism occurs by the same lysosomal degradation responsible for catabolism of naked Abs into their substituent amino acids. Further supporting the idea of lysosomal metabolism, Sutherland et al. have demonstrated that an auristatin-conjugated ADC based on a chimeric anti-CD30 Ab, cAC10, trafficked to the lysosome, and cysteine protease metabolism was implicated to play a role in drug release (146).

Metabolism studies have also helped to explain what we observed as "bystander effect," that is, the antigen-negative cells may be killed by administration and subsequent metabolism of an ADC. This concept was illustrated by Kovtun et al. who showed that the $N^{2'}$-deacetyl-$N^{2'}$-(3-mercapto-1-oxopropyl)-maytansine (DM1) or DC1 conjugate of anti-CanAg (cancer antigen) (huC242) killed antigen-positive cells and neighboring antigen-negative cells in culture, and that in vivo, the same conjugates effectively eradicated tumors containing both antigen-positive and antigen-negative cells (149,150). Similar conclusions were drawn by Erickson et al. at ImmunoGen, Inc. in relating the in vitro potency and in vivo activity of huC242-maytansinoid conjugates; however, in this case, the choice of linker heavily influenced the outcome (145). A disulfide-linked ADC yielded an uncharged, lipophilic drug metabolite that was found to be highly toxic to cells; however, a charged lysine-maytansinoid adduct resulting from metabolism of the thioether-linked conjugate had a very low cell-killing potency, presumably due to its inability to cross plasma membranes to exit/enter cells (145).

Summary of Antibody-Drug Conjugates

ADCs represent a special case of Ab therapeutics, combining many important concepts discussed throughout this chapter. Numerous examples of IVIVCs involving naked Abs were outlined in the preceding sections. Similar IVIVCs also apply for the Ab component of ADCs. When the cytotoxic drug is released from the ADC, IVIVC principles as established for small-molecule drugs remain applicable and will not be further discussed as they are beyond the scope of this presentation. Instead, the discussions herein address the specific details of IVIVCs for ADCs in the context of entire macromolecular ADC constructs.

ADCs are associated with additional risks compared with Abs. The highly cytotoxic nature of the free drug and the possibility of its detachment through chemical or metabolic processes are both complications that must be approached with great caution. One must constantly assess whether in vitro cytotoxicity or in vivo efficacy is due to the intact ADC or results from a metabolite/catabolite, and the same questions must also be applied to any normal organ toxicity observed in vivo. Examples described here and throughout the literature indicate that initial in vitro/in vivo studies provide critical information on identifying the determinant factors in optimizing ADCs. The behavior of these complex macromolecules is often target, antibody, linker, and drug specific; therefore, a combination of both in vitro and in vivo methods will be necessary to dissect the appropriate factors in their development.

MODELING OF ANTIBODY IN VITRO–IN VIVO CORRELATIONS

Throughout most of the currently available literature, IVIVC for mAbs are generally qualitative. Quantitative correlations of in vitro and in vivo data in a statistical sense have not been established in most cases. The primary reasons for the general lack of IVIVCs are insufficient data and the limited mechanistic understanding of the relationship between in vitro and in vivo data. Accordingly, most examples of IVIVC for Abs are primarily focused on the assessment of in vitro binding behavior to qualitatively support in vivo binding, targeting, and the related impact on PK. Such qualitative or semiquantitative comparisons with few or even single compounds relating in vitro to in vivo information do hold some value; however, for consistency, evaluations assessing the validity of either the in vitro or the in vivo assay provides more reliability. In this sense, a more systematic approach providing IVIVC driven toward mechanism-based quantitative assessments would provide far greater benefits including mechanistic insight and predictive power to estimate the outcome of future experiments. One must also realize that modeling and simulation (M&S) as a data-driven approach is a continuously evolving process in drug development; learning from prior knowledge is extracted from data and used to predict future experiments, followed by the next learning cycle. Since IVIVCs are not yet established for many Abs, most current M&S approaches are primarily focused on estimating drug specific PK/PD, efficacy, safety parameters (e.g., PK parameters,

ED50, K_d, k_{on}, k_{off}), and biological system information (e.g., target expression, turnover constants of binding targets) by using in vivo data from experiments with animal species or from clinical trials in healthy volunteers or patients. The information residing in in vitro data is only rarely used in these M&S evaluations. Retrospectively, in vivo parameter estimates are sometimes compared with those obtained from in vitro data. One commonly used modeling approach called target-mediated drug disposition modeling (TDDM) can capitalize on in vitro data by capturing the binding of a mAb to its target antigen in addition to other nonspecific distribution and clearance processes (3). In cases where relevant portions of the mAb are bound with high affinity to their cognate target antigen, the PK becomes very dependent on the binding and the downstream processing of the antigen-mAb complex (e.g., internalization into target cells and metabolism). TDDM models describe binding to the target antigen by estimating in vivo binding kinetics on the basis of the following parameters: the dissociation constant K_d, the association constant K_a, the second-order association rate constant k_{on}, and the first-order dissociation rate constant k_{off}. If bioassays allow measurement of occupied and free cell surface target antigen in addition to the unbound mAb as part of PK/PD studies, these parameters can be estimated from in vivo experiments. Consistency between these in vivo parameter estimates and binding parameters obtained from in vitro studies provides further confidence of physiologically meaningful parameters, providing a sound basis to use such models to optimize dose and dosing regimens to obtain optimal safety and efficacy in clinical settings. Excellent examples for TDDM modeling have been demonstrated with several mAbs including efaluzimab (anti-CD11a Ab) (151) and muromonab (anti-CD3 Ab) (152).

In the case of anti-CD11a in the treatment of psoriasis, mechanism-based modeling assisted the optimal clinical dose regimen selection (31). In vitro binding studies of anti-CD11a to CD11a on the surface of human T cells were conducted, and the binding affinity constant was measured. It was observed in the early clinical trials that intravenous doses of anti-CD11a higher than 0.3 mg/kg saturated CD11a binding sites, and the receptor clearance was dependent on the plasma concentration of Ab (40). A two-compartmental TDDM model was proposed to model the clearance of anti-CD11a coupled with a feedback loop of CD11a to the clearance of anti-CD11a (153). From this model, the affinity of anti-CD11a to CD11a was estimated on the basis of in vivo data and was very similar to the in vitro estimates. This model was subsequently studied both in vitro and in vivo (39,40). Collectively, these studies and modeling exercises provided a sound basis for further optimal dose selection to guide future clinical trials, eventually leading to an approved dose regimen for clinical use. Ideally, in vitro studies and modeling can be designed in such a way that more relevant in vivo studies may be conducted and the resulting data may then be related back to the in vitro studies. Combination of in vitro and in vivo studies can provide more insight to future study design, especially clinical designs, through mechanistic modeling. After the first clinical trial, more information is gathered, and the mechanical model can be modified and further refined to guide the next clinical trial design, including patient and dose selection.

CONCLUSIONS

The development and application of IVIVCs for Ab-based therapeutic drugs will continue to represent an exciting area of scientific exploration upon evaluation of numerous drug candidates. Significant breakthroughs in protein engineering technology continue to impact the availability of Ab derivatives beyond those originally identified by the murine immune system, yielding variations in IgG structures, receptor binding affinities, and other critical parameters (24). There are currently more than 100 mAbs in various phases of clinical evaluation for treatment of various disease states. Through the use of radioimmunoconjugates, the coupling of immunotherapy with molecular imaging allows confirmation of the presence of a given target prior to initiation of therapy and permits monitoring of the progress of treatment during the course of therapy (12). In addition, ADCs continue to prove efficacious with acceptable toxicities and will extend our therapeutic range from unconjugated mAbs to a broad array of highly active and specific immunoconjugates (154). Because of this potential diversity in Ab characteristics, IVIVC will prove invaluable in estimating the in vivo behavior (e.g., PK, PD) on the basis of in vitro information (e.g., Fc receptor binding affinity).

The principles involved in developing IVIVCs for Abs are inherently more complex than those needed to establish IVIVCs for SMD. The challenging nature of Ab IVIVC can be explained by the existence of unique interactions between Abs and biological molecules coupled with the structural complexity and diversity of the macromolecules themselves. Because of such complications, more progress is needed to establish predictive IVIVC tools similar to those already existing in the area of SMD.

Continual improvements in the accuracy, sensitivity, and ruggedness of in vitro assays will play an integral role in the evolution of IVIVC for Abs. A major limiting factor in the development of high-throughput assays has been the low number of candidates in preclinical evaluation; however, this number is ever increasing, and the demand for efficient in vitro measurements will likely begin to reflect this trend. In vivo tools will also help drive the use of Ab IVIVCs, especially considering the very rapid growth of the molecular imaging field within the past decade (35). Imaging is especially well suited to clinical applications owing to its noninvasive nature and compatibility with extended dynamic measurements. Specifically, both PET and SPECT may be powerful tools in preclinical drug development by (*i*) providing therapeutic rationale, (*ii*) enabling collection of data for rational drug dosing, (*iii*) permitting verification of binding of a drug to its receptor, and (*iv*) allowing evaluation of a drug's mechanism of action (155). To summarize, it is likely that the anticipated growth of three fields—Ab engineering, in vitro assay technologies, and in vivo detection methods—will influence the establishment of new IVIVCs for Abs in the near term. The resulting IVIVC data will be valuable in providing mechanistic insight into the relationships that exist between Ab PK, PD, and, ultimately, clinical efficacy.

REFERENCES

1. Whitebread S, Hamon J, Bojanic D, et al. Keynote review: in vitro safety pharmacology profiling: an essential tool for successful drug development. Drug Discov Today 2005; 10(21):1421–1433.
2. Li AP. Screening for human ADME/Tox drug properties in drug discovery. Drug Discov Today 2001; 6(7):357–366.
3. Mager DE. Target-mediated drug disposition and dynamics. Biochem Pharmacol 2006; 72(1):1–10.

4. Theil FP, Guentert TW, Haddad S, et al. Utility of physiologically based pharmacokinetic models to drug development and rational drug discovery candidate selection. Toxicol Lett 2003; 138(1–2):29–49.
5. Lave T, Parrott N, Grimm HP, et al. Challenges and opportunities with modelling and simulation in drug discovery and drug development. Xenobiotica 2007; 37(10–11):1295–1310.
6. Day ED, Rigsbee LC, Rosenthal JT, et al. Adsorption properties in vitro and in vivo of antibodies raised against Rat brain blood vessels. J Immunol 1974; 112(2):607–616.
7. Mire-Sluis AR. Progress in the use of biological assays during the development of biotechnology products. Pharm Res 2001; 18(9):1239–1246.
8. Gunaratna C. Drug metabolism and pharmacokinetics in drug discovery: a primer for bioanalytical chemists, part I. Curr Separations 2000; 19(1):17–23.
9. Gunaratna C. Drug metabolism and pharmacokinetics in drug discovery: a primer for bioanalytical chemists, part II. Curr Separations 2001; 19(3):87–92.
10. Leveque D, Wisniewski S, Jehl F. Pharmacokinetics of therapeutic monoclonal antibodies used in oncology. Anticancer Res 2005; 25(3c):2327–2343.
11. Lobo ED, Hansen RJ, Balthasar JP. Antibody pharmacokinetics and pharmacodynamics. J Pharma Sci 2004; 93(11):2645–2668.
12. Boswell CA, Brechbiel MW. Development of radioimmunotherapeutic and diagnostic antibodies: an inside-out view. Nucl Med Biol 2007; 34(7):757–778.
13. Milenic DE, Brady ED, Brechbiel MW. Antibody-targeted radiation cancer therapy. Nat Rev Drug Discov 2004; 3(6):488–499.
14. Walsh G. Biopharmaceutical benchmarks 2006. Nat Biotechnol 2006; 24(7):769–776.
15. Epstein AL, Khawli LA. Tumor biology and monoclonal antibodies: overview of basic principles and clinical considerations. Antibody Immunoconj and Radiopharmacol 1991; 4:373–383.
16. Epstein AL, Khawli LA. Tumor necrosis therapy of cancer: new methods of antibody targeting. In: Henkin RE, Bova D, Dillehay GL, et al. eds. Nuclear Medicine: Principles & Practice. 2nd ed. Philadelpha: Mosby-Elsevier, 2006.
17. Roskos LK, Davis CG, Schwab GM. The clinical pharmacology of therapeutic monoclonal antibodies. Drug Dev Res 2004; 61(3):108–120.
18. Scallon BJ, Snyder LA, Anderson GM, et al. A review of antibody therapeutics and antibody-related technologies for oncology. J Immunother 2006; 29(4):351–364.
19. The PyMOL Molecular Graphics System. DeLano Scientific, 2002. Available at: http://www.pymol.org.
20. Ghetie V. The neonatal Fc receptor is a regulator of the homeostasis of IgG. Curr Trends Immunol 2006; 7:31–46.
21. Christiansen J, Rajasekaran AK. Biological impediments to monoclonal antibody-based cancer immunotherapy. Mol Cancer Ther 2004; 3(11):1493–1501.
22. Kohler G, Milstein C. Continuous cultures of fused cells secreting antibody of predefined specificity. Nature 1975; 256:495–497.
23. Khazaeli MB, Conry RM, LoBuglio AF. Human immune response to monoclonal antibodies. J Immunother Emphasis Tumor Immunol 1994; 15(1):42–52.
24. Oldham RK, Dillman RO. Monoclonal antibodies in cancer therapy: 25 years of progress. J Clin Oncol 2008; 26(11):1774–1777.
25. Kim SJ, Park Y, Hong HJ. Antibody engineering for the development of therapeutic antibodies. Mol Cells 2005; 20(1):17–29.
26. Holliger P, Hudson PJ. Engineered antibody fragments and the rise of single domains. Nat Biotechnol 2005; 23(9):1126–1136.
27. Holliger P, Prospero T, Winter G. "Diabodies": small bivalent and bispecific antibody fragments. Proc Natl Acad Sci U S A 1993; 90(14):6444–6448.
28. Hudson PJ, Kortt AA. High avidity scFv multimers; diabodies and triabodies. J Immunol Methods 1999; 231(1–2):177–189.
29. Presta L. Antibody engineering for therapeutics. Curr Opin Struct Biol 2003; 13(4):519–525.
30. Albrecht H, DeNardo SJ. Recombinant antibodies: from the laboratory to the clinic. Cancer Biother Radiopharm 2006; 21(4):285–304.

31. Wu AM, Senter PD. Arming antibodies: prospects and challenges for immuno-conjugates. Nat Biotechnol 2005; 23(9):1137–1146.
32. Shen WC, Persiani S, Srivastava K. Chemical linkages in drug-antibody conjugation. BioPharm 1990; 3(1):16–22.
33. Vaidyanathan G, Zalutsky MR. Preparation of N-succinimidyl 3-[*I]iodobenzoate: an agent for the indirect radioiodination of proteins. Nat Protoc 2006; 1(2):707–713.
34. Larson SM. Radiolabeled monoclonal anti-tumor antibodies in diagnosis and therapy. J Nucl Med 1985; 26(5):538–545.
35. van Dongen GA, Visser GW, Lub-de Hooge MN, et al. Immuno-PET: a navigator in monoclonal antibody development and applications. Oncologist 2007; 12(12): 1379–1389.
36. Keller L, Boswell CA, Milenic DE, et al. Monoclonal antibody targeted radiation cancer therapy. In: Boehncke WH, Radeke HH, eds. Biologics in General Medicine. Berlin: Springer, 2007:50–58.
37. Beckman RA, Weiner LM, Davis HM. Antibody constructs in cancer therapy: protein engineering strategies to improve exposure in solid tumors. Cancer 2007; 109(2):170–179.
38. Joshi A, Bauer R, Kuebler P, et al. An overview of the pharmacokinetics and pharmacodynamics of efalizumab: a monoclonal antibody approved for use in psoriasis. J Clin Pharmacol 2006; 46(1):10–20.
39. Coffey GP, Stefanich E, Palmieri S, et al. In vitro internalization, intracellular transport, and clearance of an anti-CD11a antibody (Raptiva) by human T-cells. J Pharmacol Exp Ther 2004; 310(3):896–904.
40. Coffey GP, Fox JA, Pippig S, et al. Tissue distribution and receptor-mediated clearance of anti-CD11a antibody in mice. Drug Metab Dispos 2005; 33(5):623–629.
41. Sautes-Fridman C, Cassard L, Cohen-Solal J, et al. Fc gamma receptors: a magic link with the outside world. In: ASHI Quarterly: American Society for Histocompatibility and Immunogenetics, 2003:148–151.
42. Shields RL, Namenuk AK, Hong K, et al. High resolution mapping of the binding site on human IgG1 for Fc gamma RI, Fc gamma RII, Fc gamma RIII, and FcRn and design of IgG1 variants with improved binding to the Fc gamma R. J Biol Chem 2001; 276(9):6591–6604.
43. Gillies SD, Lan Y, Lo KM, et al. Improving the efficacy of antibody-interleukin 2 fusion proteins by reducing their interaction with Fc receptors. Cancer Res 1999; 59(9):2159–2166.
44. Hutchins JT, Kull FC Jr., Bynum J, et al. Improved biodistribution, tumor targeting, and reduced immunogenicity in mice with a gamma 4 variant of Campath-1H. Proc Natl Acad Sci U S A 1995; 92(26):11980–11984.
45. Sharma A, Davis CB, Tobia LA, et al. Comparative pharmacodynamics of keliximab and clenoliximab in transgenic mice bearing human CD4. J Pharmacol Exp Ther 2000; 293(1):33–41.
46. Kanda Y, Yamada T, Mori K, et al. Comparison of biological activity among nonfucosylated therapeutic IgG1 antibodies with three different N-linked Fc oligosaccharides: the high-mannose, hybrid, and complex types. Glycobiology 2007; 17(1):104–118.
47. Mannik M, Arend MP, Hall AP, et al. Studies on antigen-antibody complexes. I. Elimination of soluble complexes from rabbit circulation. J Exp Med 1971; 133(4): 713–739.
48. Ghetie V, Hubbard JG, Kim JK, et al. Abnormally short serum half-lives of IgG in beta 2-microglobulin-deficient mice. Eur J Immunol 1996; 26(3):690–696.
49. Junghans RP, Anderson CL. The protection receptor for IgG catabolism is the beta2-microglobulin-containing neonatal intestinal transport receptor. Proc Natl Acad Sci U S A 1996; 93(11):5512–5516.
50. Junghans RP. Finally! The Brambell receptor (FcRB). Mediator of transmission of immunity and protection from catabolism for IgG. Immunol Res 1997; 16(1):29–57.
51. Brambell F, Hemmings W, Morris I. A theoretical model of gamma-globulin catabloism. Nature 1964; 203:1352–1355.

52. Simister NE, Mostov KE. Cloning and expression of the neonatal rat intestinal Fc receptor, a major histocompatibility complex class I antigen homolog. Cold Spring Harb Symp Quant Biol 1989; 54(pt 1):571–580.

53. Simister NE, Mostov KE. An Fc receptor structurally related to MHC class I antigens. Nature 1989; 337(6203):184–187.

54. Roopenian DC, Christianson GJ, Sproule TJ, et al. The MHC class I-like IgG receptor controls perinatal IgG transport, IgG homeostasis, and fate of IgG-Fc-coupled drugs. J Immunol 2003; 170(7):3528–3533.

55. Medesan C, Matesoi D, Radu C, et al. Delineation of the amino acid residues involved in transcytosis and catabolism of mouse IgG1. J Immunol 1997; 158(5): 2211–2217.

56. Ober RJ, Radu CG, Ghetie V, et al. Differences in promiscuity for antibody-FcRn interactions across species: implications for therapeutic antibodies. Int Immunol 2001; 13(12):1551–1559.

57. Petkova SB, Akilesh S, Sproule TJ, et al. Enhanced half-life of genetically engineered human IgG1 antibodies in a humanized FcRn mouse model: potential application in humorally mediated autoimmune disease. Int Immunol 2006; 18(12):1759–1769.

58. Jaggi JS, Carrasquillo JA, Seshan SV, et al. Improved tumor imaging and therapy via i.v. IgG-mediated time-sequential modulation of neonatal Fc receptor. J Clin Invest 2007; 117(9):2422–2430.

59. Zhou J, Johnson JE, Ghetie V, et al. Generation of mutated variants of the human form of the MHC class I-related receptor, FcRn, with increased affinity for mouse immunoglobulin G. J Mol Biol 2003; 332(4):901–913.

60. Vaccaro C, Bawdon R, Wanjie S, et al. Divergent activities of an engineered antibody in murine and human systems have implications for therapeutic antibodies. Proc Natl Acad Sci U S A 2006; 103(49):18709–18714.

61. Hinton PR, Xiong JM, Johlfs MG, et al. An engineered human IgG1 antibody with longer serum half-life. J Immunol 2006; 176(1):346–356.

62. Datta-Mannan A, Witcher DR, Tang Y, et al. Monoclonal antibody clearance. Impact of modulating the interaction of IgG with the neonatal Fc receptor. J Biol Chem 2007; 282(3):1709–1717.

63. Datta-Mannan A, Witcher DR, Tang Y, et al. Humanized IgG1 variants with differential binding properties to the neonatal Fc receptor: relationship to pharmacokinetics in mice and primates. Drug Metab Dispos 2007; 35(1):86–94.

64. Yeung YA, Leabman MK, Marvin JS, et al. Engineering Human IgG1 Affinity to Human Neonatal Fc Receptor: Impact of Affinity Improvement on Pharmacokinetics in Primates. J Immunol 2009; 182:7663–7671.

65. Dall'Acqua WF, Woods RM, Ward ES, et al. Increasing the affinity of a human IgG1 for the neonatal Fc receptor: biological consequences. J Immunol 2002; 169(9):5171–5180.

66. Dall'Acqua WF, Kiener PA, Wu H. Properties of human IgG1s engineered for enhanced binding to the neonatal Fc receptor (FcRn). J Biol Chem 2006; 281(33): 23514–23524.

67. Nicholson JP, Wolmarans MR, Park GR. The role of albumin in critical illness. Br J Anaesth 2000; 85(4):599–610.

68. Kratz F, Muller-Driver R, Hofmann I, et al. A novel macromolecular prodrug concept exploiting endogenous serum albumin as a drug carrier for cancer chemotherapy. J Med Chem 2000; 43(7):1253–1256.

69. Kurtzhals P, Havelund S, Jonassen I, et al. Albumin binding of insulins acylated with fatty acids: characterization of the ligand-protein interaction and correlation between binding affinity and timing of the insulin effect in vivo. Biochem J 1995; 312(pt 3):725–731.

70. Dennis MS, Jin H, Dugger D, et al. Imaging tumors with an albumin-binding Fab, a novel tumor-targeting agent. Cancer Res 2007; 67(1):254–261.

71. Dennis MS, Zhang M, Meng YG, et al. Albumin binding as a general strategy for improving the pharmacokinetics of proteins. J Biol Chem 2002, 277(38), 35035–35043.

72. Nguyen A, Reyes AE II, Zhang M, et al. The pharmacokinetics of an albumin-binding Fab (AB.Fab) can be modulated as a function of affinity for albumin. Protein Eng Des Sel 2006; 19(7):291–297.

73. Anderson CL, Chaudhury C, Kim J, et al. Perspective—FcRn transports albumin: relevance to immunology and medicine. Trends Immunol 2006; 27(7):343–348.
74. Khawli LA, Chen FM, Alauddin MM, et al. Radioiodinated monoclonal antibody conjugates: synthesis and comparative evaluation. Antibody Immunoconj Radiopharm 1991; 4:163–182.
75. Khawli LA, Glasky MS, Alauddin MM, et al. Improved tumor localization and radioimaging with chemically modified monoclonal antibodies. Cancer Biother Radiopharm 1996; 11(3):203–215.
76. Khawli LA, Mizokami MM, Sharifi J, et al. Pharmacokinetic characteristics and biodistribution of radioiodinated chimeric TNT-1, -2, and -3 monoclonal antibodies after chemical modification with biotin. Cancer Biother Radiopharm 2002; 17(4):359–370.
77. Chen S, Yu L, Jiang C, et al. Pivotal study of iodine-131-labeled chimeric tumor necrosis treatment radioimmunotherapy in patients with advanced lung cancer. J Clin Oncol 2005; 23(7):1538–1547.
78. Yu L, Ju DW, Chen W, et al. 131I-chTNT radioimmunotherapy of 43 patients with advanced lung cancer. Cancer Biother Radiopharm 2006; 21(1):5–14.
79. Silva Filho FC, Santos AB, de Carvalho TM, et al. Surface charge of resident, elicited, and activated mouse peritoneal macrophages. J Leukoc Biol 1987; 41(2):143–149.
80. Lee HJ, Pardridge WM. Monoclonal antibody radiopharmaceuticals: cationization, pegylation, radiometal chelation, pharmacokinetics, and tumor imaging. Bioconjug Chem 2003; 14(3):546–553.
81. Khawli LA, Biela B, Hu P, et al. Comparison of recombinant derivatives of chimeric TNT-3 antibody for the radioimaging of solid tumors. Hybrid Hybridomics 2003; 22(1):1–9.
82. Khawli LA, Biela BH, Hu P, et al. Stable, genetically engineered F(ab')(2) fragments of chimeric TNT-3 expressed in mammalian cells. Hybrid Hybridomics 2002; 21(1):11–18.
83. Moin K, McIntyre OJ, Matrisian LM, et al. Fluorescent imaging of tumors. In: Shields AF, Price P, eds. In Vivo Imaging of Cancer Therapy. Totowa, NJ: Humana Press, Inc., 2007:281–302.
84. Shockley TR, Lin K, Sung C, et al. A quantitative analysis of tumor specific monoclonal antibody uptake by human melanoma xenografts: effects of antibody immunological properties and tumor antigen expression levels. Cancer Res 1992; 52(2):357–366.
85. Jain RK. Determinants of tumor blood flow: a review. Cancer Res 1988; 48(10):2641–2658.
86. Sung C, Youle RJ, Dedrick RL. Pharmacokinetic analysis of immunotoxin uptake in solid tumors: role of plasma kinetics, capillary permeability, and binding. Cancer Res 1990; 50(22):7382–7392.
87. Blumenthal RD, Osorio L, Ochakovskaya R, et al. Regulation of tumour drug delivery by blood flow chronobiology. Eur J Cancer 2000; 36(14):1876–1884.
88. Lewis MR, Boswell CA, Laforest R, et al. Conjugation of monoclonal antibodies with TETA using activated esters: biological comparison of 64Cu-TETA-1A3 with 64Cu-BAT-2IT-1A3. Cancer Biother Radiopharm 2001; 16(6):483–494.
89. Rodwell JD, Alvarez VL, Lee C, et al. Site-specific covalent modification of monoclonal antibodies: in vitro and in vivo evaluations. Proc Natl Acad Sci U S A 1986; 83(8):2632–2636.
90. Burvenich I, Schoonooghe S, Cornelissen B, et al. In vitro and in vivo targeting properties of iodine-123- or iodine-131-labeled monoclonal antibody 14C5 in a non-small cell lung cancer and colon carcinoma model. Clin Cancer Res 2005; 11(20):7288–7296.
91. Zuckier LS, Berkowitz EZ, Sattenberg RJ, et al. Influence of affinity and antigen density on antibody localization in a modifiable tumor targeting model. Cancer Res 2000; 60(24):7008–7013.
92. Sakahara H, Endo K, Koizumi M, et al. Relationship between in vitro binding activity and in vivo tumor accumulation of radiolabeled monoclonal antibodies. J Nucl Med 1988; 29(2):235–240.

93. Ng CM, Stefanich E, Anand BS, et al. Pharmacokinetics/pharmacodynamics of nondepleting anti-CD4 monoclonal antibody (TRX1) in healthy human volunteers. Pharm Res 2006; 23(1):95–103.

94. Worn A, Auf der Maur A, Escher D, et al. Correlation between in vitro stability and in vivo performance of anti-GCN4 intrabodies as cytoplasmic inhibitors. J Biol Chem 2000; 275(4):2795–2803.

95. Hochhaus G, Brookman L, Fox H, et al. Pharmacodynamics of omalizumab: implications for optimised dosing strategies and clinical efficacy in the treatment of allergic asthma. Curr Med Res Opin 2003; 19(6):491–498.

96. Liu J, Lester P, Builder S, et al. Characterization of complex formation by humanized anti-IgE monoclonal antibody and monoclonal human IgE. Biochemistry 1995; 34(33): 10474–10482.

97. Fox JA, Hotaling TE, Struble C, et al. Tissue distribution and complex formation with IgE of an anti-IgE antibody after intravenous administration in cynomolgus monkeys. J Pharmacol Exp Ther 1996; 279(2):1000–1008.

98. Schulman ES. Development of a monoclonal anti-immunoglobulin E antibody (omalizumab) for the treatment of allergic respiratory disorders. Am J Respir Crit Care Med 2001; 164(8 pt 2):S6–S11.

99. Putnam WS, Li J, Haggstrom J, et al. Use of quantitative pharmacology in the development of HAE1, a high-affinity anti-IgE monoclonal antibody. AAPS J 2008; 10(2):425–430.

100. Clarke J, Leach W, Pippig S, et al. Evaluation of a surrogate antibody for preclinical safety testing of an anti-CD11a monoclonal antibody. Regul Toxicol Pharmacol 2004; 40(3):219–226.

101. Smith MR. Rituximab (monoclonal anti-CD20 antibody): mechanisms of action and resistance. Oncogene 2003; 22(47):7359–7368.

102. Daniel D, Yang B, Lawrence DA, et al. Cooperation of the proapoptotic receptor agonist rhApo2L/TRAIL with the CD20 antibody rituximab against non-Hodgkin lymphoma xenografts. Blood 2007; 110(12):4037–4046.

103. Zhang N, Khawli LA, Hu P, et al. Generation of rituximab polymer may cause hyper-cross-linking-induced apoptosis in non-Hodgkin's lymphomas. Clin Cancer Res 2005; 11(16):5971–5980.

104. Byrd JC, Kitada S, Flinn IW, et al. The mechanism of tumor cell clearance by rituximab in vivo in patients with B-cell chronic lymphocytic leukemia: evidence of caspase activation and apoptosis induction. Blood 2002; 99(3):1038–1043.

105. Borchmann P, Treml JF, Hansen H, et al. The human anti-CD30 antibody 5F11 shows in vitro and in vivo activity against malignant lymphoma. Blood 2003; 102(10):3737–3742.

106. Kelley SK, Gelzleichter T, Xie D, et al. Preclinical pharmacokinetics, pharmacodynamics, and activity of a humanized anti-CD40 antibody (SGN-40) in rodents and non-human primates. Br J Pharmacol 2006; 148(8):1116–1123.

107. Baselga J, Perez EA, Pienkowski T, et al. Adjuvant trastuzumab: a milestone in the treatment of HER-2-positive early breast cancer. Oncologist 2006; 11(suppl 1):4–12.

108. Mann M, Sheng H, Shao J, et al. Targeting cyclooxygenase 2 and HER-2/neu pathways inhibits colorectal carcinoma growth. Gastroenterology 2001; 120(7):1713–1719.

109. Agus DB, Akita RW, Fox WD, et al. Targeting ligand-activated ErbB2 signaling inhibits breast and prostate tumor growth. Cancer Cell 2002; 2(2):127–137.

110. Lin BC, Wang M, Blackmore C, et al. Liver-specific activities of FGF19 require Klotho beta. J Biol Chem 2007; 282(37):27277–27284.

111. Desnoyers LR, Pai R, Ferrando RE, et al. Targeting FGF19 inhibits tumor growth in colon cancer xenograft and FGF19 transgenic hepatocellular carcinoma models. Oncogene 2008; 27(1):85–97.

112. Bell SJ, Kamm MA. Review article: the clinical role of anti-TNFalpha antibody treatment in Crohn's disease. Aliment Pharmacol Ther 2000; 14(5):501–514.

113. Balkwill F. Tumor necrosis factor or tumor promoting factor? Cytokine Growth Factor Rev 2002; 13(2):135–141.

114. Mpofu S, Fatima F, Moots RJ. Anti-TNF-alpha therapies: they are all the same (aren't they?). Rheumatology (Oxford) 2005; 44(3):271–273.

115. Egberts JH, Cloosters V, Noack A, et al. Anti-tumor necrosis factor therapy inhibits pancreatic tumor growth and metastasis. Cancer Res 2008; 68(5):1443–1450.
116. Teng MN, Park BH, Koeppen HK, et al. Long-term inhibition of tumor growth by tumor necrosis factor in the absence of cachexia or T-cell immunity. Proc Natl Acad Sci U S A 1991; 88(9):3535–3539.
117. Bromberg JS, Chavin KD, Kunkel SL. Anti-tumor necrosis factor antibodies suppress cell-mediated immunity in vivo. J Immunol 1992; 148(11):3412–3417.
118. Waterston AM, Salway F, Andreakos E, et al. TNF autovaccination induces self anti-TNF antibodies and inhibits metastasis in a murine melanoma model. Br J Cancer 2004; 90(6):1279–1284.
119. Ferrara N, Hillan KJ, Gerber HP, et al. Discovery and development of bevacizumab, an anti-VEGF antibody for treating cancer. Nat Rev Drug Discov 2004; 3(5): 391–400.
120. Bhaskara A, Eng C. Bevacizumab in the treatment of a patient with metastatic colorectal carcinoma with brain metastases. Clin Colorectal Cancer 2008; 7(1):65–68.
121. Scott LJ. Bevacizumab: in first-line treatment of metastatic breast cancer. Drugs 2007; 67(12):1793–1799.
122. Manegold C. Bevacizumab for the treatment of advanced non-small-cell lung cancer. Expert Rev Anticancer Ther 2008; 8(5):689–699.
123. Gerber HP, Wu X, Yu L, et al. Mice expressing a humanized form of VEGF-A may provide insights into the safety and efficacy of anti-VEGF antibodies. Proc Natl Acad Sci U S A 2007; 104(9):3478–3483.
124. Gaudreault J, Fei D, Rusit J, et al. Preclinical pharmacokinetics of Ranibizumab (rhuFabV2) after a single intravitreal administration. Invest Ophthalmol Vis Sci 2005; 46(2):726–733.
125. Lowe J, Araujo J, Yang J, et al. Ranibizumab inhibits multiple forms of biologically active vascular endothelial growth factor in vitro and in vivo. Exp Eye Res 2007; 85(4):425–430.
126. Sands H, Jones PL. Methods for the study of the metabolism of radiolabeled monoclonal antibodies by liver and tumor. J Nucl Med 1987; 28(3):390–398.
127. Rogers BE, Franano FN, Duncan JR, et al. Identification of metabolites of 111In-diethylenetriaminepentaacetic acid-monoclonal antibodies and antibody fragments in vivo. Cancer Res 1995; 55(23 suppl):5714s–5720s.
128. Tsai SW, Li L, Williams LE, et al. Metabolism and renal clearance of 111In-labeled DOTA-conjugated antibody fragments. Bioconjug Chem 2001; 12(2):264–270.
129. Schwartz RS. Paul Ehrlich's magic bullets. N Engl J Med 2004; 350(11):1079–1080.
130. Schrama D, Reisfeld RA, Becker JC. Antibody targeted drugs as cancer therapeutics. Nat Rev Drug Discov 2006; 5(2):147–159.
131. McCarron PA, Olwill SA, Marouf WM, et al. Antibody conjugates and therapeutic strategies. Mol Interv 2005; 5(6):368–380.
132. Garnett MC. Targeted drug conjugates: principles and progress. Adv Drug Deliv Rev 2001; 53(2):171–216.
133. Maxfield FR, McGraw TE. Endocytic recycling. Nat Rev Mol Cell Biol 2004; 5(2): 121–132.
134. Yelton DE, Scharff MD. Monoclonal antibodies: a powerful new tool in biology and medicine. Annu Rev Biochem 1981; 50:657–680.
135. Beeram M. A phase i study of trastuzumab-DM1, a first-in-class HER2 antibody-drug conjugate (ADC), in patients with advanced HER2-positive breast cancer (abstract 1028). 44th Annual Meeting of the American Society of Clinical Oncology (ASCO), 2008, Chicago, IL.
136. Doronina SO, Mendelsohn BA, Bovee TD, et al. Enhanced activity of monomethylauristatin F through monoclonal antibody delivery: effects of linker technology on efficacy and toxicity. Bioconjug Chem 2006; 17(1):114–124.
137. Guillemard V, Saragovi HU. Novel approaches for targeted cancer therapy. Curr Cancer Drug Targets 2004; 4(4):313–326.
138. Reff M, Braslawsky G, Hanna N. Future approaches for treating hematologic disease. Curr Pharm Biotechnol 2001; 2(4):369–382.

139. Berek JS, Thomas GM, Ozols RF. Ovarian cancer. In: Holland JF, Frei E, Bast RC, eds. Oncology: Principles and Practice. 2nd ed. Philadelphia: Lippincott, 1996.

140. Chen Y, Clark S, Wong T, et al. Armed antibodies targeting the mucin repeats of the ovarian cancer antigen, MUC16, are highly efficacious in animal tumor models. Cancer Res 2007; 67(10):4924–4932.

141. Ross S, Spencer SD, Holcomb I, et al. Prostate stem cell antigen as therapy target: tissue expression and in vivo efficacy of an immunoconjugate. Cancer Res 2002; 62(9):2546–2553.

142. Mao W, Luis E, Ross S, et al. EphB2 as a therapeutic antibody drug target for the treatment of colorectal cancer. Cancer Res 2004; 64(3):781–788.

143. Polson AG, Yu SF, Elkins K, et al. Antibody-drug conjugates targeted to CD79 for the treatment of non-Hodgkin lymphoma. Blood 2007; 110(2):616–623.

144. Hamblett KJ, Senter PD, Chace DF, et al. Effects of drug loading on the antitumor activity of a monoclonal antibody drug conjugate. Clin Cancer Res 2004; 10(20): 7063–7070.

145. Erickson HK, Park PU, Widdison WC, et al. Antibody-maytansinoid conjugates are activated in targeted cancer cells by lysosomal degradation and linker-dependent intracellular processing. Cancer Res 2006; 66(8):4426–4433.

146. Sutherland MS, Sanderson RJ, Gordon KA, et al. Lysosomal trafficking and cysteine protease metabolism confer target-specific cytotoxicity by peptide-linked anti-CD30-auristatin conjugates. J Biol Chem 2006; 281(15):10540–10547.

147. Alley SC, Benjamin DR, Jeffrey SC, et al. Contribution of linker stability to the activities of anticancer immunoconjugates. Bioconjug Chem 2008; 19(3):759–765.

148. Doronina SO, Toki BE, Torgov MY, et al. Development of potent monoclonal antibody auristatin conjugates for cancer therapy. Nat Biotechnol 2003; 21(7):778–784.

149. Kovtun YV, Audette CA, Ye Y, et al. Antibody-drug conjugates designed to eradicate tumors with homogeneous and heterogeneous expression of the target antigen. Cancer Res 2006; 66(6):3214–3221.

150. Kovtun YV, Goldmacher VS. Cell killing by antibody-drug conjugates. Cancer Lett 2007; 255(2):232–240.

151. Ng CM, Joshi A, Dedrick RL, et al. Pharmacokinetic-pharmacodynamic-efficacy analysis of efalizumab in patients with moderate to severe psoriasis. Pharm Res 2005; 22(7):1088–1100.

152. Meijer RT, Koopmans RP, ten Berge IJ, et al. Pharmacokinetics of murine anti-human CD3 antibodies in man are determined by the disappearance of target antigen. J Pharmacol Exp Ther 2002; 300(1):346–353.

153. Bauer RJ, Dedrick RL, White ML, et al. Population pharmacokinetics and pharmacodynamics of the anti-CD11a antibody hu1124 in human subjects with psoriasis. J Pharmacokinet Biopharm 1999; 27(4):397–420.

154. Ricart AD, Tolcher AW. Technology insight: cytotoxic drug immunoconjugates for cancer therapy. Nat Clin Pract Oncol 2007; 4(4):245–255.

155. Wong DF. Imaging in drug discovery, preclinical, and early clinical development. J Nucl Med 2008; 49(6):26N–28N.

3

In Vitro/In Vivo Correlations of Pharmacokinetics, Pharmacodynamics, and Metabolism for Extracellular Matrix–Binding Growth Factors

Matthew A. Nugent
Departments of Biochemistry and Ophthalmology, Boston University School of Medicine, and Department of Biomedical Engineering, Boston University, Boston, Massachusetts, U.S.A.

Maria Mitsi and Jean L. Spencer
Departments of Biochemistry and Ophthalmology, Boston University School of Medicine, Boston, Massachusetts, U.S.A.

INTRODUCTION

Growth factors are a large class of extracellular proteins that regulate nearly all aspects of cell function. While they were originally defined on the basis of their ability to stimulate the growth of cells, it has become clear that their range of activities is much broader and includes the ability to regulate cell survival, differentiation, movement, metabolic state, and biosynthesis. With the original discovery of growth factors by Rita Levi-Montalcini and Stanley Cohen in the 1950s came the prospects of using these small proteins to stimulate the repair and regeneration of injured tissues and organs in humans. After half a century, and with the discovery of well over one hundred growth factors, the great clinical potential for growth factor therapies remains unrealized. Indeed, the development of growth factor therapies has been met with the sobering reality that their potent activities in well-controlled in vitro systems do not often translate smoothly to the more complex in vivo setting. Thus, most large-scale growth factor clinical trials have produced disappointing outcomes (1,2). While there remains considerable excitement for the potential wide clinical use of growth factors, it has become clear that the complex mechanisms in place controlling growth factor activity need to be understood to design effective means of administration. In particular, it has become clear that a large number of growth factors are controlled locally within tissues by interactions with components of the extracellular matrix (ECM) (3–7). Consequently, this chapter will focus on the subclass of growth factors that bind to the ECM, will review the mechanisms that lead to their deposition and release from the ECM, and will discuss the consequences on growth factor pharmacokinetics, pharmacodynamics, and metabolism (Fig. 1). Unfortunately, the magnitude of this field of research and the space limitations of this chapter will prevent us from comprehensively citing many of the primary references for the findings that are described.

EXTRACELLULAR MATRIX–BINDING GROWTH FACTORS

The initial discovery of epidermal growth factor within the salivary gland of mice and the later demonstration of high growth factor levels within mammalian (e.g., bovine) serum used to stimulate the growth of cultured cells suggested

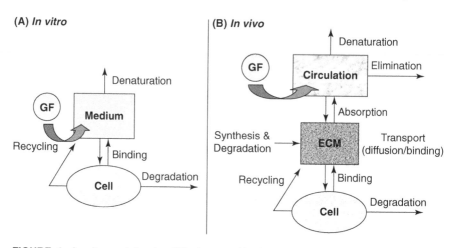

FIGURE 1 In vitro and in vivo GF pharmacokinetic and pharmacodynamic considerations. (**A**) GF interactions with cells in in vitro culture models can be effectively modeled using a relatively simple set of parameters that describe the physical stability of the protein in the culture medium (denaturation), the reversible binding of the GF with cell surface binding sites (binding), cell signaling and internalization, and the intracellular fate of the GF that includes degradation (degradation) or released from the cell in intact or partially degraded form (recycling). Models that capture these steps can be correlated to biological response. (**B**) In vivo models need to consider additional components including active clearance (elimination) and interactions with the ECM. ECM absorption, transport, and release of GFs are controlled by the composition of the ECM, which is in constant flux as a result of synthesis and degradation. The homeostasis of the ECM can be modified dramatically by tissue injury, disease, and inflammation. Hence, models describing the pharmacokinetics and pharmacodynamics of ECM-binding GFs in vivo need to include parameters describing the binding, transport, and release from the ECM. *Abbreviations*: GF, growth factor; ECM, extracellular matrix.

that this class of bioactive proteins might represent a new type of systemic hormone. However, it has become clear that growth factors, in contrast to traditional hormones, do not generally act systemically as circulating proteins but are instead produced or released at local target sites. For example, the aggregation and degranulation of blood platelets at sites of injury lead to the release of a range of growth factors, including platelet-derived growth factor (PDGF) and transforming growth factor β (TGF-β), accounting for the high levels of these growth factors in the serum of clotted blood. In this way, blood platelets store growth factors for local release upon demand. Similarly, activated mast cells are known to release a range of growth factors including fibroblast growth factor 2 (FGF-2) vascular endothelial growth factor (VEGF), and nerve growth factor (8–13). Thus, eventual clinical applications of growth factors might be most effective if they are able to mimic the endogenous localized control of these potent proteins.

Within tissue environments, a number of growth factors are produced by resident cells that act via autocrine or paracrine mechanisms over relatively short distances. Indeed, the diffusive range of many growth factors appears quite limited in tissues even when the resident cells do not appear to be binding and responding to the growth factors present (14,15). This phenomenon is, in large part, a consequence of the fact that many growth factors physically bind to components of the ECM (5,16,17). Interactions between growth factors and their ECM-binding sites can lead to prolonged retention of growth factors in a latent state and can thus constitute a stored form of growth factor that can be activated upon release. Release of growth factors from ECM storage sites can be activated by degradation of the ECM by proteases and polysaccharide-degrading enzymes released by inflammatory cells or produced locally, or by alterations in the mass-action balance of growth factor–binding sites on cell surfaces and the ECM (14,18–22). Considerable research effort has been aimed at elucidating this sophisticated process of growth factor control, and results of these studies have begun to inform the design of the most appropriate means for controlled growth factor delivery to produce clinical benefit. ECM binding will likely prove to be a significant issue for most growth factors. However, it is clear that this issue is of major significance for the FGF and VEGF families where disappointing results from several clinical trials likely reflect the lack of appreciation of the consequences of ECM binding in modulating these growth factors (23–25).

Fibroblast Growth Factors

The FGF family represents a large class (23 members) of structurally related proteins that share the ability to bind to heparin with high affinity (26–28). In addition, some members of the FGF family (i.e., FGF-2) have been shown to be expressed as multiple forms from a single mRNA through the use of alternate translation start sites (29). FGF-1 to FGF-10 bind and activate the four-member family of fibroblast growth factor receptors (FGFRs), which are single-pass transmembrane receptor tyrosine kinases (30). Alternate mRNA splicing gives rise to a large number of FGFR forms providing an additional level of ligand selectivity. The other FGF family members either do not bind and activate the canonical FGFR members or have yet to be fully characterized.

FGFs have been demonstrated to play critical roles in tissue morphogenesis and repair in a number of organisms. While an extensive range of biological activities have been attributed to FGFs, the well-characterized ability of FGF-1 and FGF-2 to stimulate the growth of new blood vessels from preexisting ones (angiogenesis) has produced considerable interest in their use to stimulate wound healing and relieve myocardial and peripheral ischemia. Indeed, initial small-scale trials using the direct administration of recombinant FGF-2 protein, naked DNA, and viral expression vectors produced promising results. However, the results of larger, randomized, placebo-controlled trials have revealed less impressive clinical benefit (31). The disconnect between the larger double-blind trials and the smaller initial studies may reflect the significant placebo effect observed. Even so, there also appear to be considerable pharmacological issues that may be partly responsible. For example, studies with FGF-2 in rats revealed an extremely short half-life (~3 minutes) for intravenously administered protein that reflected both fast clearance and significant tissue deposition (15). This complex pharmacokinetics most likely relates to the interaction of FGF-2 with the heparan sulfate (HS) class of complex polysaccharides, which can, somewhat paradoxically, enhance clearance, increase stability, and prolong tissue retention. In a clinical evaluation of the pharmacokinetics of a single dose of FGF-2 in 66 subjects after intravenous or intracoronary administration, biphasic elimination kinetics were observed with mean half-times of 21 minutes and 7.6 hours, and no differences were noted between the two routes of administration (32). Greater systemic exposure and slower elimination were noted when heparin was administered close to FGF-2 infusion. Heparin and HS can also modify cell receptor binding, uptake, and metabolism of FGF-2 in vitro and in vivo (7,29,33). As such, the modulation of endogenous HS interactions needs to be considered when interpreting the effects of heparin on FGF-2 pharmacokinetics and pharmacodynamics. The consequences of FGF-2 interactions with HS defy simple categorization and will need to be appreciated in greater detail if FGF-2 is to be effectively delivered to produce a predictable and controlled response. HS binding is an important aspect of FGF-2-ECM interactions, yet the influence of HS on FGF-2 is not restricted to the ECM, as described later in this chapter.

Vascular Endothelial Growth Factors

VEGF-A, originally identified and isolated on the basis of its ability to stimulate permeability in blood vessels, is a critical factor for normal vascular development and angiogenesis (34). VEGF-A is a member of the PDGF family of dimeric cysteine-knot growth factors that also includes the other VEGF members: VEGF-B, VEGF-C, VEGF-D, and placental growth factor (35). Multiple forms of VEGF-A are produced through alternate mRNA splicing to generate at least four isoforms in humans: VEGF121, VEGF165, VEGF189, and VEGF206 (36). All of the major VEGF-A isoforms are secreted from cells as homodimers, and all except VEGF121 contain a consensus heparin-binding domain in the C-terminal domain that is comprised of 15 basic amino acids. The binding of some VEGF-A isoforms to heparin is believed to reflect important interactions with HS proteoglycans within the ECM. VEGF-A binds to major receptor tyrosine kinases VEGFR-1 (also known as Flt-1) and VEGFR-2 (also known as Flk-1 and KDR) (34,37,38), and to the non-tyrosine kinase transmembrane proteins neuropilin-1 and neuropilin-2 (39). In endothelial cells, VEGF-A is believed to transduce its biological activities mainly through VEGFR-2. However, many open questions remain regarding the role of other VEGF receptors and binding sites on endothelial and other cell types. For example, there is evidence that VEGFR-1 acts as a decoy receptor that sequesters VEGF and attenuates the activity of VEGFR-2 (37,40), and there is evidence that VEGFR-1 signals directly in response to VEGF-A to stimulate monocyte migration (41). Clearly, VEGF-A is critical to many aspects of tissue growth and repair, and the activity of this growth factor is controlled in a complex manner. The complexity is even more daunting when one considers the potential roles of the various co-receptors on the cell surface and ECM-binding sites, which not only interact with VEGF-A but also with VEGF receptors. Thus, it is clear that useful pharmacodynamic models will need to consider a wide array of factors present on and near the target cell surface.

VEGF-A expression is regulated at the transcriptional, translational, and posttranslational levels. Of particular interest is the regulation of VEGF expression by hypoxia-induced factor (HIF), a transcription factor that is stabilized at low oxygen pressure (42). Through this mechanism, increased VEGF-A expression is thought to drive the development of new blood vessels to hypoxic tissue sites. It is also interesting to note that the deposition of VEGF-A within the ECM might additionally be modulated by the local tissue environment. Indeed,

VEGF-A binding to HS and fibronectin, two major ECM components, is enhanced at acidic pH, indicating a possible biological mechanism for VEGF-A deposition at sites of tissue injury where local hypoxia would lead to acidification of the extracellular space (43–45). This pH-regulated growth factor deposition might, in turn, be co-opted for the development of regulated growth factor release systems. For example, incorporation of VEGF-A into synthetic ECM-like devices at low pH could provide a local depot of growth factor that would progressively release VEGF-A as the tissue becomes vascularized and the local pH returns to neutrality. Thus, the regulation of VEGF-A by interactions with the ECM will be a critical consideration in the design of approaches to deliver VEGF-A in a therapeutically effective manner.

THE EXTRACELLULAR MATRIX

The ECM is a complex fibrillar mesh comprised of proteins, proteoglycans, and glycosaminoglycans. The ECM provides the structural support for cells to organize into tissues, and its properties vary widely depending on the particular functional demands placed on it. For example, the fibrillar collagen network and elastic fibers of the lung provide the lung with its unique ability to expand and contract to allow for gas exchange, whereas the mineralized collagen within bone imparts the structural strength required to support the body. While these classic structural roles for the ECM have been appreciated for some time, more recent research has revealed a wide array of activities for the ECM. Indeed, the specific composition of the ECM can directly influence cells through matrix receptors on the cell surface for ECM components that when activated lead to intracellular signaling and altered metabolism and synthesis of the ECM itself. In this way, a sophisticated signaling loop can lead to the development, maintenance, and repair of complex tissues. In addition to these direct ECM-cell interactions, it has also become recognized that the ECM is the medium of communication between cells. Hence, all soluble messenger molecules are regulated to some degree as they encounter the ECM through a range of interactions between the soluble messenger molecules (i.e., growth factors) and the components of the ECM. Thus, the high density of high-affinity binding sites for growth factors within the ECM creates a situation where growth factors can be considered as "transient" components of the ECM. As such, the local pharmacodynamics and metabolism of growth factors can vary quite extensively at the surface of cells on the basis of the local composition of the ECM.

Heparan Sulfate Proteoglycans

While there are reports of growth factors binding to a wide range of ECM components, the large number of growth factors that have been shown to bind to heparin, coupled with the structural complexity of HS, indicates that these polysaccharides play particularly important roles in controlling growth factors in vivo. Heparan sulfate proteoglycans (HSPGs) represent a class of complex macromolecules that are comprised of a protein core and at least one covalently linked HS chain. HS chains are linear polysaccharides characterized by repeating disaccharide units of alternating N-substituted glucosamine and hexuronic (glucuronic or iduronic) acid residues (46). HS chains are subject to extensive modification during biosynthesis, including sulfation at the N-position as well as at the C-6 and C-3 O-positions of the glucosamine and at the C-2 O-position of the uronic acid (47,48). Thus, there are theoretically as many as 48 potential disaccharide units that together make this class of compounds one of the most information dense in biology. In addition, HS has domain structures made up of extended sequences of high or low levels of sulfation, making its overall structure and sequence considerably complex (49). The high degree of structural complexity likely underlies the ability of heparin and HS to interact with a wide array of proteins as a means to modulate their biodistribution and activity (4,47,48,50–52).

HSPGs are physically positioned on the cell surface, within the ECM, or in soluble form. Approximately 20 HSPG core proteins have been identified. They include the syndecans 1 to 4 and the glypicans 1 to 6, which are widely expressed in nearly all mammalian cell types and tissues where they are linked to plasma membrane surfaces either through a transmembrane core protein (syndecans) or by a glycosylphosphatidylinositol tail (glypicans) (53,54). In addition, the ectodomain of the syndecans, which contains the attached HS chains, can be shed from the cell surface through the action of extracellular proteases to produce "soluble" HSPG fragments (HSPGfs) (55–58). In contrast to the cell surface HSPGs, several HSPGs, such as perlecan, are localized to basement membranes and other extracellular matrices where their HS chains modulate matrix structure and molecular transport (3). In addition to the traditional role of HSPG within the extracellular space, evidence has suggested roles for these molecules within cells (59–64). Indeed, the principal role of the highly sulfated HS produced by mast cells (heparin) is to store mast cell proteases in an inactive form within secretory granules (65). Consequently, the physical location and specific HS structure allow HSPG to alternatively inhibit or promote cellular responses through direct and indirect mechanisms (29,66,67).

Growth Factor–Binding Proteoglycans

Since the original observation that members of the FGF family bind heparin, many growth factors have been demonstrated to bind to heparin, and it is now accepted that this interaction is indicative of important regulatory activities of HS chains on HSPG. The importance of these interactions for the prototypic heparin-binding growth factor, FGF-2, was originally revealed in cells that were rendered HS deficient through either genetic mutation or treatment with heparinase or sulfation inhibitor chlorate (68,69). In these instances, it was observed that FGF-2 binding to its receptor was significantly reduced in the absence of HS. The data indicate that HS participates in forming a ternary complex so that binding of FGF-2 is stabilized through the simultaneous binding to both its receptor and HS (70,71). Hence, kinetic analysis of these binding events suggested that the principal impact of HS is to reduce the rate of FGF-2 dissociation from its receptor without significantly altering the association rate (70,72,73). While the majority of the studies on the co-receptor activity of HS have focused on FGF-2, high-affinity binding of a number of growth factors for their receptors has similarly been shown to depend on HS. For example, receptor binding of VEGF is reduced in endothelial cells treated with heparinase or chlorate (43,74–77), and the binding and activity of heparin-binding epidermal growth factor–like growth factor to EGF receptors on vascular smooth muscle cells are reduced by chlorate treatment (78). Thus, HS likely plays a general role in enhancing the sensitivity of receptors for heparin-binding growth factors through its ability to increase the observed affinity for binding.

As one considers the classic principles of pharmacology, which tend to consider parameters such as expression of receptors on target cells along with traditional elimination and tissue adsorption constants, it is clear that additional components will need to be included to model growth factor responsiveness. Moreover, since the therapeutic window of growth factors in vivo is likely very narrow with the potential for toxicity or unwanted responses at high doses, approaches that simply aim to maximize growth factor concentration at the target site for as long as possible will not be effective. In vitro studies have revealed a complex concentration-response relationship for FGF-2 that is further modulated by the relative presence of HSPGs on the cell surface (Fig. 2). Toward this end, we have defined the relevant kinetic parameters that define FGF-2

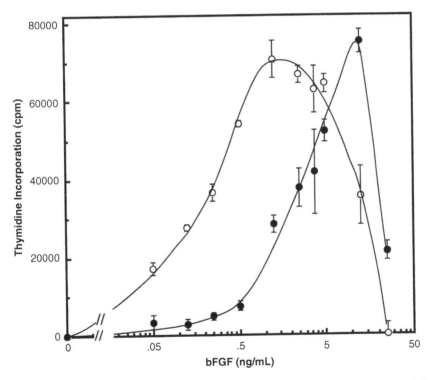

FIGURE 2 FGF-2 activity shows a narrow therapeutic window that is modulated by HSPG in vitro. Balb/c3T3 cells were treated with (●) or without (o) chlorate (50 mM) for 72 hours to inhibit the production of sulfated HSPGs. Nonsulfated HSPGs produced by chlorate-treated cells cannot bind to FGF-2 or enhance FGF-2 binding to its receptors. FGF-2 was added along with [3H] thymidine, and the cells were incubated for 36 hours. Newly synthesized DNA (^{3}H-DNA) was quantitated and presented on the *y*-axis as "Thymidine Incorporation." Maximal response to FGF-2 was similar in these cells with and without HSPG present. However, the therapeutic window was shifted approximately 10-fold. *Abbreviations*: HSPG, heparan sulfate proteoglycan; FGF, fibroblast growth factor; FGF-2 or bFGF, basic FGF. *Source*: From Ref. 79.

binding, internalization, and degradation by cells in culture in the presence and absence of HSPGs and have developed comprehensive mathematical models of these processes (80–82). These models are based on a series of ordinary differential equations that can effectively capture and predict the activation of FGF receptors on cells over time. However, even in these very controlled in vitro cell culture systems, a clear and direct link between receptor complex formation and biological response has been elusive. Indeed, studies by our laboratories and others have demonstrated that FGF-2 can elicit an array of cellular responses that are related to period and extent of activation, some of which may be mediated directly by FGF-2 binding to syndecan 4 independent of FGF receptors (83–85). Continued studies in vitro should allow models of growth factor response to become more sophisticated so that they will provide tools to accurately predict response. Extending these types of dynamic models to the more complex in vivo environment will require more complete understanding of the full range of processes that influence growth factor elimination, degradation, and metabolism in cells.

Mechanism of Co-Receptor Function of HSPGs

The recognition of the importance of HSPGs as co-receptors for FGF-2 has led to proposed approaches for FGF-2 applications that consider this HSPG function. For example, approaches involving the codelivery of heparin, HS, or HSPG, or the use of HS mimetics are being considered. Hence, a detailed understanding of the underlying mechanism of the HSPG co-receptor process is required. The chemical and physical mechanisms of this process have received considerable attention over the past 10 to 15 years. The use of oligosaccharides and chemically modified forms of heparin and HS has demonstrated that a pentasaccharide with N-sulfated glucosamine residues and at least one iduronic acid containing a sulfate in the 2-O-position is the minimal heparin unit that can bind FGF-2 (86,87). However, longer oligosaccharides (dodecasaccharides) containing 6-O sulfate groups in addition to 2-O and N sulfation are required to facilitate FGF-2 binding to its receptor (86,88). The additional requirements for ternary complex formation seemingly reflect the 6-O sulfated glucosamine–dependent binding of HS to FGF receptors (71,86,89,90). Thus, the HS–FGF-2-receptor complex likely consists of two FGF-2 molecules bound to two FGF receptors with a sufficiently long HS chain that makes contacts with all the protein components to effectively stabilize the complex. However, the exact stoichiometry and physical orientation of the various components remain an area of open investigation in spite of the availability of several FGF receptor crystal structures (91–96). The distinct HS structural requirements for binding to FGF-2 versus those for enhanced receptor binding provide an explanation for the fact that some HS species facilitate FGF-2 activity while others act in an inhibitory manner. Indeed, short heparin-derived oligosaccharides and 6-O desulfated heparin have been shown to be potent inhibitors of FGF-2 through the ability to sequester FGF-2 from its receptor (86,97). In some instances, even heparin- and HS-containing sequences capable of binding both FGF-2 and its receptor can inhibit receptor binding and activation. This is generally the case under conditions where heparin or HS molecules are not able to associate with the cell surface and are thus unable to come into proximity of the receptors (67,82,98,99). Thus, HS structure and physical localization, as well as the state of the cell surface of the target cell, are all likely to contribute to the ability of particular HS species to modulate growth factor–receptor binding. An additional consequence of this is that the state of the ECM, in particular HSPGs, will be an important determining factor in how it will interact with and modulate growth factors such as FGF-2.

Fibronectin

Another major modulator of molecular interactions within the ECM is fibronectin. Fibronectin is a high-molecular-weight glycoprotein found in a soluble form in plasma at high concentrations (300 µg/mL) as well as in an insoluble fibrillar form within the ECM of a variety of tissues (100). Fibronectin molecules are disulfide-linked dimers of approximately 500,000 kDa. The two chains ($\sim$250,000 kDa each) are similar in sequence but not identical. The diverse forms of fibronectin (at least 10 distinct polypeptide chains) are generated by alternative splicing of a common mRNA precursor and through differential posttranslational modifications (101–104). Although soluble plasma fibronectin is found almost exclusively as a dimer, it can form covalently linked multimers when incorporated in the ECM through both intermolecular disulfide bonds and other cross-linking events (105,106). Sequence analysis has shown that fibronectin is a modular protein containing three different types of repeats, known as homology types I, II, and III (100,107). Later structural studies have confirmed the hypothesis that these repeats correspond to individually folded modules linked via short flexible regions (108,109).

Fibronectin as a Site of Integration of Multiple Interactions

Individual modules in fibronectin combine to form proteolytically resistant domains that exhibit distinct binding properties (110). Numerous studies employing sequential proteolysis using several proteases have led to our current understanding regarding the structure of fibronectin (111–115). Fibrin-binding sites have been localized on the 29-kDa N-terminal and the 19-kDa C-terminal fragments, which consist exclusively of type I modules (116,117). The central 110-kDa fragment (type III modules) contains the cell-binding sites, including the arginine-glycine-aspartate (RGD) sequence, as well as several cryptic sites involved in fibronectin fibrillogenesis. The C-terminal 40-kDa fragment, containing three or four type III modules depending on splicing, contains a high-affinity heparin-binding site (Hep II) (118–121), although there are some studies that show weak affinity for heparin within the 29-kDa N-terminal fragment (122). Fibronectin contains only a pair of type II modules that can be released as part of a 40-kDa fragment, located between the 29-kDa N-terminal and 110-kDa central fragments (110). Type II modules are important for collagen/gelatin binding to fibronectin (123–125).

Several growth factors can interact with fibronectin. Connective tissue growth factor (CTGF) binds to type I modules of fibronectin and enhances the affinity of fibronectin for fibrin, with important implications for cell adhesion and migration during wound healing (126,127). Hepatocyte growth factor (HGF) also interacts with fibronectin, and these binding events modulate the angiogenic activity of HGF (128,129). In addition fibronectin has been linked to the insulin-like growth factor (IGF) system by its ability to interact with IGF-binding protein 3 and IGF-binding protein 5, thus regulating the transfer, availability, and biological activity of IGFs (130,131). The interactions between VEGF and fibronectin and their implications for angiogenesis will be discussed below.

Fibronectin Interactions with Heparin/HSPGs

As mentioned above, fibronectin contains at least two heparin/HS-binding sites (121,132), but there is evidence that the interaction is dominated by the C-terminal 40-kDa Hep II domain (type III repeats 12–14) (119,133). During cell adhesion to fibronectin, binding of HS chains of syndecan 4 to the Hep II domain is necessary for focal adhesion formation (134). It is thought that multiple interactions between fibronectin and HS chains lead to oligomerization of cell surface syndecans with consequent activation of protein kinase C (PKC) and focal adhesion formation (53,135).

Structural studies indicate that interactions of fibronectin with heparin and HS chains cause conformational alterations in fibronectin (45,136) associated with exposure of otherwise cryptic binding sites, as well as self-association sites, which play important roles in fibronectin fibrillogenesis and ECM assembly (137–139). Indeed, it is known that heparin stimulates fibronectin fibrillogenesis (140,141). Moreover, cells impaired in HS synthesis or sulfation show reduced fibronectin matrix assembly (142,143).

Fibronectin Interactions with VEGF

Several studies have shown that fibronectin possesses VEGF-binding sites (44,144,145). These binding sites are cryptic and can be exposed upon treatment of fibronectin with heparin (45). They have been localized on the Hep II domain of fibronectin (146). The interactions between fibronectin and VEGF are more pronounced at acidic pH (44). Acidic pH can be considered as a characteristic of hypoxic tissue, since lack of oxygenation leads to anaerobic metabolism as well as an O_2/CO_2 imbalance and, consequently, a drop in the extracellular pH (147–150). It is known that hypoxia regulates VEGF at the gene expression and mRNA levels (151–153), and it is possible that it can modulate the biological function of VEGF at higher levels (44). Deposition of VEGF within the ECM through interactions with fibronectin could generate a gradient that would direct the growth of new blood vessels. The formation of such a gradient would be dictated by an underlying pH gradient generated within the hypoxic tissue, with the pH decreasing with distance from the existing vasculature. Such regulation of VEGF-fibronectin interactions by the tissue microenvironment may be an important aspect of the control mechanisms operating during wound healing and tissue repair.

GROWTH FACTOR–ECM INTERACTION MODELS
AND APPLICATIONS

The interaction of growth factors with the ECM can obviously impact growth factor distribution and activity in vivo. The original observation that many growth factors bind to heparin suggested that HS within the ECM might function as a storage depot for these potent growth-regulatory proteins (21,154,155). Moreover, as additional binding sites for growth factors within the ECM have been identified, it has become clear that the ECM would likely play a critical role in defining growth factor pharmacokinetics and pharmacodynamics. For the FGF family members, it appears that HS interactions constitute the major sites of interaction within the ECM. Indeed, several studies have confirmed that members of the FGF family are endogenously deposited within the basement membrane bound to HS (33,156–159). The interaction of FGFs and VEGF with

HS has also been demonstrated to protect these proteins against proteases and physical denaturation (160–163). Furthermore, ECM, containing stored growth factors, has been demonstrated to function as a prolonged source of active growth factor, suggesting that the extracellular matrices in vivo play important roles as reservoirs of active growth factor (22,155,164,165). Several studies have also demonstrated that degradation of HSPG, or ECM in general, by heparanase or inflammatory proteases can result in the release of active growth factor that might participate in the tissue response to injury (16,58,166–168). In a sense, the ECM may be viewed as an endogenous matrix that mediates the controlled release of growth factors. The clinical application of this regulated storage and release system has already begun to be realized with the development of synthetic heparin-based controlled drug delivery systems, which are currently being evaluated in humans (169–172).

The storage and release of growth factors from ECM are likely controlled at many levels. Obviously, alterations in the local synthesis and deposition of ECM components and growth factors would alter the dynamics of this process, indicating that the "stored" growth factors are not static but in constant dynamic equilibrium with matrix-binding sites (14,70,72,173). Consequently, changes in the structure and density of growth factor–binding sites within the ECM, for example, through the action of extracellular enzymes such as the 6-O sulfatases, heparanases, or matrix metalloproteases, could modulate the binding kinetics of growth factors to the resident HS chains and could alter the structure of HS-modulated proteins such as fibronectin, leading to changes in ECM-growth factor binding and release (174–178). In addition, the interactions of growth factors with HSPG within the ECM might also be modulated by changes in the extracellular environment. As a specific example of this type of process, VEGF binding to HS is dramatically stabilized at low pH, suggesting that VEGF deposition in ECM would be controlled by factors such as hypoxia, which lead to decreased local pH (43,44). Together, these myriad actions have been demonstrated to provide a means for the ECM to modulate the molecular transport of growth factors and, in turn, the formation of growth factor and morphogen gradients required for coordinated tissue development (51,177–181). An appreciation of how this physiological process can modulate growth factor action is needed as growth factor treatment regimes are designed.

Mathematical Model of HSPG Influence on Growth Factor Movement

While the in vivo pharmacokinetics and pharmacodynamics of growth factors will relate to a large number of processes within complex whole-organism systems, it is clear that the role of ECM interactions will need to be understood quantitatively. Hence, a system to measure growth factor transport through isolated basement membranes has been developed (14,182), and subsequent studies revealed that FGF-2 binding to HS within the basement membrane was the major determinant of growth factor transport rate. Consequently, mathematical models of this transport process have been generated so that the full range of this effect can be evaluated under a variety of conditions that might relate to various tissues and target sites for growth factor treatment. The initial approach treats the transport of growth factors through the ECM as a diffusion-with-reaction process where the reversible binding of growth factor to HSPG

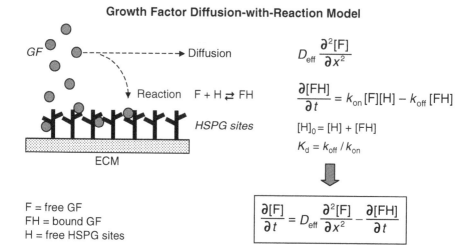

FIGURE 3 Schematic description of GF diffusion and reaction with ECM. The movement of heparin-binding GF was modeled as being governed by diffusion and reversible binding to HSPG sites within the ECM. Binding of GF to HSPGs is described by mass action reversible binding where K_d represents the equilibrium dissociation binding constant. The concentration of unbound (free) GF at any time is described by the equation in the box where D_{eff} represents the effective diffusion coefficient and x the distance from the initial GF source. *Abbreviations*: HSPG, heparan sulfate proteoglycan; GF, growth factor; ECM, extracellular matrix.

resident within the ECM is controlled by the affinity of this interaction and the numbers of binding sites (Fig. 3). An example of how these kinds of models can be developed and utilized is described below.

System Description

To develop a mathematical model to examine the effect of HSPG on growth factor movement through the ECM, some simplifying assumptions were made about the physical system. The ECM was considered to be a large volume of isotropic material with a homogenous distribution of stationary HSPG sites. A region of tissue where growth factor is introduced was represented as a plane surface of finite width that is placed in the ECM. At time $t = 0$, growth factor is instantaneously released from this region and diffuses through the surrounding medium. As free growth factor (F) moves through the matrix, reversible binding occurs with HSPG sites (H), resulting in fixed and nondiffusing growth factor (FH). The importance of this interaction on growth factor movement can be evaluated by knowing how the concentration of diffusing growth factor varies in response to changes in HSPG binding. In mathematical terms, this means that the concentration of free growth factor must be defined in terms of position and time through the ECM.

Mathematical Development

The simplified physical system described in the preceding section was simulated by using a mathematical model based on one-dimensional diffusion with reaction in an infinite slab (see relevant equations in Appendix 1) (183,184). This model is derived by writing mass balances for both free growth factor and bound growth factor. Each mass balance relates the accumulation of the component in the volume to the net amount that diffuses in through the surface plus the net amount that is produced by reaction within the volume. After some manipulations, substituting Fick's law for the diffusive term and assuming a constant diffusion coefficient, expressions are obtained for the change in each component with respect to time. In equation (1) for free growth factor (Appendix 1), [F], is the concentration, D_{eff} is the effective diffusion coefficient for diffusion in a porous medium, r_F is the rate of production per volume, x is the distance, and t is the time. Similarly, in equation (2) for bound growth factor, [FH] is the concentration and r_{FH} is the rate of production per volume. It is obvious that equation (2) does not have a diffusive term because the component is bound to the ECM and does not diffuse.

The reaction terms in equations (1) and (2) are evaluated by writing the reversible binding reaction between free growth factor and each HSPG site [equation (3), Appendix 1], where k_{on} and k_{off} are the rate constants for association and dissociation, respectively. The binding sites are assumed to be homogeneous, with 1:1 binding and no cooperativity. They are constrained by the mass balance in equation (4), which states that the initial concentration of sites $[H]_0$ is equal to the concentration of unoccupied sites [H] plus the concentration of occupied sites represented by the concentration of bound growth factor [FH]. From the binding reaction in equation (3), the value of r_F is set equal to the rate of dissociation minus the rate of association, and after substitution for [H] from equation (4), the expression for r_F becomes equation (5). Since $r_{FH} = -r_F$ [equation (6)], appropriate substitution of these expressions into equations (1) and (2) gives the working equations for the model in the form of equations (7) and (8). These equations are subject to the initial conditions and boundary conditions specified by equations (9) and (10), respectively. The initial conditions [equation (9)] indicate that the instantaneous pulse of free growth factor is located in a region of width $2w$ with an initial concentration of $[F]_0$. There are no HSPG sites within this region. The concentration of bound growth factor is zero everywhere. The boundary conditions [equation (10)] impose an infinite sink on the system, forcing both components to zero at the extremes. Equations (7) to (10) form the mathematical basis for the model of this system, and the solution gives the distributions of free growth factor and bound growth factor through the matrix as a function of time.

Numerical Implementation

The solution to the system defined by equations (7) to (10) is obtained by using a numerical method in which the partial derivatives are replaced by finite differences (185). In this case, the second-order partial derivative is substituted with a central-difference approximation, and the first-order partial derivatives are replaced with forward-difference approximations [equations (11–13)]. As a result of these transformations, equations (14) and (15) become the new working equations for the model. These equations allow for marching along the axes of a grid in the x, t plane, where i represents a node on the x-axis (distance) in increments of Δx and m represents a node on the t-axis (time) in increments of Δt. At time $t = 0$, the values of [F] and [FH] are known at all x_i from the initial conditions. At time $t = \Delta t$, the first increment of time, new values of [F] are calculated implicitly from equation (14) by generating a set of linear equations for all nodes along the x-axis on the basis of values from the previous time step and the boundary conditions. If these equations are written in matrix-vector form, the coefficient matrix is tridiagonal and diagonally dominant and the equations can be solved by Gaussian elimination. For [FH] at this time step, new values are determined explicitly from equation (15) using values from the previous time step. By repeating this procedure for as many time steps as desired, concentration profiles for [F] and [FH] are produced as a function of time. Because both implicit and explicit formulas are used in the approximation of the concentrations, this technique is called a semi-implicit finite difference method. The numerical implementation of this method can be efficiently carried out by computer (e.g., a program written in Fortran 90/95 using a double-precision IMSL (International Mathematical and Statistical Libraries) routine called DLSLTR for solving a tridiagonal linear system by Gaussian elimination). Results from the semi-implicit finite difference method are in good agreement with the analytical solution to the simpler problem of one-dimensional diffusion in an infinite slab.

Model Results

Values of model parameters. Figure 4 shows the output of the diffusion-with-reaction model for various conditions of growth factor diffusion through the ECM. Many of the values of the model parameters are based on information from previous studies (14,70,81,186). For Figure 4A and B, the y-axis is centered on a 20-μm-wide region that contains growth factor at an $[F]_0$ of 26 μM, and the concentration profiles $[F]/[F]_0$ appear as symmetric curves around this region after instantaneous release. The x-axis represents the perpendicular distance from the central plane of this region outward through the ECM. HSPG sites are at an $[H]_0$ of 10 μM in the ECM. Growth factor diffuses with a D_{eff} of 1×10^{-7} cm^2/sec and binds to HSPG with a K_d of 0.1 nM. Because binding is much faster than diffusion, components of free growth factor, HSPG, and bound growth factor are at local equilibrium.

Effect of HSPG on growth factor diffusion. Comparison of Figure 4A and B clearly shows the effect of the presence of HSPG in the ECM on the dissipation of growth factor. The curves in each figure represent the concentration distributions of free growth factor at successive times (0.05, 0.15, 0.5, and 5 minutes) from the initial release. In Figure 4A, there is no binding to HSPG in the matrix, and the results reflect typical distributions for diffusion with no reaction. After five minutes, there is still more than 4% of the initial concentration of growth factor within 100 μm of the release site. If a receptor for this growth factor has a K_d on the order of 10^{-11} M, then a functional response (50% receptor occupancy) is possible from a cell that is as far out as 400 μm. When HSPG binding is added to the matrix (Fig. 4B), growth factor binds to the proteoglycan as it diffuses outward, reducing the local concentration of free growth factor and providing the drive for more to diffuse into the area. As a result, when HSPG binding is present, growth factor is cleared more rapidly from the release site and at the same time prevented from establishing a wide range of influence. For example, after 30 seconds, a 4% concentration of free growth factor reaches its greatest extent at about 24 μm from the release site, and after five minutes, the concentration drops to less than 0.1% everywhere. A cell with receptors for this growth factor ($K_d \approx 10^{-11}$ M) is unlikely to produce a functional response at a distance greater than about 37 μm.

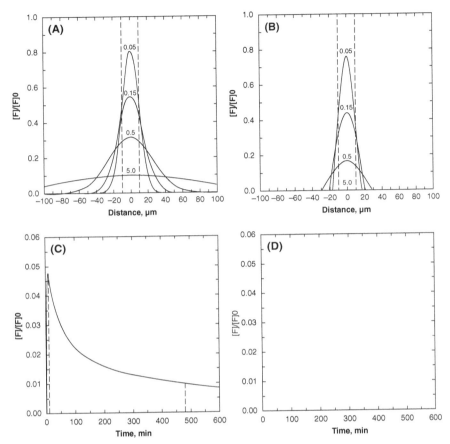

FIGURE 4 Simulation results for one-dimensional diffusion-with-reaction model of GF movement through ECM after instantaneous GF release from region of width 2*w*. (**A**) Concentration-distance curves for nonbinding GF at 0.05, 0.15, 0.5, and 5.0 minutes. (**B**) Concentration-distance curves for HSPG-binding GF at 0.05, 0.15, 0.5, and 5.0 minutes. (**C**) Concentration-time curve for nonbinding GF at 100 μm from release site. (**D**) Concentration-time curve for HSPG-binding GF at 100 μm from release site. Model parameters: $[F]/[F]_0$ = relative concentration of GF; $[F]_0 = 26$ μM; $[H]_0 = 10$ μM; $D_{eff} = 1 \times 10^{-7}$ cm^2/sec; $K_d = 0.1$ nM; $k_{off} = 0.01$/sec; $k_{on} = 1 \times 10^8$/M·sec; $w = 10$ μm; $\Delta x = 2$ μm; $\Delta t = 0.0001$ to 0.001 second. *Abbreviations*: HSPG, heparan sulfate proteoglycan; GF, growth factor; ECM, extracellular matrix.

Another way to view this effect is illustrated in Figure 4C for a location in the ECM that is 100 μm from the site of growth factor release. Without HSPG binding in the intervening space, a substantial pulse of growth factor is experienced at this location, peaking at about 5% after 10 minutes and delivering at least 1% for eight hours. With the addition of HSPG binding to the space, a strong pulse does not form, and growth factor cannot be detected at this location (Fig. 4D). Thus, the presence of HSPG has enormous consequences on the signaling potential of growth factors in the ECM.

Modification of growth factor diffusion through ECM. The base model describing growth factor movement through ECM can be used to evaluate a wide range of parameters (Fig. 5). For example, growth factor transport can be related to the variable levels of HSPG in the ECM in various tissues and in disease states or to the affinity of the specific growth factor (or engineered mutant) for HSPG. These molecules can also be used to evaluate how soluble heparin, HS, or HSPGfs might alter growth factor distribution. This is especially interesting as codelivery or cotreatment with heparin is often considered for growth factor–based angiogenic therapies. It is also important to recognize that areas of tissue injury or inflammation generally contain high levels of extracellular proteases, which would lead to the release of HSPGfs from the ECM.

In simulating the presence of HSPGfs, the central region of the model represented an area injured by protease attack possibly associated with inflammation or disease (Fig. 5B). The local damage to the ECM would result in the generation of HSPGfs, which would retain their ability to bind growth factor. These growth factor–HSPGf complexes would then be able to diffuse from the site of growth factor administration. Because growth factor–HSPGf complexes are larger and less globular than growth factor itself, the value of D_{eff} was assumed to be smaller by a factor of 10 (10^{-8} cm^2/sec vs. 10^{-7} cm^2/sec for growth factor only), and the complexes diffuse at a slower rate than growth factor alone. However, since the growth factor is bound to a proteoglycan fragment, it is less likely to bind to HSPG residing in the uninjured ECM as it diffuses away from the site of administration. As a result of these combined factors, the concentration profiles of growth factor–HSPGf complexes begin to approach those observed for nonbinding growth factor through ECM. On the basis of these results, the model suggests that growth

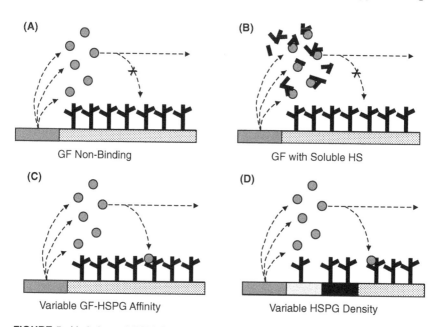

FIGURE 5 Variations of ECM-GF interactions that lead to altered transport. The movement of heparin-binding GFs through ECM can be manipulated by (**A**) the loss of heparin binding, (**B**) the presence of soluble heparin, HS, or HSPG fragments, (**C**) variations in the heparin-binding affinity, and (**D**) variations in the density of HSPGs within the ECM. *Abbreviations*: HSPG, heparan sulfate proteoglycan; GF, growth factor; HS, heparan sulfate; ECM, extracellular matrix.

factor bound to HSPGfs can be kept in the vicinity of an injured site for a much longer time than free growth factor, providing a large reservoir of growth factor to aid in the repair of tissue. These model predictions have begun to suggest interesting possibilities regarding approaches for administering growth factors in vivo. While it is clear that these models remain an oversimplification of the complex process of growth factor interactions with and movement through ECM, they have begun to provide critical insight that might help drive the development of more optimal growth factor therapies.

SUMMARY AND PROSPECTIVE

The range of potent biological activities that have been attributed to growth factors presents a compelling argument for their use in a wide range of clinical settings. FGF-2 and VEGF-A have especially been the subject of considerable attention, and a number of small and large clinical trials have evaluated the effectiveness of administering these proteins and their cognate gene sequences (1,2,24,31,187,188). However, the majority of these trials have produced little evidence of clear clinical benefit. The lack of success might reflect the fact that the design of these trials did not effectively account for the complexity of the endogenous mechanisms in place to control the action of these growth factors. In particular, the interactions of both FGF-2 and VEGF-A with the ECM are likely a major factor controlling tissue response. Thus, in this chapter, we have focused on how the interaction of these growth factors with HSPGs within the ECM might begin to be appreciated in a quantitative manner. The initial models that we have developed reveal a range of ways in which HSPGs might influence growth factor access to target cells. The interactions of growth factors with ECM might represent a process that could be exploited in the design of the most effective and specific therapeutic modalities. For example, modifications in growth factor binding to HSPGs via alterations in growth factor sequence and structure, or codelivery of heparin-like compounds might be used to effectively alter the local pharmacokinetics and pharmacodynamics of ECM-binding growth factors. Alternatively, increased knowledge about the specific composition of the ECM in various tissues and in particular disease states might allow for the targeting of growth factors on the basis of their particular ECM-binding characteristics. However, it is important to note that the interactions of growth factors with HSPGs that we have described in detail above are likely only one of the many important interactions that will impact growth factor activity and distribution. Indeed, the binding of VEGF-A to fibronectin likely represents the major interaction within the ECM for this growth factor. Thus, considerably more research is needed to fully identify the various elements involved in controlling growth factor function normally to develop comprehensive models that will allow the design of rational treatment regimes. In this way, the full potential of specific growth factor–based therapies may be realized.

REFERENCES

1. de Muinck ED, Simons M. Re-evaluating therapeutic neovascularization. J Mol Cell Cardiol 2004; 36(1):25–32.
2. Simons M. Angiogenesis: where do we stand now? Circulation 2005; 111(12):1556–1566.
3. Iozzo RV. Matrix proteoglycans: from molecular design to cellular function. Annu Rev Biochem 1998; 67:609–652.

4. Iozzo RV, San Antonio JD. Heparan sulfate proteoglycans: heavy hitters in the angiogenesis arena. J Clin Invest 2001; 108(3):349–355.
5. Vlodavsky I, Fuks Z, Ishai-Michaeli R, et al. Extracellular matrix-resident basic fibroblast growth factor: implication for the control of angiogenesis. J Cell Biochem 1991; 45(2):167–176.
6. Hausser H, Groning A, Hasilik A, et al. Selective inactivity of TGF-beta/decorin complexes. FEBS Lett 1994; 353(3):243–245.
7. Nugent MA, Forsten-Williams K, Karnovsky MJ, et al. Mechanisms of cell growth regulation by heparin and heparan sulfate. In: Garg HG, ed. Chemistry and Biology of Heparin and Heparan Sulfate. Oxford: Elsevier Limited, 2005:533–570.
8. Leon A, Buriani A, Dal Toso R, et al. Mast cells synthesize, store, and release nerve growth factor. Proc Natl Acad Sci U S A 1994; 91(9):3739–3743.
9. Metcalfe DD, Baram D, Mekori YA. Mast cells. Physiol Rev 1997; 77(4):1033–1079.
10. Boesiger J, Tsai M, Maurer M, et al. Mast cells can secrete vascular permeability factor/vascular endothelial cell growth factor and exhibit enhanced release after immunoglobulin E-dependent upregulation of fc epsilon receptor I expression. J Exp Med 1998; 188(6):1135–1145.
11. Grutzkau A, Kruger-Krasagakes S, Baumeister H, et al. Synthesis, storage, and release of vascular endothelial growth factor/vascular permeability factor (VEGF/ VPF) by human mast cells: implications for the biological significance of VEGF206. Mol Biol Cell 1998; 9(4):875–884.
12. Norrby K. Mast cells and angiogenesis. APMIS 2002; 110(5):355–371.
13. Qu Z, Liebler JM, Powers MR, et al. Mast cells are a major source of basic fibroblast growth factor in chronic inflammation and cutaneous hemangioma. Am J Pathol 1995; 147(3):564–573.
14. Dowd CJ, Cooney CL, Nugent MA. Heparan sulfate mediates bFGF transport through basement membrane by diffusion with rapid reversible binding. J Biol Chem 1999; 274(8):5236–5244.
15. Edelman ER, Nugent MA, Karnovsky MJ. Perivascular and intravenous administration of basic fibroblast growth factor: vascular and solid organ deposition. Proc Natl Acad Sci U S A 1993; 90:1513–1517.
16. Vlodavsky I, Korner G, Ishai-Michaeli R, et al. Extracellular matrix-resident growth factors and enzymes: possible involvement in tumor metastasis and angiogenesis. Cancer Metastasis Rev 1990; 9:203–226.
17. Baird A, Ling N. Fibroblast growth factors are present in the extracellular matrix produced by endothelial cells in vitro: implications for a role of heparinase-like enzymes in the neovascular response. Biochem Biophys Res Commun 1987; 142:428–435.
18. Dabin I, Courtois Y. In vitro kinetics of basic fibroblast growth factor diffusion across a reconstituted corneal endothelium. J Cell Physiol 1991; 147:396–402.
19. Flaumenhaft R, Moscatelli D, Saksela O, et al. Role of extracellular matrix in the action of basic fibroblast growth factor: matrix as a source of growth factor for long-term stimulation of plasminogen activator production and DNA synthesis. J Cell Physiol 1989; 140:75–81.
20. Flaumenhaft R, Moscatelli D, Rifkin DB. Heparin and heparan sulfate increase the radius of diffusion and action of basic fibroblast growth factor. J Cell Biol 1990; 111:1651–1659.
21. Folkman J, Klagsbrun M, Sasse J, et al. Heparin-binding angiogenic protein-basic fibroblast growth factor-is stored within basement membrane. Am J Pathol 1988; 130:393–400.
22. Nugent MA, Edelman ER. Transforming growth factor β1 stimulates the production of basic fibroblast growth factor binding proteoglycans in Balb/c3T3 cells. J Biol Chem 1992; 267:21256–21264.
23. Chen RR, Silva EA, Yuen WW, et al. Integrated approach to designing growth factor delivery systems. FASEB J 2007; 21(14):3896–3903.
24. Annex BH, Simons M. Growth factor-induced therapeutic angiogenesis in the heart: protein therapy. Cardiovasc Res 2005; 65(3):649–655.

25. Rhodes JM, Simons M. The extracellular matrix and blood vessel formation: not just a scaffold. J Cell Mol Med 2007; 11(2):176–205.
26. Presta M, Dell'era P, Mitola S, et al. Fibroblast growth factor/fibroblast growth factor receptor system in angiogenesis. Cytokine Growth Factor Rev 2005; 16(2): 159–178.
27. Rusnati M, Presta M. Fibroblast growth factors/fibroblast growth factor receptors as targets for the development of anti-angiogenesis strategies. Curr Pharm Des 2007; 13 (20):2025–2044.
28. Ornitz DM, Itoh N. Fibroblast growth factors. Genome Biol 2001; 2(3):reviews3005.
29. Nugent MA, Iozzo RV. Fibroblast growth factor-2. Int J Biochem Cell Biol 2000; 32:115–120.
30. Eswarakumar VP, Lax I, Schlessinger J. Cellular signaling by fibroblast growth factor receptors. Cytokine Growth Factor Rev 2005; 16(2):139–149.
31. Khurana R, Simons M. Insights from angiogenesis trials using fibroblast growth factor for advanced arteriosclerotic disease. Trends Cardiovasc Med 2003; 13(3): 116–122.
32. Bush MA, Samara E, Whitehouse MJ, et al. Pharmacokinetics and pharmacodynamics of recombinant FGF-2 in a phase I trial in coronary artery disease. J Clin Pharmacol 2001; 41(4):378–385.
33. Friedl A, Filla M, Rapraeger AC. Tissue-specific binding by FGF and FGF receptors to endogenous heparan sulfates. Methods Mol Biol 2001; 171:535–546.
34. Ferrara N, Gerber HP, LeCouter J. The biology of VEGF and its receptors. Nat Med 2003; 9(6):669–676.
35. Muller YA, Christinger HW, Keyt BA, et al. The crystal structure of vascular endothelial growth factor (VEGF) refined to 1.93 A resolution: multiple copy flexibility and receptor binding. Structure 1997; 5(10):1325–1338.
36. Robinson CJ, Stringer SE. The splice variants of vascular endothelial growth factor (VEGF) and their receptors. J Cell Sci 2001; 114(pt 5):853–865.
37. Rahimi N. VEGFR-1 and VEGFR-2: two non-identical twins with a unique physiognomy. Front Biosci 2006; 11:818–829.
38. Cross MJ, Dixelius J, Matsumoto T, et al. VEGF-receptor signal transduction. Trends Biochem Sci 2003; 28(9):488–494.
39. Neufeld G, Kessler O, Herzog Y. The interaction of Neuropilin-1 and Neuropilin-2 with tyrosine-kinase receptors for VEGF. Adv Exp Med Biol 2002; 515:81–90.
40. Rahimi N, Dayanir V, Lashkari K. Receptor chimeras indicate that the vascular endothelial growth factor receptor-1 (VEGFR-1) modulates mitogenic activity of VEGFR-2 in endothelial cells. J Biol Chem 2000; 275(22):16986–16992.
41. Barleon B, Sozzani S, Zhou D, et al. Migration of human monocytes in response to vascular endothelial growth factor (VEGF) is mediated via the VEGF receptor flt-1. Blood 1996; 87(8):3336–3343.
42. Detmar M, Brown LF, Berse B, et al. Hypoxia regulates the expression of vascular permeability factor/vascular endothelial growth factor (VPF/VEGF) and its receptors in human skin. J Invest Dermatol 1997; 108(3):263–268.
43. Goerges AL, Nugent MA. Regulation of vascular endothelial growth factor binding and activity by extracellular pH. J Biol Chem 2003; 278(21):19518–19525.
44. Goerges AL, Nugent MA. pH regulates vascular endothelial growth factor binding to fibronectin: a mechanism for control of extracellular matrix storage and release. J Biol Chem 2004; 279(3):2307–2315.
45. Mitsi M, Hong Z, Costello CE, et al. Heparin-mediated conformational changes in fibronectin expose vascular endothelial growth factor binding sites. Biochemistry 2006; 45(34):10319–10328.
46. Rabenstein DL. Heparin and heparan sulfate: structure and function. Nat Prod Rep 2002; 19(3):312–331.
47. Esko JD, Selleck SB. Order out of chaos: assembly of ligand binding sites in heparan sulfate. Annu Rev Biochem 2002; 71:435–471.
48. Turnbull J, Powell A, Guimond S. Heparan sulfate: decoding a dynamic multifunctional cell regulator. Trends Cell Biol 2001; 11(2):75–82.

49. Gallagher JT, Turnbull JE, Lyon M. Heparan sulphate proteoglycans: molecular organisation of membrane—associated species and an approach to polysaccharide sequence analysis. Adv Exp Med Biol 1992; 313:49–57.

50. Shriver Z, Liu D, Sasisekharan R. Emerging views of heparan sulfate glycosaminoglycan structure/activity relationships modulating dynamic biological functions. Trends Cardiovasc Med 2002; 12(2):71–77.

51. Perrimon N, Bernfield M. Specificities of heparan sulphate proteoglycans in developmental processes. Nature 2000; 404(6779):725–728.

52. Gallagher JT. Heparan sulfate: growth control with a restricted sequence menu. J Clin Invest 2001; 108(3):357–361.

53. Bernfield M, Gotte M, Park PW, et al. Functions of cell surface heparan sulfate proteoglycans. Annu Rev Biochem 1999; 68:729–777.

54. Park PW, Reizes O, Bernfield M. Cell surface heparan sulfate proteoglycans: selective regulators of ligand-receptor encounters. J Biol Chem 2000; 275(39):29923–29926.

55. Subramanian SV, Fitzgerald ML, Bernfield M. Regulated shedding of syndecan-1 and -4 ectodomains by thrombin and growth factor receptor activation. J Biol Chem 1997; 272(23):14713–14720.

56. Rapraeger A, Bernfield M. Cell surface proteoglycan of mammary epithelial cells. Protease releases a heparan sulfate-rich ectodomain from a putative membrane-anchored domain. J Biol Chem 1985; 260(7):4103–4109.

57. Fitzgerald ML, Wang Z, Park PW, et al. Shedding of syndecan-1 and -4 ectodomains is regulated by multiple signaling pathways and mediated by a TIMP-3-sensitive metalloproteinase. J Cell Biol 2000; 148(4):811–824.

58. Buczek-Thomas JA, Nugent MA. Elastase-mediated release of heparan sulfate proteoglycans from pulmonary fibroblast cultures. A mechanism for basic fibroblast growth factor (bFGF) release and attenuation of bfgf binding following elastase-induced injury. J Biol Chem 1999; 274(35):25167–25172.

59. Richardson TP, Trinkaus-Randall V, Nugent MA. Regulation of heparan sulfate proteoglycan nuclear localization by fibronectin. J Cell Sci 2001; 114(pt 9):1613–1623.

60. Hsia E, Richardson TP, Nugent MA. Nuclear localization of basic fibroblast growth factor is mediated by heparan sulfate proteoglycans through protein kinase C signaling. J Cell Biochem 2003; 88(6):1214–1225.

61. Ishihara M, Fedarko NS, Conrad HE. Transport of heparan sulfate into the nuclei of hepatocytes. J Biol Chem 1986; 261(29):13575–13580.

62. Fedarko NS, Conrad HE. A unique heparan sulfate in the nuclei of hepatocytes: structural changes with the growth state of the cells. J Cell Biol 1986; 102(2):587–599.

63. Kovalszky I, Dudas J, Olah-Nagy J, et al. Inhibition of DNA topoisomerase I activity by heparan sulfate and modulation by basic fibroblast growth factor. Mol Cell Biochem 1998; 183(1–2):11–23.

64. Kolset SO, Prydz K, Pejler G. Intracellular proteoglycans. Biochem J 2004; 379(pt 2):217–227.

65. Humphries DE, Wong GW, Friend DS, et al. Heparin is essential for the storage of specific granule proteases in mast cells. Nature 1999; 400(6746):769–772.

66. Segev A, Nili N, Strauss BH. The role of perlecan in arterial injury and angiogenesis. Cardiovasc Res 2004; 63(4):603–610.

67. Fannon M, Forsten KE, Nugent MA. Potentiation and inhibition of bFGF binding by heparin: a model for regulation of cellular response. Biochemistry 2000; 39: 1434–1445.

68. Yayon A, Klagsbrun M, Esko JD, et al. Cell surface, heparin-like molecules are required for binding basic fibroblast growth factor to its high affinity receptor. Cell 1991; 64:841–848.

69. Rapraeger A, Krufka A, Olwin B. Requirement of heparan sulfate for bFGF-mediated fibroblast growth and myoblast differentiation. Science 1991; 252:1705–1708.

70. Nugent MA, Edelman ER. Kinetics of basic fibroblast growth factor binding to its receptor and heparan sulfate proteoglycan: a mechanism for cooperativity. Biochemistry 1992; 31:8876–8883.

71. Pantoliano MW, Horlick RA, Springer BA, et al. Multivalent ligand-receptor binding interactions in the fibroblast growth factor system produce a cooperative growth factor and heparin mechanism for receptor dimerization. Biochemistry 1994; 33 (34):10229–10248.

72. Ibrahimi OA, Zhang F, Hrstka SC, et al. Kinetic model for FGF, FGFR, and proteoglycan signal transduction complex assembly. Biochemistry 2004; 43(16):4724–4730.

73. Sperinde GV, Nugent MA. Heparan sulfate proteoglycans control bFGF processing in vascular smooth muscle cells. Biochemistry 1998; 37:13153–13164.

74. Gitay-Goren H, Soker S, Vlodavsky I, et al. The binding of vascular endothelial growth factor to its receptors is dependent on cell surface-associated heparin-like molecules. J Biol Chem 1992; 267:6093–6098.

75. Cohen T, Gitay-Goren H, Sharon R, et al. VEGF121, a vascular endothelial growth factor (VEGF) isoform lacking heparin binding ability, requires cell-surface heparan sulfates for efficient binding to the VEGF receptors of human melanoma cells. J Biol Chem 1995; 270(19):11322–11326.

76. Neufeld G, Cohen T, Gengrinovitch S, et al. Vascular endothelial growth factor (VEGF) and its receptors. FASEB J 1999; 13(1):9–22.

77. Gengrinovitch S, Berman B, David G, et al. Glypican-1 is a VEGF165 binding proteoglycan that acts as an extracellular chaperone for VEGF165. J Biol Chem 1999; 274 (16):10816–10822.

78. Higashiyama S, Abraham JA, Klagsbrun M. Heparin-binding EGF-like growth factor stimulation of smooth muscle cell migration: dependence on interactions with cell surface heparan sulfate. J Cell Biol 1993; 122:933–940.

79. Fannon M, Nugent MA. FGF binds its receptors, is internalized and stimulates DNA synthesis in Balb/c3T3 cells in the absence of heparan sulfate. J Biol Chem 1996; 271:17949–17956.

80. Gopalakrishnan M, Forsten-Williams K, Nugent MA, et al. Effects of receptor clustering on ligand dissociation kinetics: theory and simulations. Biophys J 2005; 89 (6):3686–3700.

81. Forsten-Williams K, Chua CC, Nugent MA. The kinetics of FGF-2 binding to heparan sulfate proteoglycans and MAP kinase signaling. J Theor Biol 2005; 233(4):483–499.

82. Forsten KE, Fannon M, Nugent MA. Potential mechanisms for the regulation of growth factor binding by heparin. J Theor Biol 2000; 205(2):215–230.

83. Chua CC, Rahimi N, Forsten-Williams K, et al. Heparan sulfate proteoglycans function as receptors for fibroblast growth factor-2 activation of extracellular signal-regulated kinases 1 and 2. Circ Res 2004; 94(3):316–323.

84. Zhang Y, Li J, Partovian C, et al. Syndecan-4 modulates basic fibroblast growth factor 2 signaling in vivo. Am J Physiol Heart Circ Physiol 2003; 284(6):H2078–H2082.

85. Tkachenko E, Rhodes JM, Simons M. Syndecans: new kids on the signaling block. Circ Res 2005; 96(5):488–500.

86. Guimond S, Maccarana M, Olwin BB, et al. Activating and inhibitory heparin sequences for FGF-2 (basic FGF). Distinct requirements for FGF-1, FGF-2, and FGF-4. J Biol Chem 1993; 268:23906–23914.

87. Turnbull JE, Fernig DG, Ke Y, et al. Identification of the basic fibroblast growth factor binding sequence in fibroblast heparan sulfate. J Biol Chem 1992; 267(15): 10337–10341.

88. Pye DA, Vives RR, Turnbull JE, et al. Heparan sulfate oligosaccharides require 6-O-sulfation for promotion of basic fibroblast growth factor mitogenic activity. J Biol Chem 1998; 273(36):22936–22942.

89. Kan M, Wang F, Xu J, et al. An essential heparin-binding domain in the fibroblast growth factor receptor kinase. Science 1993; 259(26):1918–1921.

90. Loo BM, Kreuger J, Jalkanen M, et al. Binding of heparin/heparan sulfate to fibroblast growth factor receptor 4. J Biol Chem 2001; 276(20):16868–16876.

91. Stauber DJ, DiGabriele AD, Hendrickson WA. Structural interactions of fibroblast growth factor receptor with its ligands. Proc Natl Acad Sci U S A 2000; 97(1):49–54.

92. Schlessinger J, Plotnikov AN, Ibrahimi OA, et al. Crystal structure of a ternary FGF-FGFR-heparin complex reveals a dual role for heparin in FGFR binding and dimerization. Mol Cell 2000; 6(3):743–750.

93. Plotnikov AN, Hubbard SR, Schlessinger J, et al. Crystal structures of two FGF-FGFR complexes reveal the determinants of ligand-receptor specificity. Cell 2000; 101(4):413–424.

94. Plotnikov AN, Schlessinger J, Hubbard SR, et al. Structural basis for FGF receptor dimerization and activation. Cell 1999; 98(5):641–650.

95. Pellegrini L. Role of heparan sulfate in fibroblast growth factor signalling: a structural view. Curr Opin Struct Biol 2001; 11(5):629–634.

96. Pellegrini L, Burke DF, von Delft F, et al. Crystal structure of fibroblast growth factor receptor ectodomain bound to ligand and heparin. Nature 2000; 407(6807):1029–1034.

97. Lundin L, Larsson H, Kreuger J, et al. Selectively desulfated heparin inhibits fibroblast growth factor-induced mitogenicity and angiogenesis. J Biol Chem 2000; 275(32):24653–24660.

98. Fannon M, Forsten-Williams K, Dowd CJ, et al. Binding inhibition of angiogenic factors by heparan sulfate proteoglycans in aqueous humor: potential mechanism for maintenance of an avascular environment. FASEB J 2003; 17(8):902–904.

99. Forsten KE, Courant NA, Nugent MA. Endothelial proteoglycans inhibit bFGF binding and mitogenesis. J Cell Physiol 1997; 172:209–220.

100. Mosher DF, ed. Fibronectin. San Diego: Academic Press Inc., 1989.

101. Paul JI, Hynes RO. Multiple fibronectin subunits and their post-translational modifications. J Biol Chem 1984; 259(21):13477–13487.

102. Paul JI, Schwarzbauer JE, Tamkun JW, et al. Cell-type-specific fibronectin subunits generated by alternative splicing. J Biol Chem 1986; 261(26):12258–12265.

103. Schwarzbauer JE, Paul JI, Hynes RO. On the origin of species of fibronectin. Proc Natl Acad Sci U S A 1985; 82(5):1424–1428.

104. Kornblihtt AR, Pesce CG, Alonso CR, et al. The fibronectin gene as a model for splicing and transcription studies. FASEB J 1996; 10(2):248–257.

105. Morla A, Zhang Z, Ruoslahti E. Superfibronectin is a functionally distinct form of fibronectin. Nature 1994; 367(6459):193–196.

106. Chen R, Gao B, Huang C, et al. Transglutaminase-mediated fibronectin multimerization in lung endothelial matrix in response to TNF-alpha. Am J Physiol Lung Cell Mol Physiol 2000; 279(1):L161–L174.

107. Odermatt E, Tamkun JW, Hynes RO. Repeating modular structure of the fibronectin gene: relationship to protein structure and subunit variation. Proc Natl Acad Sci U S A 1985; 82(19):6571–6575.

108. Sticht H, Pickford AR, Potts JR, et al. Solution structure of the glycosylated second type 2 module of fibronectin. J Mol Biol 1998; 276(1):177–187.

109. Constantine KL, Brew SA, Ingham KC, et al. 1H-n.m.r. studies of the fibronectin 13 kDa collagen-binding fragment. Evidence for autonomous conserved type I and type II domain folds. Biochem J 1992; 283(pt 1):247–254.

110. Pankov R, Yamada KM. Fibronectin at a glance. J Cell Sci 2002; 115(pt 20):3861–3863.

111. Sekiguchi K, Hakomori S. Domain structure of human plasma fibronectin. Differences and similarities between human and hamster fibronectins. J Biol Chem 1983; 258(6):3967–3973.

112. Sekiguchi K, Hakomori S. Topological arrangement of four functionally distinct domains in hamster plasma fibronectin: a study with combination of S-cyanylation and limited proteolysis. Biochemistry 1983; 22(6):1415–1422.

113. Ruoslahti E, Hayman EG, Engvall E, et al. Alignment of biologically active domains in the fibronectin molecule. J Biol Chem 1981; 256(14):7277–7281.

114. Richter H, Hormann H. Early and late cathepsin D-derived fragments of fibronectin containing the C-terminal interchain disulfide cross-link. Hoppe Seylers Z Physiol Chem 1982; 363(4):351–364.

115. Hayashi M, Yamada KM. Domain structure of the carboxyl-terminal half of human plasma fibronectin. J Biol Chem 1983; 258(5):3332–3340.

116. Makogonenko E, Tsurupa G, Ingham K, et al. Interaction of fibrin(ogen) with fibronectin: further characterization and localization of the fibronectin-binding site. Biochemistry 2002; 41(25):7907–7913.
117. Rostagno A, Williams MJ, Baron M, et al. Further characterization of the NH2-terminal fibrin-binding site on fibronectin. J Biol Chem 1994; 269(50):31938–31945.
118. Hayashi M, Schlesinger DH, Kennedy DW, et al. Isolation and characterization of a heparin-binding domain of cellular fibronectin. J Biol Chem 1980; 255(21):10017–10020.
119. Benecky MJ, Kolvenbach CG, Amrani DL, et al. Evidence that binding to the carboxyl-terminal heparin-binding domain (Hep II) dominates the interaction between plasma fibronectin and heparin. Biochemistry 1988; 27(19):7565–7571.
120. Ingham KC, Brew SA, Migliorini MM, et al. Binding of heparin by type III domains and peptides from the carboxy terminal hep-2 region of fibronectin. Biochemistry 1993; 32(46):12548–12553.
121. Sekiguchi K, Hakomori S, Funahashi M, et al. Binding of fibronectin and its proteolytic fragments to glycosaminoglycans. Exposure of cryptic glycosaminoglycan-binding domains upon limited proteolysis. J Biol Chem 1983; 258(23):14359–14365.
122. Gold LI, Frangione B, Pearlstein E. Biochemical and immunological characterization of three binding sites on human plasma fibronectin with different affinities for heparin. Biochemistry 1983; 22(17):4113–4119.
123. Ingham KC, Brew SA, Migliorini MM. Further localization of the gelatin-binding determinants within fibronectin. Active fragments devoid of type II homologous repeat modules. J Biol Chem 1989; 264(29):16977–16980.
124. Ingham KC, Brew SA, Miekka SI. Interaction of plasma fibronectin with gelatin and complement C1q. Mol Immunol 1983; 20(3):287–295.
125. Ingham KC, Brew SA, Isaacs BS. Interaction of fibronectin and its gelatin-binding domains with fluorescent-labeled chains of type I collagen. J Biol Chem 1988; 263(10):4624–4628.
126. Yoshida K, Munakata H. Connective tissue growth factor binds to fibronectin through the type I repeat modules and enhances the affinity of fibronectin to fibrin. Biochim Biophys Acta 2007; 1770(4):672–680.
127. Pi L, Ding X, Jorgensen M, et al. Connective tissue growth factor with a novel fibronectin binding site promotes cell adhesion and migration during rat oval cell activation. Hepatology 2008; 47(3):996–1004.
128. Rahman S, Patel Y, Murray J, et al. Novel hepatocyte growth factor (HGF) binding domains on fibronectin and vitronectin coordinate a distinct and amplified Met-integrin induced signalling pathway in endothelial cells. BMC Cell Biol 2005; 6(1):8.
129. Huang SD, Liu XH, Bai CG, et al. Synergistic effect of fibronectin and hepatocyte growth factor on stable cell-matrix adhesion, re-endothelialization, and reconstitution in developing tissue-engineered heart valves. Heart Vessels 2007; 22(2):116–122.
130. Gui Y, Murphy LJ. Insulin-like growth factor (IGF)-binding protein-3 (IGFBP-3) binds to fibronectin (FN): demonstration of IGF-I/IGFBP-3/fn ternary complexes in human plasma. J Clin Endocrinol Metab 2001; 86(5):2104–2110.
131. Xu Q, Yan B, Li S, et al. Fibronectin binds insulin-like growth factor-binding protein 5 and abolishes its ligand-dependent action on cell migration. J Biol Chem 2004; 279(6):4269–4277.
132. Yamada KM, Kennedy DW, Kimata K, et al. Characterization of fibronectin interactions with glycosaminoglycans and identification of active proteolytic fragments. J Biol Chem 1980; 255(13):6055–6063.
133. Ingham KC, Brew SA, Migliorini M. An unusual heparin-binding peptide from the carboxy-terminal hep-2 region of fibronectin. Arch Biochem Biophys 1994; 314(1):242–246.
134. Woods A, Longley RL, Tumova S, et al. Syndecan-4 binding to the high affinity heparin-binding domain of fibronectin drives focal adhesion formation in fibroblasts. Arch Biochem Biophys 2000; 374(1):66–72.

135. Mahalingam Y, Gallagher JT, Couchman JR. Cellular adhesion responses to the heparin-binding (HepII) domain of fibronectin require heparan sulfate with specific properties. J Biol Chem 2007; 282(5):3221–3230.
136. Osterlund E, Eronen I, Osterlund K, et al. Secondary structure of human plasma fibronectin: conformational change induced by calf alveolar heparan sulfates. Biochemistry 1985; 24(11):2661–2667.
137. Schwarzbauer JE. Identification of the fibronectin sequences required for assembly of a fibrillar matrix. J Cell Biol 1991; 113(6):1463–1473.
138. Ingham KC, Brew SA, Huff S, et al. Cryptic self-association sites in type III modules of fibronectin. J Biol Chem 1997; 272(3):1718–1724.
139. Watanabe K, Takahashi H, Habu Y, et al. Interaction with heparin and matrix metalloproteinase 2 cleavage expose a cryptic anti-adhesive site of fibronectin. Biochemistry 2000; 39(24):7138–7144.
140. Bultmann H, Santas AJ, Peters DM. Fibronectin fibrillogenesis involves the heparin II binding domain of fibronectin. J Biol Chem 1998; 273(5):2601–2609.
141. Richter H, Wendt C, Hormann H. Aggregation and fibril formation of plasma fibronectin by heparin. Biol Chem Hoppe Seyler 1985; 366(5):509–514.
142. Chung CY, Erickson HP. Glycosaminoglycans modulate fibronectin matrix assembly and are essential for matrix incorporation of tenascin-C. J Cell Sci 1997; 110(pt 12): 1413–1419.
143. Galante LL, Schwarzbauer JE. Requirements for sulfate transport and the diastrophic dysplasia sulfate transporter in fibronectin matrix assembly. J Cell Biol 2007; 179(5):999–1009.
144. Wijelath ES, Murray J, Rahman S, et al. Novel vascular endothelial growth factor binding domains of fibronectin enhance vascular endothelial growth factor biological activity. Circ Res 2002; 91(1):25–31.
145. Wijelath ES, Rahman S, Murray J, et al. Fibronectin promotes VEGF-induced CD34 cell differentiation into endothelial cells. J Vasc Surg 2004; 39(3):655–660.
146. Mitsi M. Heparin Catalyzes Conformational Changes in Fibronectin that Enhance Vascular Endothelial Growth Factor Binding [Ph.D.]. Boston: Boston University School of Medicine, 2008.
147. Xu L, Fukumura D, Jain RK. Acidic extracellular pH induces vascular endothelial growth factor (VEGF) in human glioblastoma cells via ERK1/2 MAPK signaling pathway: mechanism of low pH-induced VEGF. J Biol Chem 2002; 277(13):11368–11374.
148. Mousa SA, Lorelli W, Campochiaro PA. Role of hypoxia and extracellular matrix-integrin binding in the modulation of angiogenic growth factors secretion by retinal pigmented epithelial cells. J Cell Biochem 1999; 74(1):135–143.
149. D'Arcangelo D, Facchiano F, Barlucchi LM, et al. Acidosis inhibits endothelial cell apoptosis and function and induces basic fibroblast growth factor and vascular endothelial growth factor expression. Circ Res 2000; 86(3):312–318.
150. D'Arcangelo D, Gaetano C, Capogrossi MC. Acidification prevents endothelial cell apoptosis by Axl activation. Circ Res 2002; 91(7):e4–e12.
151. Shweiki D, Neeman M, Itin A, et al. Induction of vascular endothelial growth factor expression by hypoxia and by glucose deficiency in multicell spheroids: implications for tumor angiogenesis. Proc Natl Acad Sci U S A 1995; 92(3):768–772.
152. Stein I, Neeman M, Shweiki D, et al. Stabilization of vascular endothelial growth factor mRNA by hypoxia and hypoglycemia and coregulation with other ischemia-induced genes. Mol Cell Biol 1995; 15(10):5363–5368.
153. Shweiki D, Itin A, Soffer D, Keshet E. Vascular endothelial growth factor induced by hypoxia may mediate hypoxia-initiated angiogenesis. Nature 1992; 359(6398): 843–845.
154. Folkman J, Klagsbrun M. Angiogenic factors. Science 1985; 235:442.
155. Bashkin P, Doctrow S, Klagsbrun M, et al. Basic fibroblast growth factor binds to subendothelial extracellular matrix and is released by heparitinase and heparin-like molecules. Biochemistry 1989; 28(4):1737–1743.

156. Folkman J, Klagsbrun M, Sasse J, et al. A heparin-binding angiogenic protein–basic fibroblast growth factor–is stored within basement membrane. Am J Pathol 1988; 130(2):393–400.

157. Vlodavsky I, Folkman J, Sullivan R, et al. Endothelial cell-derived basic fibroblast growth factor: synthesis and deposition into the subendothelial extracellular matrix. Proc Natl Acad Sci U S A 1987; 84:2292–2296.

158. Whalen GF, Shing Y, Folkman J. The fate of intravenously administered bFGF and the effect of heparin. Growth Factors 1989; 1(2):157–164.

159. Friedl A, Chang Z, Tierney A, et al. Differential binding of fibroblast growth factor-2 and -7 to basement membrane heparan sulfate: comparison of normal and abnormal human tissues. Am J Pathol 1997; 150(4):1443–1455.

160. Gospodarowicz D, Cheng J. Heparin protects basic and acidic FGF from inactivation. J Cell Physiol 1986; 128:475–484.

161. Edelman ER, Nugent MA. Controlled release of basic fibroblast growth factor. Drug News Perspect 1991; 4(6):352–357.

162. Sommer A, Rifkin DB. Interaction of heparin with human basic fibroblast growth factor: protection of the angiogenic protein from proteolytic degradation by a glycosaminoglycan. J Cell Physiol 1989; 138:215–220.

163. Saksela O, Moscatelli D, Sommer A, et al. Endothelial cell-derived heparan sulfate binds basic fibroblast growth factor and protects it from proteolytic degradation. J Cell Biol 1988; 107:743–751.

164. Rogelj S, Klagsbrun M, Atzmon R, et al. Basic fibroblast growth factor is an extracellular matrix component required for supporting the proliferation of vascular endothelial cells and the differentiation of PC12 cells. J Cell Biol 1989; 109(2):823–831.

165. Presta M, Maier JAM, Rusnati M, et al. Basic fibroblast growth factor is released from endothelial extracellular matrix in a biologically active form. J Cell Physiol 1989; 140:68–74.

166. Rich CB, Nugent MA, Stone P, et al. Elastase release of basic fibroblast growth factor in pulmonary fibroblast cultures results in down-regulation of elastin gene transcription. A role for basic fibroblast growth factor in regulating lung repair. J Biol Chem 1996; 271(38):23043–23048.

167. Liu J, Rich CB, Buczek-Thomas JA, et al. Heparin-binding EGF-like growth factor regulates elastin and FGF-2 expression in pulmonary fibroblasts. Am J Physiol Lung Cell Mol Physiol 2003; 285(5):L1106–L1115.

168. Buczek-Thomas JA, Lucey EC, Stone PJ, et al. Elastase mediates the release of growth factors from lung in vivo. Am J Respir Cell Mol Biol 2004; 31(3):344–350.

169. Sellke FW, Laham RJ, Edelman ER, et al. Therapeutic angiogenesis with basic fibroblast growth factor: technique and early results. Ann Thorac Surg 1998; 65:1540–1544.

170. Laham RJ, Sellke FW, Edelman ER, et al. Local perivascular delivery of basic fibroblast growth factor in patients undergoing coronary bypass surgery: results of a phase I randomized, double-blind, placebo-controlled trial. Circulation 1999; 100 (18):1865–1871.

171. Edelman ER, Mathiowitz E, Langer R, et al. Controlled and modulated release of basic fibroblast growth factor. Biomaterials 1991; 12:619–626.

172. Nugent MA, Chen OS, Edelman ER. Controlled release of fibroblast growth factor: activity in cell culture. Mat Res Soc Symp Proc 1992; 252:273–284.

173. Moscatelli D. Basic fibroblast growth factor (bFGF) dissociates rapidly from heparan sulfate but slowly from receptors. J Biol Chem 1992; 267:25803–25809.

174. Wang S, Ai X, Freeman SD, et al. QSulf1, a heparan sulfate 6-O-endosulfatase, inhibits fibroblast growth factor signaling in mesoderm induction and angiogenesis. Proc Natl Acad Sci U S A 2004; 101(14):4833–4838.

175. Ai X, Do AT, Lozynska O, et al. QSulf1 remodels the 6-O sulfation states of cell surface heparan sulfate proteoglycans to promote Wnt signaling. J Cell Biol 2003; 162(2):341–351.

176. Dhoot GK, Gustafsson MK, Ai X, et al. Regulation of Wnt signaling and embryo patterning by an extracellular sulfatase. Science 2001; 293(5535):1663–1666.

177. Izvolsky KI, Zhong L, Wei L, et al. Heparan sulfates expressed in the distal lung are required for Fgf10 binding to the epithelium and for airway branching. Am J Physiol Lung Cell Mol Physiol 2003; 285(4):L838–L846.
178. Izvolsky KI, Shoykhet D, Yang Y, et al. Heparan sulfate-FGF10 interactions during lung morphogenesis. Dev Biol 2003; 258(1):185–200.
179. Lin X, Buff EM, Perrimon N, et al. Heparan sulfate proteoglycans are essential for FGF receptor signaling during Drosophila embryonic development. Development 1999; 126(17):3715–3723.
180. Lin X, Perrimon N. Developmental roles of heparan sulfate proteoglycans in Drosophila. Glycoconj J 2002; 19(4–5):363–368.
181. Lander AD, Nie Q, Wan FY. Do morphogen gradients arise by diffusion? Dev Cell 2002; 2(6):785–796.
182. Fannon M, Forsten-Williams K, Nugent MA, et al. Sucrose octasulfate regulates fibroblast growth factor-2 binding, transport, and activity: potential for regulation of tumor growth. J Cell Physiol 2008; 215(2):434–441.
183. Crank J. The Mathematics of Diffusion, 2nd ed. Oxford: Oxford University Press, 2004.
184. Cussler EL. Diffusion: Mass Transfer in Fluid Systems, 2nd ed. Cambridge: Cambridge University Press, 2003.
185. Golub GH, Ortega JM. Scientific Computing and Differential Equations: An Introduction to Numerical Methods. San Diego: Academic Press, 1992.
186. Sperinde GV, Nugent MA. Mechanisms of FGF-2 intracellular processing: a kinetic analysis of the role of heparan sulfate proteoglycans. Biochemistry 2000; 39:3788–3796.
187. Baklanov D, Simons M. Arteriogenesis: lessons learned from clinical trials. Endothelium 2003; 10(4–5):217–223.
188. Simons M, Ware JA. Therapeutic angiogenesis in cardiovascular disease. Nat Rev Drug Discov 2003; 2(11):863–871.

APPENDIX 1. EQUATIONS FOR DIFFUSION-WITH-REACTION MODEL

$$\frac{\partial [F]}{\partial t} = D_{\text{eff}} \frac{\partial^2 [F]}{\partial x^2} + r_F \tag{1}$$

$$\frac{\partial [FH]}{\partial t} = r_{FH} \tag{2}$$

$$F + H \underset{k_{\text{off}}}{\overset{k_{\text{on}}}{\rightleftharpoons}} FH \tag{3}$$

$$[H]_0 = [H] + [FH] \tag{4}$$

$$r_F = -k_{\text{on}}[F]([H]_0 - [FH]) + k_{\text{off}}[FH] \tag{5}$$

$$r_{FH} = -r_F = k_{\text{on}}[F]([H]_0 - [FH]) - k_{\text{off}}[FH] \tag{6}$$

$$\frac{\partial [F]}{\partial t} = D_{\text{eff}} \frac{\partial^2 [F]}{\partial x^2} - k_{\text{on}}[F]([H]_0 - [FH]) + k_{\text{off}}[FH] \tag{7}$$

$$\frac{\partial [FH]}{\partial t} = k_{\text{on}}[F]([H]_0 - [FH]) - k_{\text{off}}[FH] \tag{8}$$

$$t = 0 \quad \begin{matrix} -w < x < +w \\ -w > x > +w \end{matrix} \quad \begin{matrix} [F] = [F]_0 \\ [F] = 0 \end{matrix} \quad \begin{matrix} [FH] = 0 \\ [FH] = 0 \end{matrix} \quad \begin{matrix} [H] = 0 \\ [H] = [H]_0 \end{matrix} \tag{9}$$

$$t > 0 \quad x = \pm\infty \quad [F] = [FH] = 0 \tag{10}$$

$$\frac{\partial^2 [F]}{\partial x^2} = \frac{[F]_{i-1}^{m+1} - 2[F]_i^{m+1} + [F]_{i+1}^{m+1}}{\Delta x^2} \tag{11}$$

$$\frac{\partial [F]}{\partial t} = \frac{[F]_i^{m+1} - [F]_i^m}{\Delta t} \tag{12}$$

$$\frac{\partial [FH]}{\partial t} = \frac{[FH]_i^{m+1} - [FH]_i^m}{\Delta t} \tag{13}$$

$$\left(-\frac{D_{\text{eff}}\Delta t}{\Delta x^2} \right) [F]_{i-1}^{m+1} + \left(1 + \frac{2D_{\text{eff}}\Delta t}{\Delta x^2} \right) [F]_i^{m+1} + \left(-\frac{D_{\text{eff}}\Delta t}{\Delta x^2} \right) [F]_{i+1}^{m+1}$$
$$= [F]_i^m + \Delta t (k_{\text{on}} [F]_i^m [FH]_i^m - k_{\text{on}} [H]_0 [F]_i^m + k_{\text{off}} [FH]_i^m) \tag{14}$$

$$[FH]_i^{m+1} = [FH]_i^m + \Delta t (k_{\text{on}} [H]_0 [F]_i^m - k_{\text{on}} [F]_i^m [FH]_i^m - k_{\text{off}} [FH]_i^m) \tag{15}$$

4

Metabolic Processes at Injection Sites Affecting Pharmacokinetics, Pharmacodynamics, and Metabolism of Protein and Peptide Therapeutics

Randall J. Mrsny

Department of Pharmacy and Pharmacology, University of Bath, Bath, U.K.

INTRODUCTION

The advent of recombinant biotechnology and efficient synthetic processes has facilitated the development of a wide variety of protein and peptide therapeutics. Although a plethora of delivery routes and methods have been examined for the administration of these materials, to date the administration of these molecules is dominated by injection. Of these injected molecules, only a few, mostly high-dose antibodies, are administered by intravenous (IV) infusion; the vast majority being delivered by subcutaneous (SC) injection. Further, efforts are ongoing to identify formulation strategies to allow delivery of high-dose biopharmaceuticals such as antibodies that can also be administered by SC injection. Although SC injection is a prominent delivery route for many biopharmaceuticals, relatively little is known about specific events that might occur to these therapeutic agents following their injection into a SC site that could affect their pharmacokinetic (PK) properties, pharmacodynamic (PD) outcome(s), and metabolic fate.

The marketing of protein and peptide therapeutics has focused on SC injections as this route of administration can be performed at home by patients with minimal training. This strategy of limiting the involvement of health care professionals for the administration of therapeutic protein and peptide drugs has proven to be a critical component for widespread use of biotherapeutics. Further, the pharmaceutical industry has strived to make SC injections as pain-free, fast, and simple as possible to increase patient compliance and maximize the consistency of delivery for self-administered SC injections. This is particularly important for some protein or peptide therapeutics where, because of their PK and PD characteristics, patients perform multiple injections every day.

Presently, one of the few assessments made regarding the SC injection of a protein or peptide therapeutic is to compare its PK profile and overall bioavailability to that observed following an IV infusion. In some ways, this comparison is an important indicator of potential events that might occur at the SC injection site. These SC to IV comparisons, however, provide little, if any, specific information about events at the SC injection site. This chapter will examine some potential events that could occur to a protein or peptide therapeutic following its SC injection that might affect its PK profile, PD outcome, and metabolic fate. The goal of this examination is to highlight potential mechanisms that could result in undesirable outcomes for injected protein and peptide therapeutics.

BACKGROUND

Before examining the various potential issues that might affect PK, PD, and metabolic parameters of SC injected proteins and peptides, one should ask the basic question: why. In general, SC injections of protein and peptide therapeutics have proven to be particularly successful for a wide range of diseases and conditions. One must consider, however, that any list of marketed (successful) products does not include the failures that have occurred, and indeed there have been failures for a number of promising protein and peptide therapeutics following their SC injection. Additionally, there is the common perception that outcomes for SC injections at various sites of the body (typically performed at the posterior surface of the upper arm, anterior surface of the thigh, buttocks, and abdominal regions that exclude the navel and waistline) provide comparable PK, PD, and metabolism outcomes. The reality, however, is that SC injection at these various sites are not always the same. That diabetic patients must recalibrate their insulin dose on the basis of the anatomical site used for injection (1) while the same patient population does not need to do this for SC injections of Exenatide (2) suggests that the inherent properties of a protein or peptide therapeutic can dictate its PK, PD, and metabolism outcomes following SC injection at various anatomical sites. Thus, SC injections are not necessarily a simple and straightforward form of administration.

Variations in uptake from an SC injection can also vary from patient to patient and from one day to the next in the same patient. In one study where the same operator was used to inject insulin on three consecutive days in the same site and at the same depth into the SC space in the same individual, there was approximately 20% variability in PK outcomes (3). Under less-controlled conditions, the coefficient of variation of PD parameters was in the range of 20% to 40% (4). In clinical practice, these PD variations can be even greater (5). Clearly, such differences in SC injection site outcomes cannot be deconvoluted by the standard SC to IV comparison of PK or PD outcomes without an understanding of events occurring at the SC injection site. It is also important to consider that even though the above-described studies focus on insulin, the information gathered for this patient population may provide a general sense for SC delivery variability of protein or peptide therapeutics without such an extensive historical database.

THE SUBCUTANEOUS INJECTION SITE

To provide a framework to examine potential PK, PD, and metabolism issues for a protein or peptide administered by SC injection, it is important to appreciate the architecture of this site and some of the events that occur because of accessing this site. The physical, chemical, and biological characteristics of the SC injection site can provide a more or less hostile environment depending on the nature of the injected protein or peptide therapeutic. Because the SC injection site environment will change over time through tissue repair events initiated by the trauma produced by an injection, the SC injection site should be considered dynamic. Further, the nature of injected materials (including formulation elements) and their residence time at an injection site will potentially affect these dynamic SC injection site events. Thus, the use of a needle to access the SC space, the elements within the SC space that are resident and/or drawn to this site as a result of the injection, and the time course of clearance from injected materials from the SC site could all potentially alter PK, PD, and metabolic outcomes for an injected protein or peptide therapeutic.

Architecture of the SC Injection Site

Skin is composed of two layers: the dermis and an overlying epidermis (Fig. 1). Beneath this structure and superficial to underlying structures of muscle, tendon, or bone is a region termed the SC space, a site used for the injection of therapeutic proteins and peptides. The SC space is typically accessed via a needle that penetrates first the epidermis and then the dermis to reach a layer of loose connective tissue composed of fibroblasts, fat cells, and capillaries; its anatomical design allows slippage between the overlying skin and underlying structures. Further, the SC site is considered a potential space as its design to allow this lateral slippage can also accommodate vertical movement. Indeed, the loose nature of this potential space can expand significantly through the introduction of fluids (swelling) that can occur during a local inflammatory response or because of dysfunction of outflowing lymphatic vessels (7,8). If an SC site contains a large amount of fat, the connective tissue attachments become more obvious and the skin cannot move as easily, giving rise to an appearance to the skin that is commonly described as cellulite (9). Importantly, fat is a highly vascularized tissue. While this property is critical for rapid utilization of fat stores upon demand, it also makes the SC site a good place for efficient vascular uptake of an injected therapeutic protein or peptide.

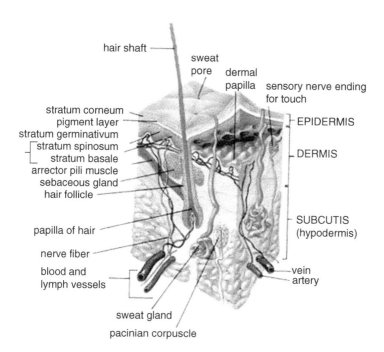

FIGURE 1 Diagram of the layers and human skin demonstrating a composite of structures and their relative locations within the epidermis, dermis, and subcutaneous space. *Source*: From Ref. 6.

The nature of the SC space and its accessibility can vary depending on the functional properties of its location. In most areas of the body, the epidermis is only 35 to 50 μm thick. In the area around the eye, it is only about 20 μm thick. On the palms of the hands and soles of the feet, it is usually much thicker, up to several millimeters, and the dermis can be as much as 3000 μm thick. The soles of the feet, palms of the hands, and postauricular region all have little or no fat in the SC space, while the inner arm, inner thigh, abdomen, and buttocks can have varying degrees of fat in this SC space. Further, females tend to distribute a slightly higher proportion of their fat composition than males to SC sites around the body (10). Thus, the depth and nature of this SC layer can differ from one person to another and can vary from one anatomical region to another for an individual.

SC Site Components and Their Properties

Accessing the SC site is typically achieved using a needle to penetrate the epidermis and dermis; as it does so, the needle has the potential to come in contact with a number of anatomical elements present within the skin that include vessels and nerves (Fig. 1). Cells of the epidermis originate from a single layer of basal cells, called the basal layer, which lie immediately superficial to an acellular basement membrane. Beneath the epidermis lies the dermis, a complex arrangement of proteins secreted by resident fibroblasts and referred to as the extracellular matrix (ECM). The ECM is composed of an interlocking mesh of fibrous proteins and glycosaminoglycans (GAGs). GAGs are carbohydrate polymers attached to ECM proteins to form proteoglycans (PGs) [hyaluronic acid (HA) is a notable exception, discussed below]. Having a net negative charge, PGs attract water molecules to maintain a level of hydration in the ECM. The negative charge density of PGs also helps to trap and store growth factors within the ECM.

ECM proteins serve many functions; including providing support and anchorage for cells and regulating intercellular communication through their ability to sequester a wide range of cellular growth factors to stabilize them within the ECM (11). Heparan sulfate (HS), a linear polysaccharide that occurs as a PG, is one example of an ECM protein that can bind growth factors (12). ECM multidomain proteins perlecan, agrin, and collagen XVIII are the main proteins to which HS is attached. The non-PG matrix component HA, or "hyaluronan," is a polysaccharide consisting of alternative residues of D-glucuronic acid and N-acetylglucosamine. Unlike other GAGs, HA is not found as a PG. HA in the SC site confers an ability to resist compression by providing a counteracting turgor (swelling) force through its ability to absorb water. Thus, HA is found in load-bearing joints, is a chief component of the interstitial space, and can regulate cell behavior during embryonic development, healing processes, inflammation, and tumor development through its specific interaction with the transmembrane receptor CD44 (13).

Collagens, fibronectins, elastins, and laminins are also present in the SC space, being organized into a complex matrix within the ECM. Collagen accounts for the majority of these proteins and is exocytosed in a precursor form known as procollagen and processed by proteinases to facilitate extracellular assembly. There are at least 10 different isoforms of collagen, the selective distribution of these various isoforms being organized around tissue and cellular functions (14). Fibronectins are proteins that facilitate cellular contacts with collagen fibers within the ECM by binding both collagen and cell surface protein complexes known as integrins. These fibronectin-mediated interactions are critical for cell movement (migration) through the ECM. Fibronectins are secreted by cells in an unfolded, inactive form, and their binding to integrins induces dimeric associations into dimers so that they can function properly.

Elastins, in contrast to collagens, give elasticity to tissues, allowing them to stretch when needed and then return to their original state. While collagens account for up to 75% by weight of the dermis, elastins account for less than 5%. Elastins are highly insoluble; tropoelastins are secreted as a complex with other proteins to stabilize them prior to their contact with a fiber of mature elastin. Tropoelastins are deaminated within the ECM to allow their incorporation into the elastin strand. Unlike collagen-containing fibers, laminins form networks of weblike structures that resist tensile forces and assist in cell adhesion. Laminins bind other ECM components such as collagens, nidogens, and entactins (15). Cell surface integrins associate not only with fibronectin but also with laminin. Thus, there is a wide range of protein elements that make up the ECM of the SC site environment that have the potential to interact with each other as well as cell surface components. Delivery of a protein or peptide therapeutic into this environment provides the opportunity for these bioactive agents to interact with or modulate the function of these ECM materials.

Nerves and Vessels of the SC Space

While the epidermis lacks blood vessels, it does contain nerves that can respond to mechanical, thermal, or noxious stimulation. In contrast, the dermis contains blood vessels, lymph structures, hair roots, sweat glands, and a variety of nerves. These nerves are associated with nerve endings that function in response to vibrations (Pacinian and Meissner corpuscles), pressure (Ruffini endings), and alterations in hair follicle position (16). The deeper part of the dermis contains fewer blood vessels than the more superficial layers. Rather than being straight, the junction between the epidermis and the dermis undulates; these undulations, known as rete pegs, are more marked in some areas of the body than in others. Rete pegs increase the area of contact between the layers of skin and help to prevent the epidermis from being sheared off from the dermis. Interestingly, rete pegs are not present in newborn skin but rapidly develop after birth, become very noticeable in a young person's skin, and become smaller and flatter as skin ages. Older skin also has reduced collagen production, a loss of elasticity, decreases in SC fat, reduced HA production, and a reduced ability to retain water. Indeed, collagen, HA, and other materials have been injected into this SC space to recover the loss of these components that come with age; such injections can have the cosmetic outcome of reducing wrinkles (17).

The network of blood vessels running through rete pegs provides a critical blood source to the epidermis and can provide an external indicator of the status of blood flow in the skin. Without sufficient oxygenated blood in these vessels, the skin will take on a bluish hue. Cold temperatures can also produce a bluish hue as this network of vessels contract (shunting blood to the central core compartment), reducing the amount of oxygenated blood flowing through rete vessels. Oppositely, increased temperatures and ample oxygenation will produce a more pinkish hue to the skin as rete vessels are more dilated. Excessive dilation of these blood vessels and a reddening of the skin would potentially indicate an inflammatory response. (Obviously, the extent of perception of these color differences depends on the inherent coloration of each person's skin.) Vessel leakage due to excessive dilation or even vessel disruption would produce local swelling. The important issue here is that skin hue and swelling can provide an important external indicator of events occurring within the SC space following the injection of a protein or peptide therapeutic.

Interstitial Fluid Composition

Because of arteriole-to-venule pressure differences, there is a physical force for serum fluid and solute components to exit capillary beds and enter the SC space (18). Active and passive uptake mechanisms present in venules recover most, but not all, of what is lost from these capillary beds. The absorptive function of lymphatic structures is critical in maintaining the overall volume influx/efflux equilibrium within the SC space. Proper organization of vascular and lymphatic elements within the SC space is maintained by a series of growth factors that are regulated by systems that sense signaling agents such as oxygen and nitrous oxide (19). In this way, the fluid composition of the interstitial space under normal conditions is maintained within fairly defined, steady-state limits.

On the basis of this intimate interaction with vascular and lymphatic vessels, interstitial fluid elements are comparable to those of blood serum; maintained by constant regulated replenishment and uptake. Thus, the environment experienced by a protein or peptide therapeutic injected into the SC space would have a pH and ionic composition similar to serum. Under normal conditions, the interstitial space is relatively free from cells that are normally present in the blood and lymph: erythrocytes, lymphocytes, as well as leukocytes (e.g., neutrophils, eosinophils, and basophils). Platelets are also generally absent from the environment of the interstitial space. The primary cells of residence are adipocytes, fibroblasts, and macrophages. Resident mast cells, present at approximately $7000/cm^3$, can also play an important role in the biology of the SC space through their attraction by and response to potential allergens that might be present in the injected material (20).

TISSUE RESPONSES AT A SUBCUTANEOUS INJECTION SITE
Events Stimulated by Needle Penetration at Injection

Following withdrawal, large-diameter needles can produce a wound that is not immediately closed by the elastic nature of the epidermis; use of large-diameter needles can result in the loss of some material leaving the SC injection site because of back pressure. This is one reason small gauge needles are preferred for SC injections. Another reason is that larger needles are likely to come in contact with a greater number of pain-registering nerves than a smaller needle. Further, needle lengths used for SC injections are specifically selected to restrict accidental delivery to sites deep to this loose connective tissue compartment. Overall, small gauge needles of specific lengths are the standards for delivering protein and peptide therapeutics to the SC site to reduce the extent of pain on injection and minimize the chance of missing the SC space.

Under nonpathological conditions, the SC space is relatively devoid of blood-derived cells and protease/peptidase activities. The act of delivering a protein or peptide via an injection to the SC space amounts to introduction of a transient wound in the skin. Unless a very large needle is used, the opening in the epidermis and dermis produced by the needle tract closes within seconds. The act of needle penetration, however, initiates processes associated with wound repair, the resolution of which involves activation of cell- and protease-based mechanisms. Although this discussion is not intended to focus on the complexities of events involved in wound repair mechanisms, a brief summary highlighting some aspects of these events is helpful in appreciating potential implications on PK, PD, and metabolism of an injected protein or peptide.

Typically, the actions of serum proteases and peptidases that leak from capillary beds into the SC space are restrained through the presence of inhibitory elements; pathological conditions can arise from an imbalance of these inhibitor/protease functions (21). Although a number of proteases and peptidases can be present in the SC space, additional activities may enter or become activated within this space by an SC injection. For example, in response to changes in physiological conditions, cells of the ECM can secrete proteases, such as serine and threonine proteases as well as matrix metalloproteinases, that induce the local release of growth factors from an ECM depot (22). Thus, although the SC space is not normally enriched in proteolytic activities, this can change in response to events associated with the injection of a protein or peptide therapeutic.

Further, a number of specialized cells (discussed below) associated with the immune system are resident within the SC space; these cells guard against pathogen entry that might occur as a result of skin wounding. Once activated by signals associated with skin wounding, these resident cells could also secrete a variety of protease activities that function as part of the innate immune response. Thus, the repertoire of protease activities in the SC space can be shifted and augmented within the interstitial space by inflammatory processes (23). Additionally, it is possible that other factors released by the cells associated with the innate immune response might affect vascular permeability (discussed in detail later). Together, these factors and their affects on protease and peptidase activities could affect PK and PD of a protein or peptide therapeutic injected into the SC space through metabolic events.

Responses to the Injected Protein or Peptide

While most proteins and peptides delivered by SC injection do not stimulate local effects that could affect their delivery, some proteins can have local actions that might complicate an SC injection. It should also be remembered that most particular therapeutic protein or peptide delivered to an SC site would never normally be present at that site in such high quantities or concentrations. For example, chemokines that mediate pain nerve responses could make an SC injection potentially unbearable for the patient and possibly enhance the potential for innate immune outcomes (24). Further, within the epidermis and dermis, there is an extensive network of dendritic cells (DCs); those in the epidermis are known as Langerhans cells (LCs). Although their exact role in antigen processing is still unclear, LCs are relatively long lived and have several unique properties that define their immunological role in skin (25,26). One of these unique properties is defined by an activating role by transforming growth factor β (TGF-β) (27). Thus, SC injection of TGF-β might affect local tissue responses and immune outcomes that could alter PK, PD, and/or metabolism outcomes for this cytokine or a co-administered therapeutic protein or peptide.

With regard to peptide and protein immunogenicity following SC injection, review of this literature points to the interesting fact that many injected proteins and peptides result in antibody responses (28). The majority of these events appear to involve nonneutralizing antibodies (Table 1), but a few are more significant with neutralizing events that can result in serious clinical issues (39). As endogenous proteins and peptides of the body should be poorly immunogenic because of previously established tolerance, it is usually unclear why and how this self-tolerance is broken and alloimmunization occurs (40). In general, autoimmune outcomes can be observed following years of SC injection where a therapeutic protein or peptide is administered in much higher than normal concentrations. Some studies have suggested that acute, supraphysiological levels of a protein therapeutic, in this case erythropoietin, can generate neutralizing antibodies to the endogenous molecule (41).

It is possible that high concentrations of a protein or peptide therapeutic injected into the SC space might precipitate, leading to increased scrutiny by the

TABLE 1 Immunogenicity of Some Human Therapeutic Proteins

Protein	Indication	Immunogenicity	References
Insulin	Insulin-dependent diabetes	1. 78% ($n = 9$) of nondiabetics not previously exposed to insulin produced antibodies to semisynthetic insulin.	29
		2. 100% ($n = 7$) nondiabetics not previously exposed to insulin produced antibodies to human insulin.	30
IFN-β1a	Relapsing/remitting multiple sclerosis	24% ($n = 560$) patients receiving 22 µg IFN-β1a produced neutralizing antibodies.	31
IFN-α	Hairy cell leukemia	51% ($n = 31$) patients produced antibodies, which neutralized recombinant IFN-α in vitro. 37% of these patients showed clinical resistance to treatment.	32
IFN α	Chronic myeloid leukemia	22% ($n = 67$) patients produced neutralizing antibodies, of whom 73% were unresponsive to treatment. Of the 78% not producing antibodies, only 21% were unresponsive to treatment.	33
GM-CSF	Metastatic colorectal carcinoma	95% ($n = 20$) patients produced GM-CSF-binding antibodies. 47% of these individuals produced antibodies, which neutralized biological activity of GM-CSF.	34
EPO	Anemia associated with chronic kidney disease	Pure red cell aplasia was observed to sharply increase from 1999 to 2002 in patients treated with EPO; events correlated with a formulation change involving replacement of human serum albumin with Tween 80.	35
TPO	Thrombocytopenia associated with chemotherapy	Synthetic TPO induced antibodies that cross-reacted with endogenous TPO, resulting in severe, persistent thrombocytopenia in 4% of healthy volunteers and 0.6% of oncology patients who received intensive chemotherapy.	36–38

Abbreviations: IFN, interferon; GM-CSF, granulocyte-macrophage colony-stimulating factor; EPO, erythropoietin-α; TPO, thrombopoietin.

immune system and possible recognition as a foreign (even pathogenic) material (42). Similarly, the injection of a denatured or conformationally damaged protein or peptide might result in an immune response. In other cases such as insulin, nonhuman versions of the protein were injected by patients for years; it is not so surprising that an antibody response was observed and that common determinants present on both human and nonhuman insulin preparations could ultimately become the focus of an immune response (40). Such an outcome might have been involved in the immune responses observed in patients dosed with thrombopoietin (36–38,43,44). Determining the reason(s) for these alloimmune events will require improved characterization of the type and duration of these responses; recent improvements in these areas have been described (45).

Impact of Fluid-Induced Distention

One obvious consequence resulting from the SC injection of a protein or peptide therapeutic is that some finite volume of displacement must occur at the site of injection. In general, best efforts are made to minimize this injection volume as excessive distention can cause discomfort, pain, and even sufficient back pressure to reflux drug through the needle track used to penetrate the epidermis/dermis. SC injection of a 200-µL formulation is frequently the goal, but physical/chemical properties of protein and peptide therapeutics and the dose required sometimes demand larger volumes to be injected. If the injection site is constrained by septa associated with adipose of the SC site, a discrete, distended fluid pocket is formed. Some locations of the deeper SC space can be connected to large continuous areas that function to separate deeper fatty and muscular structures; these continuous areas are referred to as fascial planes (46). Thus, if a volume is injected into an SC site that is associated with a fascial plane, the fluid will fan out in a manner such that discrete borders of the injected volume are not discernable and no discrete, distended fluid pocket will be formed. It is not difficult to imagine the potential for different PK, PD, and metabolism outcomes for a protein or peptide therapeutic delivered into a fluid-filled pocket versus that distributed into a fascial plane.

Fluid injected into an SC site can be absorbed into both vascular and lymphatic capillary beds. The rate of fluid absorption from an SC injection site can be controlled by a number of factors, but in general, fluid from a 200-µL injection volume can be completely cleared in a few hours. In the absence of its removal, the hydrodynamic pressure exerted by fluid introduced into an SC injection site could disrupt associations between cells and ECM components. Such an outcome would likely trigger cellular responses associated with tissue remodeling, resulting in the local release of factors from resident macrophages and mast cells (e.g., histamine). Accordingly, the SC injection site will respond according to how long that excess fluid volume exists. SC injections typically use iso-osmotic formulations, on the basis of the belief that hypo- or hyperosmotic preparations will incite a painful injection. Iso-osmotic (or slightly hypo-osmotic) formulations will also minimize the amount of water drawn into the injection site because of an imbalance in local osmotic pressure; hyper-osmotic formulation would draw water into the injection site. Whatever the nature of the formulation, the fluid volume injected can initiate a series of responses that can affect the PK of the protein or peptide being delivered.

In most cases, insignificant tissue damage results from an SC injection, and thus, there is little to trigger tissue remodeling events. As efforts are made, however, to increase the volume injected into an SC space, the potential for tissue damage and the initiation of tissue remodeling events are increased. To reduce this outcome, slow infusion methods into the SC space have been described (47). Additionally, it is possible to alleviate some issues associated with the impact of fluid-induced distention at an SC injection site by increasing the effective volume of an SC injection site. This can be achieved by altering the physical nature of this potential space by cleaving ECM elements by injection of specific enzymes, such as hyaluronidase. In doing so, the cross-linking elements that constrain the effective volume at a site of SC injection can be expanded (48,49).

Dilution of Interstitial Fluid Composition and Return to Steady-State Conditions

Although the concentration of endogenous interstitial fluid elements is typically overwhelmed at the site of an SC injection, the body sets in motion a series of processes to ensure the rapid recovery of its steady-state composition. Various factors, in particular the composition of the injected formulation and the injection volume, will drive the rate of protein or peptide therapeutic absorption. One can envision that at the time of SC injection, there is a transition from an initial nonhomogeneous condition to some sort of overall more homogeneous condition with initial absorptive events starting to occur (Fig. 2). During this initial absorption phase, the rate of protein or peptide uptake from an SC injection site would be influenced by several changing parameters: pH, ionic composition, fluid flux, etc. Once equilibrium is achieved, the rate of protein or peptide uptake would become constant and remain constant for as long as dominating parameters of this equilibrium are maintained within a critical range. When these parameters move beyond some threshold, the rate of protein and peptide uptake shifts away from some constant rate and would continue to change as these parameters continue to move to a state of recovery. Ultimately, recovery moves to resolution as the cellular and acellular (e.g., protease) changes return to that of their preinjection status. The amount of material and the volume injected should modulate each of these stages, as exemplified in the case of

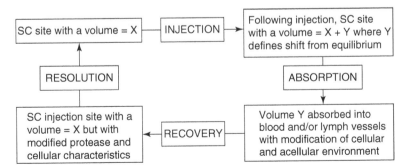

FIGURE 2 Diagram describing the dynamic changes and continuum of events associated with the volume changes associated with an SC injection and resolution of these changes following correction of fluid distention by blood and lymph volume uptake. *Abbreviation*: SC, subcutaneous.

insulin where changing the dose and concentration of the administered material can affect PK and PD profiles (50).

The time required for achieving an initial steady-state or equilibrium condition at an SC injection site might take seconds, minutes, or hours; the same can be said for the duration of a steady state/equilibrium and the time course for returning to preinjection conditions (Fig. 2). The rate and duration of each of these three phases provide a basis for the uptake profile of an injected protein or peptide therapeutic. For very fast absorption events, there would be little or no time for tissue responses to produce a significant alteration of the SC injection site. Injected formulations designed to establish a more chronic uptake of an injected protein or peptide, however, could result in cellular and tissue component changes at the site of injection, as in the case of proteins released from microparticle preparations (51). Since such changes would essentially follow the biological processes associated with tissue adhesion events, injected proteins or peptides that modulate inflammatory events could provide the basis for affecting the PK, PD, and metabolism of injected materials at these sites (52).

Characteristics of the protein or peptide and its formulation and/or administration to an SC site can significantly affect the PK, PD, and metabolism outcomes for the material. For example, doubling a regular insulin dose can extend the observed T_{max} by 62%, but there is much less of this effect for fast-acting form of insulin (53). It is possible that differences in the volume of formulations delivered also had some effect on these outcomes. Additionally, alteration of the protein or peptide after its delivery into an SC site could affect PK, PD, and metabolism outcomes. Formulations of insulin using glargine and detemir achieve their slower rates of protein release from the injection site by precipitation of the protein as the injected volume becomes neutralized. Formulation elements, such as protamine in some insulin preparations, might act to locally affect the vascular response at the SC injection site (54), with the potential outcome of affecting PK, PD, and metabolic outcomes.

MODIFICATIONS OF INJECTED PROTEINS AND PEPTIDES IN THE SC SPACE

For an injected protein or peptide to achieve its desired biological action(s), it must typically leave the SC injection site in a biologically active form, overcoming physical, chemical, and biological stresses experienced at the site of injection. Without an ability to directly examine the fate of a protein or peptide prior to its systemic uptake from the SC injection site, it is difficult to understand the potential for a protein or peptide to undergo modification(s) that might alter its ability to leave the SC injection site in a biologically active form. In general, the only outcome that can be monitored is the presence (and condition) of that protein or peptide in the systemic circulation. Events associated with tissue response, interaction with endogenous elements, and recovery of steady-state conditions might occur at the SC site following injection that could limit systemic uptake of a protein or peptide therapeutic.

Repair mechanisms associated with epidermal/dermal wounding that involve protease/peptidase activation derived from damaged and/or activated cells provide one possible pathway for damage of an injected protein or peptide therapeutic. Obviously, the extent and duration of exposure to such protease and peptidase activities are crucial. While such possibilities do not appear to represent an overwhelming issue for those proteins and peptide therapeutics that have become commercial products, there is no reason to assume that such metabolism will never be an issue. In vitro screening of protein and peptide therapeutics to the actions of enzymes such as collagenase, fibrin, and elastase would be one way to initially evaluate some of these potential concerns. In this regard, introducing enzyme inhibitors, formulation additives that protect the protein or peptide, or increasing the amount of substrate (injected protein or peptide) could all provide some benefit. Typically, efforts are made to reduce SC injection volumes; this increases the local concentration of the injected protein or peptide, which could act to reduce the fractional loss due to local catabolism. Although formulations at these higher than physiological concentrations may be beneficial for the successful pharmaceutical application of a therapeutic protein or peptide, compatibility with the environment of the SC injection site is critical.

Some therapeutic proteins and peptides interact with protective serum binding proteins in the body. In the absence of such protection, a purified protein or peptide may require nonphysiological pH conditions to keep it from becoming biologically inactive (55). Injecting a strong acid or base into the SC space would likely result in local tissue damage and be perceived as painful; thus, weak buffering systems are commonly used for SC injection formulations. This approach, which allows for rapid re-equilibration of the SC space to its normal, near-neutral pH, can lead to an undesirable stress on the protein if it is required to transition through its isoelectric point (pI) as the SC site pH re-equilibration occurs. For example, if a protein has a pI of 6 and is formulated in a weak buffer at pH 5, this protein will potentially undergo an increased potential for unfolding and/or aggregation as the space returns to neutral pH.

As with most, if not all, potential issues for a protein and peptide injected into the SC space, timing and environment are critical. If the injected protein or peptide transitions through potentially labile states with sufficient speed or in a sufficiently stabilized status because of its local high concentration (or other formulation elements), then little, if any, loss may occur. There is currently little information regarding such events at an SC injection site that might affect PK, PD, and metabolism of an injected protein or peptide. One exemption is what has been acquired through accumulated anecdotal information for outcomes following SC injection of insulin (56). In this case, the protein delivered is formulated in various states of solubility that can affect uptake rate and thus PK and PD profiles.

Efforts have been made to obtain accurate local pH information that could be applied to analyzing the events in the SC space after injection of a protein or peptide. Both magnetic resonance and positron emission methodologies have been described (57,58). That these methods lack sensitivity has been the rationale to introduce a doping probe such as gadolinium (Gd^{3+}) ions to function as a chelate relaxation agent (59). More recently, a method using hyperpolarized ^{13}C-labeled bicarbonate to monitor the pH environment experienced by this physiological buffering agent has been described (60). Although this most recent method looks appealing, it is not trivial to prepare this probe material, and its addition to a formulation injected into the SC space would still suffer from the axiom known as "the price of peeking." In general terms, this axiom emphasizes that there is always a price paid for any analysis; the act of disrupting the site to collect its contents or the introduction of a probe that would not normally be present alters the measurement and thus could provide spurious information.

Physical Stresses

Two frequent concerns regarding the potential for physical modification of an injected protein or peptide is aggregation or binding (to cell receptors or ECM elements). Both events can dramatically reduce the amount and rate of biologically active material leaving the SC space following injection. In the instance of aggregation, this can be a serious concern as many proteins and peptides are prone to increased contacts when formulated at high, nonphysiological concentrations. To prepare a sufficiently stable yet simple formulation, many proteins and peptides are delivered to the SC space in a mixture of salts, sugars, surfactants, and preservatives in a biologically compatible buffer system (61). Once injected into the SC space, the rapid return to physiological pH could incite conditions prone to cause protein or peptide aggregation (as discussed above).

One reason protein and peptide therapeutics have a tendency to aggregate can be attributed to the nature of the amino acids that make up their sequence; amino acid side chain structures are classified as polar, charged, nonpolar, and hydrophobic. Hydrophobic and nonpolar amino acids typically reside within a central core of a protein that is sequestered away from the surface of the protein where polar and charged residues dominate; such an arrangement helps stabilize the three-dimensional (3D) structure of proteins in an aqueous environment. Although many exceptions exist, events that alter this typical organization of hydrophobic and hydrophilic side chain structures can lead to aggregation (62). As a gross oversimplification, events that drive protein unfolding can position hydrophobic side chains at the protein surface where their potential to interact with similarly exposed amino acids of another protein is enhanced, increasing their potential for self-association through hydrophobic-hydrophobic interactions (63). The local high concentration of an injected protein increases this probability of one unfolded protein finding another. Extensive nonphysiological self-association can result in several outcomes that can be categorized whether the materials establish covalent associations or not and whether the materials remain soluble or become insoluble (64).

Insulin represents one biopharmaceutical well known for its propensity to undergo a structural transformation that results in aggregation and the formation of insoluble fibrils (65). Some protein fibrils, such as collagen, are fully functional and not immunogenic (66), but in general, fibril formation typically connotes loss of function and/or increased immunogenicity; this type of protein aggregation correlates with the development of Alzheimer's disease, Parkinson's disease, prion-associated transmissible spongiform encephalopathies, and type II diabetes (67). It has been suggested that fibril formation involves disordered globular proteins with a defined 3D structure or natively unfolded proteins in a soluble conformation (68). Additionally, amyloid fibril formation has been shown to occur in proteins and peptides that share the common structural feature of β-strands perpendicular to the fibril axis; lysozyme, transthyretin, $β_2$-microglobulin, calcitonin, and apolipoprotein AI can all form amyloid fibrils under certain conditions (69–71).

While the increase in free energy put into a protein or peptide by elevated temperatures is one way in which hydrophobic residues can become energetically driven to the surface of a protein, there are some proteins that form aggregates through low-temperature conformational changes. While purified therapeutic antibodies are frequently stored at reduced temperatures (e.g. 4°C) to increase shelf-life, it is surprising that some antibodies undergo precipitation at temperatures below 37°C. The presence of antibodies with this trait of cold-dependent insolubility is associated with pathological conditions (72). Fluorescence and circular dichroism studies have shown that these cold-susceptible antibodies may have a slightly perturbed structure (73). This fact that some antibodies, a class of proteins known to exist in serum in high mg/mL concentrations, can aggregate while others do not complicates our ability to identify rules that govern and potentially predict protein and peptide aggregation.

A variety of agents have been identified to protect proteins from aggregating in vitro. On the basis of such in vitro studies, some combination of these agents is typically incorporated into protein and peptide formulations used for SC injection, although it is unclear how useful these agents are in vivo following SC injection. These agents fall into several categories: chaotropic agents (urea, guanidinium chloride), amino acids (commonly arginine, glycine), sugars (trehalose, sucrose), polyols (mannitol), polymers (polyethylene glycol, cyclodextrins), and surfactants (polysorbates). Some of these agents have been suggested to function as "neutral crowders," acting to increase the distance between protein domains that might participate in aggregation (74,75). Given that some of these additives have the potential to affect protease and peptidase activities, they may provide added benefit to improve the stability of a protein or peptide therapeutic at an SC injection site. What should be kept in mind is that, on the basis of their size and other physical properties, most of these stabilizing agents would be anticipated to leave the SC injection site faster than the coinjected protein or peptide.

Chemical Stresses

In general, proteins and peptides are not robust molecules; they can be readily oxidized, deamidated, cleaved, and aggregated into forms with reduced biological activity when handled in vitro. All of these same events could occur following their SC injection, providing multiple potential pathways for destruction and clearance in vivo. It is important to remember that most proteins and peptides produced by cells in the body are broken down and cleared in minutes to hours; only a select few seem to last for days to weeks. Thus, the desire to identify protein and peptide formulations with multiyear stability and 100% bioavailability following SC injection is certainly a tremendous challenge. A clear picture of the SC environment and how it affects injected protein and peptide formulations does not exist at present. Having this picture would expedite efforts to find optimal formulation strategies to protect injected proteins or peptides from chemical decomposition. While it is possible to anticipate some chemical stressors confronting a protein or peptide therapeutic on the basis of in silico and in vitro studies, it is unclear how much of this information will ultimately define events in the SC space.

Proteins and peptides can also undergo a variety of amino acid–specific chemical modifications that can affect their structural and biological properties, including aggregation and/or diminished bioactivity (76). Oxidation of amino acids, in particular tryptophan, methionine, reduced cysteine, tyrosine, and phenylalanine residues, can alter structural and biological properties of protein and peptide therapeutics (77). Similarly, protein oxidation can be detrimental; excessive oxidation of serum proteins has been associated with pathological states (78). Deamidation of asparagine at neutral pH can lead to the production of an isoaspartic acid structure with backbone scission as one potential outcome (79).

More frequently, asparagine and glutamine can undergo deamidation to produce aspartic acid and glutamic acid, respectively, both in vitro and in vivo (80). As deamidation events appear to be a normal process of proteins in vivo (81), the suggestion has been made that proteins use this modification as a sort of molecular clock (82). Several disease states have been correlated with protein deamidation, including prion disease (83), Alzheimer's disease (84), and cataracts (85).

Biological Stresses

One can envisage that a large spectrum of potential protease and peptidase-mediated modifications of a protein or peptide could occur at an SC injection site. While some formulation elements, such as divalent cations, might even provide a slight activating capacity to some proteases or peptidases potentially present at an SC injection site (86), introduction of a divalent cation chelator in a formulation might act to reduce the activity of other proteases and peptidase. Neither of these possibilities has been clearly demonstrated. One formulation element that is sometimes used and that could also have an effect on protease/peptidase activities at an SC injection site is human serum albumin (HSA). Although the one benefit for incorporation of HSA is stabilization and protection from loss to container surfaces, particularly for very potent molecules dosed at very low levels, introduction of HSA at an SC injection site could also act as a potential sacrificial element for the protection of a labile therapeutic protein or peptide.

It is unlikely that proteases and peptides are highly active at the moment of an SC injection; outside of the intestinal lumen, protease and peptidase activities in the body are limited as inactive precursors or by association with dissociable inhibitors. Uncontrolled protease activities are strongly associated with conditions such as fibrotic pathologies (87). The balance of such protease inhibitor/activity relationships is regulated by a variety of cytokines (88); proteases potentially involved in remodeling the SC architecture could become activated by these molecules. The likely source of these proteases and peptidases would be resident fibroblasts, which function to produce, organize, and turn over ECM proteins (89). Thus, a protein or peptide therapeutic that attracts cells involved in tissue remodeling or vessel formation to an SC injection site could alter the protease/peptidase environment, regulating the expression of matrix metalloproteinases, clotting cascade proteases (e.g., thrombin), fibrinolytic enzymes (e.g., plasminogen), and collagenases (90).

As the PK, PD, and metabolism of protein and peptide therapeutics can be modulated by various binding processes, these same processes should come into play when the same molecule is delivered by SC injection. For example, if the body regulates the function(s) of a protein through association with specific binding elements, these same binding elements might alter the PK, PD, and metabolism of that material if these elements are present in the SC space. Vascular endothelial growth factor (VEGF) represents one such protein. VEGF contains a heparin-binding domain; heparin and heparin-like molecules are plentiful in the SC space. SC injection of VEGF results in nonlinear PK profiles resulting from retention of correlate with local binding of VEGF (91). Epidermal growth factor (EGF), basic fibroblast growth factor (bFGF), and insulin-like growth factor-1 (IGF-1) can each associate with the ECM protein vitronectin (92). Retarded release from an SC injection site as a result of these interactions may also lead to local metabolism and/or sustained local effects in preference to systemic effects.

UPTAKE OF PROTEINS AND PEPTIDES FROM AN SC INJECTION SITE
Nonspecific Vs. Specific Uptake
The rate and extent of nonspecific solute uptake into capillaries from an SC injection site is controlled by the net difference in driving forces between the interior and exterior of these vessels along with the relative permeability of the vascular and lymph membranes (93). There are three basic forces that determine the balance of nonspecific solute exit and entry: capillary pressure, interstitial fluid pressure, and oncotic pressure (94). Capillary pressure, produced by the beating heart, which forces fluid out of capillaries and into the SC space, is usually maintained within specific limits but can be altered during pathological states (95). The extent of solute flux from the capillary lumen to the SC space driven by the capillary pressure is limited by the interstitial pressure within the SC space established by the presence and composition of ECM components (96). Oncotic pressure within the capillary lumen is generated by the presence of vascular components; oncotic pressure tends to draw fluid from the interstitial space into the capillary. In consideration of these factors, an increase in capillary pressure, a decrease in plasma protein concentration, an increase in capillary permeability, or a reduction in the flow of vascular or lymph flow would result in an overall accumulation of fluid and protein in the interstitial space (97).

While nonspecific solute uptake should be comparable for a wide variety of protein and peptide therapeutics injected into the SC space, specific binding events could either enhance or diminish the rate and extent of this uptake. Specific binding interactions could involve a cognate receptor and/or a non-selective interaction. If such interactions occur at the surface of cells or with materials within the interstitial space, movement from the interstitial space into the capillary lumen could be retarded—essentially, the reverse of a drug leaving the capillary lumen to bind selectively to a target in the SC space (98). While retention at the injection site would affect the rate of systemic delivery, the overall extent of this delivery may also be reduced if this retention leads to increased metabolism by local protease/peptidase activities. Oppositely, if a protein or peptide injected into the interstitial space has an affinity for a resident receptor or binding partner present within the capillary lumen, uptake could be accelerated. In either case, nonspecific solute-binding event could affect PK and metabolism parameters, which could thus alter PD outcomes.

Tissue Damage due to Needle Introduction
Depending on the needle gauge size used to introduce a protein or peptide formulation into the SC site, varying degrees of damage will be sustained by components of the interstitial space: blood vessel breakage, disorganization of cellular and acellular structures of the SC space, etc. Consequently, events of the coagulation cascade (to stop further loss of blood from damaged vessel) and initiation of tissue repair mechanisms (to return SC tissue to its pre-trauma condition) could be activated. Disruption of blood vessels releases platelets that function in sealing vessel breaks, facilitating the generation of thrombin; events that could entrap and/or destroy an injected protein or peptide therapeutic. It is important to appreciate that platelet-driven blood clotting is a complex process (99). Collagen is one element of the basement membrane recognized by platelets, although there are a variety of signals that can activate platelets: adenosine diphosphate, serotonin, thromboxane A2, and fibrinogen receptors. Links

between platelets and extravascular structures are mediated by the glycoprotein known as von Willebrand factor. A series of clotting factors, famous for their absence in hemophiliac populations (e.g., factor VII, factor IX, etc.), convert soluble plasma fibrinogen to insoluble fibrin polymer—processes that trap additional platelets and complete the formation of a clot plug (99).

Following clot formation, three stages of wound repair mechanisms occur: inflammation, new tissue formation, and remodeling (100). Only inflammation occurs in a time frame that would likely affect the PK, PD, or metabolism of a formulation designed for rapid systemic absorption; the latter two phases would affect slow-release or depot outcomes that are beyond the scope of this chapter. Inflammation, driven by the infiltration of immune-associated cells, can potentially be augmented by the biological activity of an injected protein or peptide therapeutic. Inflammatory signals frequently increase local vascular permeability, and this can alter the overall absorption rate from the SC site through modulation of the parameters discussed above. Inflammation-associated cells (monocytes, neutrophils, etc.) can result in local increases in not only proteolytic activity but also reactive oxygen species in the extracellular SC space environment, applying physical and chemical stresses on an injected protein or peptide. Additionally, the immune cells that infiltrate the SC space are effective at nonselective phagocytosis. Thus, proteins and peptides that may be slow to leave the SC injection site may also be taken up and destroyed locally by these cells.

The overall point to be made here is that when a needle damages tissues to deliver a protein or peptide to an SC site, it is likely that some level of blood clotting processes and subsequent inflammatory events will ensue. Small gauge needles used for most SC injections result in minimal tissue damage. Thus, it is unclear how much impact these events might have on the PK, PD, and metabolism of an injected protein or peptide, particularly if large gauge needles required to deliver viscous, high-concentration formulations are used. What is known is that many of the factors and events associated with tissue damage associated with the introduction of a needle are becoming better understood and one can now anticipate some of the potential events that might affect the physical, chemical, and biological stability of a biopharmaceutical at an SC injection site.

Mechanisms of Absorption

Characteristics of the protein or peptide being absorbed can also function to direct it preferentially between lymphatic or vascular uptake (101) as well as factors present at the site of injection (e.g., formulation components). In general, most peptides and hydrophilic proteins of less than approximately 20 kDa are readily taken up by the vascular capillary bed through mechanisms outlined in previous sections of this chapter. Such mechanisms function to recover endogenous proteins and peptides lost from the vasculature at capillary beds. Larger proteins, other than those having specific reabsorption systems, tend to be taken up into lymphatic vessels.

One example of a specialized reabsorption system involves serum albumin, a molecule of 68 kDa that selectively binds to the cell surface neonatal immunoglobulin receptor (102,103). While an injected protein or peptide therapeutic could have specific interactions with insoluble ECM elements or cell surface components, interactions with soluble elements could also occur; drugs that interact strongly with albumin might show enhanced uptake from the interstitial space through mechanisms that drive albumin reuptake from the interstitial space. Thus, absorption characteristics of an injected protein or peptide can be altered on the basis of its interactions with albumin (104). Also, a lipid-modified protein or peptide that specifically interacts with the fatty acid–binding sites of serum albumins or the lipid/phospholipids transport protein α-1 lipoprotein might have enhanced uptake (105). Similarly, β-lipoprotein is capable of transferring cholesterol and triglycerides; proteins or peptides associated with such materials could have altered uptake characteristics from an SC injection site (106), and proteins or peptide therapeutics that contain or are associated with the fraction-crystallizable (Fc) domain of an antibody provide another strategy to alter PK, PD, and metabolic events at an SC injection site (107,108).

Uptake into the vascular capillary bed and clearance form an SC injection site can be relatively rapid compared with clearance by lymph-mediated mechanisms. While flow through the vascular bed is driven by a pressure gradient, fluid flow through lymphatic vessels is achieved though local compression and unidirectional valves present within the lymph vessels (109). Additionally, the lymphatic system is preferentially used by a variety of white cells to traffic through the body. As some of these white cells have surface receptors that could bind a delivered protein or peptide, this could provide a preferred route of uptake for a biopharmaceutical from the SC injection site. More likely, however, is the potential for a delivered protein or peptide therapeutic to induce white cells to secrete factors that locally affect the function of blood or lymph vessels, affecting SC injection site absorption properties.

Dose Volume Limitations

Instead of dividing the dose over several sites or delivering it more slowly using a continuous infusion system, the industry has focused on single-bolus strategies to simplify self-administration: efforts have been made to reduce the SC injection volume by increasing the concentration of the protein and peptide formulation being injected. Such a philosophy does reduce needle and syringe use, and this has a number of positive social and health care benefits. It also has the negative aspect of preparing formulations where the protein or peptide can be at a remarkably high concentration; antibody-based therapeutics can be formulated at concentrations of several hundred milligrams per milliliter. At such high concentrations, a protein solution can become so viscous that it can no longer be delivered by a needle and syringe. Several high-concentration antibody formulations have, however, been described where the issue of increased viscosity has been addressed (110).

Alternately, the concentration of a delivered protein or peptide can be increased by reducing its solubility—that is, delivering it as a suspended solid. Insulin is probably the best-known example of this approach, with crystallized and precipitated suspensions being injected by millions of diabetics (56). Methods have been identified to produce and stabilize a suspension of particles that can provide a fairly consistent delivery—essential for reproducible PK and PD outcomes. A depot effect for insulin can be achieved by the slow dissolution of insulin co-crystallization with zinc ions and the polycationic peptide protamine. While insulin is the most recognized biotherapeutic to be administered by SC injection as a particulate suspension, other proteins have been crystallized in an effort to expand this list (111,112), and some progress has been made to identify sustained-release crystalline forms of proteins other than insulin (113). In such cases, the size and characteristics of these particles and the formulation in which they are prepared must be acceptable for injection and able to pass through a needle of reasonable gauge.

Comparison of SC to Intramuscular and Intradermal Injections

Although the primary focus of this chapter was to discuss issues associated with SC injection or protein and peptide therapeutics, it is also possible to inject these molecules directly into muscle tissue or into the dermis of the skin. Intradermal (ID) injections are best known for the cosmetic application of collagen (114), while intramuscular (IM) injections are frequently used to administer vaccines. Because of unique biological outcomes, ID injections have also been used successfully for vaccination (115). Although very few direct comparisons have been made, it appears that IM may be superior to ID for vaccination outcomes that have been assessed, but for a few vaccines, ID can be the preferred delivery route (116). This type of assessment, however, is difficult as IM vaccines seem to benefit from the delivery of an adjuvant emulsion system (117) or virus-like particles (118), while the same material may be processed differently following ID delivery. These subjects are outside the scope of this chapter; what is relevant is comparing ID and IM delivery to an SC injection. IM delivery typically results in a delayed-uptake profile and is thus preferred to provide more of a depot delivery relative to an SC injection. ID delivery is perceived to provide a more direct targeting to lymph nodes than SC injection (119).

Many of the parameters listed above for SC injection could also affect PK, PD, and metabolism outcomes of protein and peptide therapeutics administered by IM injection. In the case of vaccination, very few protein or peptide subunit (roughly equivalent to therapeutics for their production, characterization, and formulation issues) vaccines are on the market or in trials (120). Most vaccines are prepared from crude (complex) preparations of nonpathogenic (either by the strain used or by the method of preparation) form of a pathogen that can be prepared cheaply for potential use in developing countries (121). The basic concept of such IM deliveries is to establish a depot where antigen-presenting cells are drawn to process and disseminate antigens. By their complex nature and from this functional paradigm, it is very difficult to appreciate any generalized issues of PK, PD, or metabolism of protein or peptide vaccine components that may be critical for a desirable vaccination outcome.

ID injection lends itself to targeted delivery to antigen-presenting cells that are resident in the dermis: LCs and DCs (122). It appears that physical trauma of the skin, via the activation of innate immune elements, assists in providing a desirable immunological outcome (123). One way to achieve ID delivery is using electroporation to drive materials into the skin (124). Microneedles can also be used to achieve protein or peptide delivery to the dermis (125). An in-depth assessment of transdermal delivery methods for therapeutic proteins and peptides is beyond the scope of this discussion.

Conclusions and Perspectives

SC injections of therapeutic proteins and peptides have played a significant role in establishing the biopharmaceutical industry. Years of preclinical and clinical studies go into the research and development, and several hundred millions of dollars are spent before a new biopharmaceutical is approved for human use; elaborate studies are performed to examine the biology, toxicology, potential drug-drug interactions, and patient populations. It is therefore somewhat surprising that so little is actually understood about the fate of a protein or peptide therapeutic after injection into the SC space. Efforts to formulate these materials are primarily focused on shelf-life stability and lack of pain upon injection. As these injected materials are endogenous (or at least composed mostly of endogenous elements), there is the perception that they will readily find their way to their intended pharmacological target once they are placed in the body. Concerns over events that might occur at an SC injection site are only of issue when the protein or peptide being administered fails to be efficacious or incites an immune response.

The academic literature describing physical, chemical, and biochemical changes experienced by protein and peptide therapeutics at an SC injection site is limited. As the biopharmaceutical industry has extensively examined protein and peptide formulations with only a small fraction of this information being published, it is possible that a great deal of information about events at the SC injection site exists but has not been published. What can be gleaned from the literature is information regarding the nature of the SC environment and events that could occur in this space following an SC injection event: local wounding, fluid distention, wound repair, etc. Piecing this information together provides a partial picture of the dynamic environment of this site and events driven by cellular and acellular components that could affect the PK, PD, and metabolism of an injected protein or peptide therapeutic. The response to and actions on an injected protein or peptide therapeutic can thus, in some ways, be partially predicted on the basis of some known physical and chemical properties. As SC injection of protein and peptide therapeutics is likely to remain the standard method of administration for the biotech industry for the near future, it would be great added value if information of events at the SC space could be better understood.

Although most protein and peptide therapeutics administered by SC injection are native to the body, there are several important aspects to these administrations that are frequently inconsistent with their biology when administered by this route. Many therapeutic proteins or peptides are produced and secreted from cells in the body under conditions and in a time course that result in local concentrations that are much lower than those required for a therapeutic SC injection. At high concentrations, such as the environment of an intracellular secretion granule, proteins and peptides are frequently associated with a complex mixture of stabilizing materials. Formulations strive to limit complexity as additional agents can complicate biological outcomes and toxicology. Further, proteins and peptides produced in the body are typically destroyed and remade every few days; SC formulations are typically prepared to have a shelf-life of years. Thus, the conditions used to stabilize therapeutic proteins and peptides can put a great deal of pressure on the physical, chemical, and biological properties of these materials. While nonnative proteins and peptides can be altered to correct some of these issues, there is typically a strong desire to maintain the composition of native proteins and peptides, requiring creative solutions to overcome these nonphysiological issues.

As highlighted above, there are multiple issues associated with preparing formulations that are sufficiently stable and acceptable for pharmaceutical applications involving an SC injection. Some solutions to successfully formulate protein and peptide therapeutics require pH, salt, and additives that are dramatically different from physiological conditions at the site of an SC injection. These differences can affect the PK, PD, and metabolism properties of an injected protein or peptide. This is not typically examined as outcomes of safety and efficacy are the ultimate issues of concern. As long as these are achieved within satisfactory limits, issues of PK, PD, and metabolism at an SC injection site are of little consequence. There can, however, be striking issues of PK, PD, and metabolism that could affect a protein or peptide following its SC injection. In general, such outcomes, typically observed during preclinical studies, result in molecules simply failing to be commercialized. As a rule, such failures are not openly discussed, and publications typically describe successful rather than dismal outcomes. It was the goal of this chapter to walk through potential issues associated with the SC injection of protein or peptide therapeutics to provide a template for understanding how physical, chemical, and biological matches and potential mismatches between the SC injection site and the material injected might affect PK, PD, and metabolism outcomes.

REFERENCES

1. Koivisto VA, Felig P. Alterations in insulin absorption and in blood glucose control associated with varying insulin injection sites in diabetic patients. Ann Intern Med 1980; 92(1):59–61.
2. Calara F, Taylor K, Han J, et al. A randomized, open-label, crossover study examining the effect of injection site on bioavailability of exenatide (synthetic exendin-4). Clin Ther 2005; 27(2):210–215.
3. Vaag A, Handberg A, Lauritzen M, et al. Variation in absorption of NPH insulin due to intramuscular injection. Diabetes Care 1990; 13(1):74–76.
4. Galloway JA, Spradlin CT, Nelson RL, et al. Factors influencing the absorption, serum insulin concentration, and blood glucose responses after injections of regular insulin and various insulin mixtures. Diabetes Care 1981; 4(3):366–376.
5. Lindstrom T, Olsson PO, Arnqvist HJ. The use of human ultralente is limited by great intraindividual variability in overnight plasma insulin profiles. Scand J Clin Lab Invest 2000; 60(5):341–347.
6. Wikimedia Commons. File:Skin.jpg. Available at: http://commons.wikimedia.org/wiki/Image:Skin.jpg.
7. Ji RC. Lymphatic endothelial cells, inflammatory lymphangiogenesis, and prospective players. Curr Med Chem 2007; 14(22):2359–2368.
8. Karpanen T, Alitalo K. Molecular biology and pathology of lymphangiogenesis. Annu Rev Pathol 2008; 3:367–397.
9. Wanner M, Avram M. An evidence-based assessment of treatments for cellulite. J Drugs Dermatol 2008; 7(4):341–345.
10. Brooks Y, Black DR, Coster DC, et al. Body mass index and percentage body fat as health indicators for young adults. Am J Health Behav 2007; 31(6):687–700.
11. Plopper GE. Extracellular matrix and cell junctions. In: Lewin B, Cassimeris L, Lingappa VR, et al., eds. Cells. Sudbury, MA: Jones & Bartlett, 2006:645–702.
12. Malavaki C, Mizumoto S, Karamanos N, et al. Recent advances in the structural study of functional chondroitin sulfate and dermatan sulfate in health and disease. Connect Tissue Res 2008; 49(3):133–139.
13. Peach RJ, Hollenbaugh D, Stamenkovic I, et al. Identification of hyaluronic acid binding sites in the extracellular domain of CD44. J Cell Biol 1993; 122(1):257–264.
14. Geutjes PJ, et al. From molecules to matrix: construction and evaluation of molecularly defined bioscaffolds. Adv Exp Med Biol 2006; 585:279–295.
15. Sugawara K, Tsuruta D, Ishii M, et al. Laminin-332 and -511 in skin. Exp Dermatol 2008; 17(6):473–480.

16. Lauria G, Lombardi R, Camozzi F, et al. Skin biopsy for the diagnosis of peripheral neuropathy. Histopathology 2009; 54(3):273–285.

17. Cohen JL. Understanding, avoiding, and managing dermal filler complications. Dermatol Surg 2008; 34(suppl 1):S92–S99.

18. Rizzoni D, Paiardi S, Rodella L, et al. Changes in extracellular matrix in subcutaneous small resistance arteries of patients with primary aldosteronism. J Clin Endocrinol Metab 2006; 91(7):2638–2642.

19. Wang D, Chabrashvili T, Borrego L, et al. Angiotensin II infusion alters vascular function in mouse resistance vessels: roles of O and endothelium. J Vasc Res 2006; 43(1):109–119.

20. Nelson HS. Allergen immunotherapy: where is it now? J Allergy Clin Immunol 2007; 119(4):769–779.

21. Komatsu N, Suga Y, Saijoh K, et al. Elevated human tissue kallikrein levels in the stratum corneum and serum of peeling skin syndrome-type B patients suggests an over-desquamation of corneocytes. J Invest Dermatol 2006; 126(10): 2338–2342.

22. Dean RA, Butler GS, Hamma-Kourbali Y, et al. Identification of candidate angiogenic inhibitors processed by matrix metalloproteinase 2 (MMP-2) in cell-based proteomic screens: disruption of vascular endothelial growth factor (VEGF)/heparin affin regulatory peptide (pleiotrophin) and VEGF/Connective tissue growth factor angiogenic inhibitory complexes by MMP-2 proteolysis. Mol Cell Biol 2007; 27(24):8454–8465.

23. Kainulainen V, Wang H, Schick C, et al. Syndecans, heparan sulfate proteoglycans, maintain the proteolytic balance of acute wound fluids. J Biol Chem 1998; 273(19): 11563–11569.

24. White FA, Wilson NM. Chemokines as pain mediators and modulators. Curr Opin Anaesthesiol 2008; 21(5):580–585.

25. Merad M, Hoffmann P, Ranheim E, et al., Depletion of host Langerhans cells before transplantation of donor alloreactive T cells prevents skin graft-versus-host disease. Nat Med 2004; 10(5):510–517.

26. Obhrai JS, Oberbarnscheidt M, Zhang N, et al. Langerhans cells are not required for efficient skin graft rejection. J Invest Dermatol 2008; 128(8):1950–1955.

27. Kwiek B, Peng WM, Allam JP, et al. Tacrolimus and TGF-beta act synergistically on the generation of Langerhans cells. J Allergy Clin Immunol 2008; 122(1):126–132, 132.e1.

28. Subramanyam M. Immunogenicity of biotherapeutics-an overview. J Immunotoxicol 2006; 3(3):151–156.

29. Ho LT, Lam HC, Wu MS, et al. A twelve month double-blind randomized study of the efficacy and immunogenicity of human and porcine insulins in non-insulin-dependent diabetics. Zhonghua Yi Xue Za Zhi (Taipei) 1991; 47(5):313–319.

30. Davis SN, Thompson CJ, Peak M, et al. Effects of human insulin on insulin binding antibody production in nondiabetic subjects. Diabetes Care 1992; 15(1):124–126.

31. Ebers GC, Dyment DA. Genetics of multiple sclerosis. Semin Neurol 1998; 18(3): 295–299.

32. Steis RG, Smith JW II, Urba WJ, et al. Resistance to recombinant interferon alfa-2a in hairy-cell leukemia associated with neutralizing anti-interferon antibodies. N Engl J Med 1988; 318(22):1409–1413.

33. Russo D, Candoni A, Zuffa E, et al. Neutralizing anti-interferon-alpha antibodies and response to treatment in patients with Ph+ chronic myeloid leukaemia sequentially treated with recombinant (alpha 2a) and lymphoblastoid interferon-alpha. Br J Haematol 1996; 94(2):300–305.

34. Wadhwa M, Skog AL, Bird C, et al. Immunogenicity of granulocyte-macrophage colony-stimulating factor (GM-CSF) products in patients undergoing combination therapy with GM-CSF. Clin Cancer Res 1999; 5(6):1353–1361.

35. Macdougall IC. Antibody-mediated pure red cell aplasia (PRCA): epidemiology, immunogenicity and risks. Nephrol Dial Transplant 2005; 20(suppl 4):iv9–iv15.

36. Kuter DJ. Future directions with platelet growth factors. Semin Hematol 2000; 37(2 suppl 4):41–49.

37. Li J, Yang C, Xia Y, et al. Thrombocytopenia caused by the development of antibodies to thrombopoietin. Blood 2001; 98(12):3241–3248.
38. Basser RL, O'Flaherty E, Green M, et al. Development of pancytopenia with neutralizing antibodies to thrombopoietin after multicycle chemotherapy supported by megakaryocyte growth and development factor. Blood 2002; 99(7):2599–2602.
39. Herzyk DJ. The immunogenicity of therapeutic cytokines. Curr Opin Mol Ther 2003; 5(2):167–171.
40. Dasgupta S, Bayry J, André S, et al. Auditing protein therapeutics management by professional APCs: toward prevention of immune responses against therapeutic proteins. J Immunol 2008; 181(3):1609–1615.
41. Chenuaud P, Larcher T, Rabinowitz JE, et al. Autoimmune anemia in macaques following erythropoietin gene therapy. Blood 2004; 103(9):3303–3304.
42. Bach JF. Infections and autoimmune diseases. J Autoimmun 2005; 25(suppl):74–80.
43. Chen W, Antonenko S, Sederstrom JM, et al. Thrombopoietin cooperates with FLT3-ligand in the generation of plasmacytoid dendritic cell precursors from human hematopoietic progenitors. Blood 2004; 103(7):2547–2553.
44. Cines DB, McMillan R. Pathogenesis of chronic immune thrombocytopenic purpura. Curr Opin Hematol 2007; 14(5):511–514.
45. Swanson SJ. Immunogenicity issues in drug development. J Immunotoxicol 2006; 3(3):165–172.
46. Kelley DE, Thaete FL, Troost F, et al. Subdivisions of subcutaneous abdominal adipose tissue and insulin resistance. Am J Physiol Endocrinol Metab 2000; 278(5): E941–E948.
47. Bertuzzi F, Verzaro R, Provenzano V, et al. Brittle type 1 diabetes mellitus. Curr Med Chem 2007; 14(16):1739–1744.
48. Bookbinder LH, Hofer A, Haller MF, et al. A recombinant human enzyme for enhanced interstitial transport of therapeutics. J Control Release 2006; 114(2):230–241.
49. Yocum RC, Kennard D, Heiner LS. Assessment and implication of the allergic sensitivity to a single dose of recombinant human hyaluronidase injection: a double-blind, placebo-controlled clinical trial. J Infus Nurs 2007; 30(5):293–299.
50. Gin H, Hanaire-Broutin H. Reproducibility and variability in the action of injected insulin. Diabetes Metab 2005; 31(1):7–13.
51. Daugherty AL, Cleland JL, Duenas EM, et al. Pharmacological modulation of the tissue response to implanted polylactic-co-glycolic acid microspheres. Eur J Pharm Biopharm 1997; 44(1637):89–102.
52. Kosaka H, Yoshimoto T, Yoshimoto T, et al. Interferon-gamma is a therapeutic target molecule for prevention of postoperative adhesion formation. Nat Med 2008; 14(4): 437–441.
53. Kaku K, Matsuda M, Urae A, et al. Pharmacokinetics and pharmacodynamics of insulin aspart, a rapid-acting analog of human insulin, in healthy Japanese volunteers. Diabetes Res Clin Pract 2000; 49(2–3):119–126.
54. Edelman ER, Pukac LA, Karnovsky MJ. Protamine and protamine-insulins exacerbate the vascular response to injury. J Clin Invest 1993; 91(5):2308–2313.
55. Capobianchi MR, Mattana P, Mercuri F, et al. Acid lability is not an intrinsic property of interferon-alpha induced by HIV-infected cells. J Interferon Res 1992; 12(6):431–438.
56. Brange J. Galenics of Insulin: the Physico-Chemical and Pharmaceutical Aspects of Insulin and Insulin Preparations. New York: Springer-Verlag, 1987.
57. Rottenberg DA, Ginos JZ, Kearfott KJ, et al. In vivo measurement of brain tumor pH using [11C]DMO and positron emission tomography. Ann Neurol 1985; 17(1):70–79.
58. Ward KM, Balaban RS. Determination of pH using water protons and chemical exchange dependent saturation transfer (CEST). Magn Reson Med 2000; 44(5):799–802.
59. Raghunand N, Zhang S, Sherry AD, et al. In vivo magnetic resonance imaging of tissue pH using a novel pH-sensitive contrast agent, GdDOTA-4AmP. Acad Radiol 2002; 9(suppl 2):S481–S483.
60. Gallagher FA, Zhang S, Sherry AD, et al. Magnetic resonance imaging of pH in vivo using hyperpolarized 13C-labelled bicarbonate. Nature 2008; 453(7197):940–943.

61. Chi EY, Krishnan S, Randolph TW, et al. Physical stability of proteins in aqueous solution: mechanism and driving forces in nonnative protein aggregation. Pharm Res 2003; 20(9):1325–1336.
62. Meredith SC. Protein denaturation and aggregation: cellular responses to denatured and aggregated proteins. Ann N Y Acad Sci 2005; 1066:181–221.
63. Kamerzell TJ, Middaugh CR. The complex inter-relationships between protein flexibility and stability. J Pharm Sci 2008; 97(9):3494–3517.
64. Pease LF III, Elliott JT, Tsai DH, et al. Determination of protein aggregation with differential mobility analysis: application to IgG antibody. Biotechnol Bioeng 2008; 101(6):1214–1222.
65. Brange J, Langkjoer L. Insulin structure and stability. Pharm Biotechnol 1993; 5: 315–350.
66. Peltonen L, Halila R, Ryhanen L. Enzymes converting procollagens to collagens. J Cell Biochem 1985; 28(1):15–21.
67. Fernandez-Busquets X, de Groot NS, Fernandez D, et al. Recent structural and computational insights into conformational diseases. Curr Med Chem 2008; 15(13): 1336–1349.
68. Uversky VN. Amyloidogenesis of natively unfolded proteins. Curr Alzheimer Res 2008; 5(3):260–287.
69. Avidan-Shpalter C, Gazit E. The early stages of amyloid formation: biophysical and structural characterization of human calcitonin pre-fibrillar assemblies. Amyloid 2006; 13(4):216–225.
70. Bellotti V, Mangione P, Stoppini M. Biological activity and pathological implications of misfolded proteins. Cell Mol Life Sci 1999; 55(6–7):977–991.
71. Trexler AJ, Nilsson MR. The formation of amyloid fibrils from proteins in the lysozyme family. Curr Protein Pept Sci 2007; 8(6):537–557.
72. Brouet JC, Clauvel JP, Danon F, et al. Biologic and clinical significance of cryoglobulins. A report of 86 cases. Am J Med 1974; 57(5):775–788.
73. Middaugh CR, Gerber-Jenson B, Hurvitz A, et al. Physicochemical characterization of six monoclonal cryoimmunoglobulins: possible basis for cold-dependent insolubility. Proc Natl Acad Sci U S A 1978; 75(7):3440–3444.
74. Baynes BM, Trout BL. Rational design of solution additives for the prevention of protein aggregation. Biophys J 2004; 87(3):1631–1639.
75. Baynes BM, Wang DI, Trout BL. Role of arginine in the stabilization of proteins against aggregation. Biochemistry 2005; 44(12):4919–4925.
76. Kertscher U, Bienert M, Krause E, et al. Spontaneous chemical degradation of substance P in the solid phase and in solution. Int J Pept Protein Res 1993; 41(3):207–211.
77. Cleland JL, Powell MF, Shire SJ. The development of stable protein formulations: a close look at protein aggregation, deamidation, and oxidation. Crit Rev Ther Drug Carrier Syst 1993; 10(4):307–377.
78. Morgan PE, Sturgess AD, Davies MJ. Increased levels of serum protein oxidation and correlation with disease activity in systemic lupus erythematosus. Arthritis Rheum 2005; 52(7):2069–2079.
79. Wakankar AA, Borchardt RT, Eigenbrot C, et al. Aspartate isomerization in the complementarity-determining regions of two closely related monoclonal antibodies. Biochemistry 2007; 46(6):1534–1544.
80. Huang L, Lu J, Wroblewski VJ, et al. In vivo deamidation characterization of monoclonal antibody by LC/MS/MS. Anal Chem 2005; 77(5):1432–1439.
81. Robinson NE, Robinson AB. Deamidation of human proteins. Proc Natl Acad Sci U S A 2001; 98(22):12409–12413.
82. Robinson NE, Robinson AB. Molecular clocks. Proc Natl Acad Sci U S A 2001; 98(3): 944–949.
83. Lindner H, Helliger W. Age-dependent deamidation of asparagine residues in proteins. Exp Gerontol 2001; 36(9):1551–1563.
84. Watanabe A, Takio K, Ihara Y. Deamidation and isoaspartate formation in smeared tau in paired helical filaments. Unusual properties of the microtubule-binding domain of tau. J Biol Chem 1999; 274(11):7368–7378.

85. Schey KL, Little M, Fowler JG, et al. Characterization of human lens major intrinsic protein structure. Invest Ophthalmol Vis Sci 2000; 41(1):175–182.
86. Tyagi SC, Kumar S, Katwa L. Differential regulation of extracellular matrix metalloproteinase and tissue inhibitor by heparin and cholesterol in fibroblast cells. J Mol Cell Cardiol 1997; 29(1):391–404.
87. Wygrecka M, Jablonska E, Guenther A, et al. Current view on alveolar coagulation and fibrinolysis in acute inflammatory and chronic interstitial lung diseases. Thromb Haemost 2008; 99(3):494–501.
88. Wahl LM, Corcoran ML. Regulation of monocyte/macrophage metalloproteinase production by cytokines. J Periodontol 1993; 64(5 suppl):467–473.
89. Kahari VM, Saarialho-Kere U. Matrix metalloproteinases in skin. Exp Dermatol 1997; 6(5):199–213.
90. Riley KN, Herman IM. Collagenase promotes the cellular responses to injury and wound healing in vivo. J Burns Wounds 2005; 4:e8.
91. Eppler SM, Combs DL, Henry TD, et al. A target-mediated model to describe the pharmacokinetics and hemodynamic effects of recombinant human vascular endothelial growth factor in humans. Clin Pharmacol Ther 2002; 72(1):20–32.
92. Hollier B, Harkin DG, Leavesley D, et al. Responses of keratinocytes to substrate-bound vitronectin: growth factor complexes. Exp Cell Res 2005; 305(1):221–232.
93. Stachowska-Pietka J, Waniewski J, Flessner MF, et al. A distributed model of bidirectional protein transport during peritoneal fluid absorption. Adv Perit Dial 2007; 23:23–27.
94. Haupt MT. The use of crystalloidal and colloidal solutions for volume replacement in hypovolemic shock. Crit Rev Clin Lab Sci 1989; 27(1):1–26.
95. Tong RT, Boucher Y, Kozin SV, et al. Vascular normalization by vascular endothelial growth factor receptor 2 blockade induces a pressure gradient across the vasculature and improves drug penetration in tumors. Cancer Res 2004; 64(11):3731–3736.
96. Pluen A, Boucher Y, Ramanujan S, et al. Role of tumor-host interactions in interstitial diffusion of macromolecules: cranial vs. subcutaneous tumors. Proc Natl Acad Sci U S A 2001; 98(8):4628–4633.
97. Flessner MF. The role of extracellular matrix in transperitoneal transport of water and solutes. Perit Dial Int 2001; 21(suppl 3):S24–S29.
98. Schweitzer AD, Rakesh V, Revskaya E, et al. Computational model predicts effective delivery of 188-Re-labeled melanin-binding antibody to metastatic melanoma tumors with wide range of melanin concentrations. Melanoma Res 2007; 17(5):291–303.
99. Furie B, Furie BC. Mechanisms of thrombus formation. N Engl J Med 2008; 359(9):938–949.
100. Gurtner GC, Werner S, Barrandon Y, et al. Wound repair and regeneration. Nature 2008; 453(7193):314–321.
101. Swart PJ, Beljaars L, Kuipers ME, et al. Homing of negatively charged albumins to the lymphatic system: general implications for drug targeting to peripheral tissues and viral reservoirs. Biochem Pharmacol 1999; 58(9):1425–1435.
102. Andersen JT, Dee Qian J, Sandlie I. The conserved histidine 166 residue of the human neonatal Fc receptor heavy chain is critical for the pH-dependent binding to albumin. Eur J Immunol 2006; 36(11):3044–3051.
103. Kim J, Bronson CL, Hayton WL, et al. Albumin turnover: FcRn-mediated recycling saves as much albumin from degradation as the liver produces. Am J Physiol Gastrointest Liver Physiol 2006; 290(2):G352–G360.
104. Kurtzhals P. How to achieve a predictable basal insulin? Diabetes Metab 2005; 31(4 pt 2):4S25–4S33.
105. Chuang VT, Kragh-Hansen U, Otagiri M. Pharmaceutical strategies utilizing recombinant human serum albumin. Pharm Res 2002; 19(5):569–577.
106. Glickson JD, Lund-Katz S, Zhou R, et al. Lipoprotein nanoplatform for targeted delivery of diagnostic and therapeutic agents. Mol Imaging 2008; 7(2):101–110.
107. Dumont JA, Low SC, Peters RT, et al. Monomeric Fc fusions: impact on pharmacokinetic and biological activity of protein therapeutics. BioDrugs 2006; 20(3):151–160.

108. Jazayeri JA, Carroll GJ. Fc-based cytokines: prospects for engineering superior therapeutics. BioDrugs 2008; 22(1):11–26.
109. Sharma R, Wendt JA, Rasmussen JC, et al. New horizons for imaging lymphatic function. Ann N Y Acad Sci 2008; 1131:13–36.
110. Dani B, Platz R, Tzannis ST. High concentration formulation feasibility of human immunoglubulin G for subcutaneous administration. J Pharm Sci 2007; 96(6): 1504–1517.
111. Basu SK, Govardhan CP, Jung CW, et al. Protein crystals for the delivery of bio-pharmaceuticals. Expert Opin Biol Ther 2004; 4(3):301–317.
112. Yang MX, Shenoy B, Disttler M, et al. Crystalline monoclonal antibodies for sub-cutaneous delivery. Proc Natl Acad Sci U S A 2003; 100(12):6934–6939.
113. Govardhan C, Khalaf N, Jung CW, et al. Novel long-acting crystal formulation of human growth hormone. Pharm Res 2005; 22(9):1461–1470.
114. Alam M, Dover JS. Management of complications and sequelae with temporary injectable fillers. Plast Reconstr Surg 2007; 120(6 suppl):98S–105S.
115. Nicolas JF, Guy B. Intradermal, epidermal and transcutaneous vaccination: from immunology to clinical practice. Expert Rev Vaccines 2008; 7(8):1201–1214.
116. Soares AP, Scriba TJ, Joseph S, et al. Bacillus Calmette-Guerin vaccination of human newborns induces T cells with complex cytokine and phenotypic profiles. J Immunol 2008; 180(5):3569–3577.
117. Schultze V, D'Agosto V, Wack A, et al. Safety of MF59 adjuvant. Vaccine 2008; 26(26):3209–3222.
118. Roy P, Noad R. Virus-like particles as a vaccine delivery system: myths and facts. Hum Vaccin 2008; 4(1):5–12.
119. Wallace AM, Hoh CK, Darrah DD, et al. Sentinel lymph node mapping of breast cancer via intradermal administration of Lymphoseek. Nucl Med Biol 2007; 34(7): 849–853.
120. Nagy G, Emody L, Pal T. Strategies for the development of vaccines conferring broad-spectrum protection. Int J Med Microbiol 2008; 298(5–6):379–395.
121. Chabalgoity JA. Paving the way for the introduction of new vaccines into devel-oping countries. Expert Rev Vaccines 2005; 4(2):147–150.
122. Lambert PH, Laurent PE. Intradermal vaccine delivery: will new delivery systems transform vaccine administration? Vaccine 2008; 26(26):3197–3208.
123. Liu L, Zhou X, Shi J, et al. Toll-like receptor-9 induced by physical trauma mediates release of cytokines following exposure to CpG motif in mouse skin. Immunology 2003; 110(3):341–347.
124. Hui SW. Overview of drug delivery and alternative methods to electroporation. Methods Mol Biol 2008; 423:91–107.
125. Vandervoort J, Ludwig A. Microneedles for transdermal drug delivery: a minire-view. Front Biosci 2008; 13:1711–1715.

Local Tissue Responses to Polymer Implants Affecting Pharmacokinetics, Pharmacodynamics, and Metabolism of Proteins and Peptides

James M. Anderson and William G. Brodbeck
Institute of Pathology, Case Western Reserve University, Cleveland, Ohio, U.S.A.

INTRODUCTION

In the research and development of implantable drug delivery systems, in vivo biocompatibility studies play an important role in determining the safety (biocompatibility) and efficacy (function) of these devices. The purpose of this review is to present perspectives on the in vivo tissue responses and biocompatibility of implantable drug delivery systems as they may affect the pharmacokinetics, pharmacodynamics, and metabolism of proteins and peptides. The evaluation of the biocompatibility of implantable delivery systems requires an understanding of the inflammatory and healing responses induced by implantable materials. For delivery systems, this includes an appreciation of the inflammatory and healing responses to degradable/resorbable systems as well as nondegradable systems. In this overview, tissue/material interactions and the foreign body reaction are viewed from the classical medical perspective of the pathologist. Tissue/material interactions are commonly referred to as the tissue response continuum, which is the series of responses that are initiated by the implantation procedure, as well as by the presence of the biomaterial, medical device, or drug delivery system. In this chapter, we divide the continuum of tissue/material responses into the early, transient tissue responses and the late, persistent tissue responses. Early, transient tissue/material responses include injury, blood/material interactions, provisional matrix formation, temporal sequence of inflammation and wound healing, acute inflammation, chronic inflammation, and granulation tissue development. These responses are usually of short duration, occurring over the first two to three weeks following implantation of a medical device or drug delivery system. Late, persistent tissue responses include macrophage interactions, foreign body giant cell (FBGC) formation and interactions, and fibrosis and fibrous encapsulation of the drug delivery system. The early, transient tissue responses form the basis for safety or biocompatibility considerations of the medical device or drug delivery system. Late, persistent tissue/material responses, while important to the safety and biocompatibility considerations, may be more important in modulating the performance characteristics of the drug delivery system.

Normal tissue responses to implanted controlled drug delivery systems may have a significant effect on the bioavailability and pharmacokinetics of proteins or peptides, that is, bioactive agent, released from a controlled drug delivery system, due to the release of degradative molecules. In addition to the carrier that functions as a permeability or diffusion barrier, normal tissue responses such as the foreign body reaction and the fibrous capsule may reduce the permeation and diffusion of the bioactive agent and thus lead to a reduction in bioavailability of the bioactive agent.

In vivo biocompatibility studies commonly utilize subcutaneous implantation. Therefore, examples of these types of studies available in the current literature will be used to illustrate important issues pertinent to the biocompatibility of implantable delivery systems. As the desired goal of delivery systems is to deliver or release a drug or therapeutic agent either locally or systemically, in vivo studies that concomitantly monitor the delivery of therapeutic agents in conjunction with the biocompatibility of the system are important. It is not our intent to provide a complete literature review of biocompatibility studies that have been carried out on degradable/resorbable and nonbiodegradable materials utilized for implantable delivery systems but rather to present fundamental perspectives on the in vivo biocompatibility of drug delivery systems.

EARLY, TRANSIENT TISSUE/MATERIAL RESPONSES
Injury and Acute Responses

The process of implantation of a biomaterial, medical device, or drug delivery system results in injury to tissues or organs (1–12). It is this injury and the subsequent perturbation of homeostatic mechanisms that lead to the inflammatory responses, foreign body reaction, and wound healing. The response to injury is dependent on multiple factors, which include the extent of injury, the loss of basement membrane structures, blood-material interactions, provisional matrix formation, the extent or degree of cellular necrosis, and the extent of the inflammatory response. These events, in turn, may affect the extent or degree of granulation tissue formation, foreign body reaction, and fibrosis or fibrous capsule development. These events are summarized in Table 1: The sequence of host reactions following implantation of medical devices and the temporal sequence of events are shown in Figure 1. The host reactions are considered to be tissue dependent, organ dependent, and species dependent. In addition, it is important to recognize that these reactions occur or are initiated early, that is, within two to three weeks of the time of implantation, and may significantly modify the pharmacokinetics, pharmacodynamics, and metabolism of proteins and peptides through cellular and noncellular interactions at the tissue/implant interface.

In considering these host reactions following injury, it is important to examine whether tissue resolution or organization occurs within the injured tissue. In situations where injury has occurred and exudative inflammation is present but no cellular necrosis or loss of basement membrane structures has occurred, the process of resolution takes place. Resolution is the restitution of

TABLE 1 Sequence of Host Reactions Following Implantation of Medical Devices

Injury and acute responses
Acute inflammation
Chronic inflammation
Granulation tissue
Foreign body reaction
Fibrous encapsulation

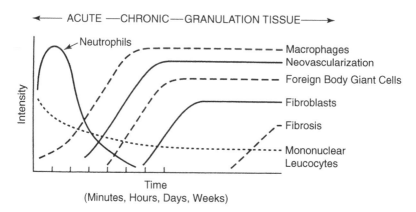

FIGURE 1   The temporal variation in the acute inflammatory response, chronic inflammatory response, granulation tissue development, and foreign body reaction to implanted biomaterials. The intensity and time variables are dependent on the extent of injury created during implantation and the size, shape, topography, and chemical and physical properties of the biomaterial, medical device, or drug delivery system.

the preexisting architecture of the tissue. On the other hand, with necrosis (cell death), granulation tissue grows into the inflammatory exudate and the process of organization with development of fibrous (scar) tissue occurs. With implants, the process of organization with development of fibrous tissue leads to the well-known fibrous capsule formation at the tissue/material interface. The proliferative capacity of cells within the tissue also plays a role in determining whether resolution or organization occurs. In general, the process of implantation in vascularized tissues leads to organization with fibrous tissue development and fibrous encapsulation.

Blood-material interactions and the inflammatory response are intimately linked, and in fact, early responses to injury involve mainly blood and the vasculature (1–5). Regardless of the tissue into which a biomaterial is implanted, the initial inflammatory response is activated by injury to vascularized connective tissue (Table 2). Because blood and its components are involved in the initial inflammatory responses, thrombus and/or blood clot also form. Thrombus formation involves activation of the extrinsic and intrinsic coagulation systems, the complement system, the fibrinolytic system, the kinin-generating system, and platelets. Thrombus or blood clot formation on the surface of a biomaterial is related to the well-known Vroman effect of protein adsorption. From a wound-healing perspective, blood protein deposition on a biomaterial surface is described as provisional matrix formation.

Immediately following injury, that is, the initiation of surgery, changes occur in vascular flow, caliber, and permeability. Fluid, proteins, and blood cells escape from the vascular system into the injured tissue in a process called exudation. Following changes in the vascular system, which also include changes induced in blood and its components, cellular events occur and characterize the inflammatory response (11–14). The effect of the injury and/or biomaterial in situ on plasma or cells can produce chemical factors that mediate

TABLE 2 Cells and Components of Vascularized Connective Tissue

Intravascular (blood) cells
  Erythrocytes (RBC)
  Neutrophils
  Monocytes
  Eosinophils
  Lymphocytes
  Basophils
  Platelets
Connective tissue cells
  Mast cells
  Fibroblasts
  Macrophages
  Lymphocytes
Extracellular matrix components
  Collagens
  Elastin
  Proteoglycans
  Fibronectin
  Laminin

many of the vascular and cellular responses of inflammation. Although injury initiates the inflammatory response, released chemicals from plasma, cells, and injured tissue mediate the response. Several important points must be noted to understand the inflammatory response and how it relates to biomaterials and drug delivery systems. First, although chemical mediators are classified on a structural or functional basis, different mediator systems interact and provide a system of checks and balances regarding their respective activities and functions. Second, chemical mediators are quickly inactivated or destroyed, suggesting that their action is predominantly local (i.e., at the implant site). Third, generally acid, lysosomal proteases, and oxygen-derived free radicals produce the most significant damage or injury. These chemical mediators may be important in the modification of released proteins and peptides and the degradation of biomaterials, for example, degradable delivery systems.

The predominant cell type present in the inflammatory response varies with the age of the injury. Neutrophils are considered professional phagocytes that can ingest or take up particulate forms of drug delivery systems in a size-dependent manner. In general, neutrophils, commonly called polymorphonuclear leukocytes or polys, predominate during the first several days following injury and are then replaced by monocytes as the predominant cell type. Three factors account for this change in cell type: (*i*) neutrophils are short lived and disintegrate and disappear after 24 to 48 hours; neutrophil emigration is of short duration because chemotactic factors for neutrophil migration are activated early in the inflammatory response; (*ii*) following emigration from the vasculature, monocytes differentiate into macrophages, and these cells are very long lived (up to months); (*iii*) monocyte emigration may continue for days to weeks, depending on the injury and implanted biomaterial, and chemotactic factors for monocytes are activated over longer periods of time.

Injury to vascularized tissue in the implantation procedure leads to immediate development of the provisional matrix at the implant site. This provisional matrix consists of fibrin, produced by activation of the coagulative and thrombosis systems, and inflammatory products released by the complement system, activated platelets, inflammatory cells, and endothelial cells (15–17). These events occur early, within minutes to hours following implantation of a medical device. Components within or released from the provisional matrix, that is, fibrin network (thrombosis or clot), initiate the resolution, reorganization, and repair processes such as inflammatory cell and fibroblast recruitment. Platelets, activated during the fibrin network formation, release platelet factor 4, platelet-derived growth factor (PDGF), and transforming growth factor β (TGF-β), which contribute to fibroblast recruitment (18,19). Monocytes and lymphocytes, upon activation, generate additional chemotactic factors including LTB_4, PDGF, and TGF-β that recruit fibroblasts.

Fibrin, the major component of the provisional matrix, has been shown to play a major role in the development of neovascularization, that is, angiogenesis. Implanted porous surfaces filled with fibrin exhibit new vessel growth within four days (20). The provisional matrix is composed of adhesive molecules such as fibronectin and thrombospondin bound to fibrin as well as platelet granule components released during platelet aggregation. The provisional matrix is stabilized by the cross-linking of fibrin by factor XIIIa.

The provisional matrix may be viewed as a naturally derived, biodegradable, sustained-release system in which mitogens, chemoattractants, cytokines, and growth factors are released to control subsequent wound-healing processes (21–26). In spite of the rapid increase in our knowledge of the provisional matrix and its capabilities, our knowledge of the control of the formation of the provisional matrix and its effect on subsequent wound-healing events is poor. In part, this lack of knowledge is due to the fact that much of our knowledge regarding the provisional matrix has been derived from in vitro studies, and there is a paucity of in vivo studies, which provide for a more complex perspective. Little is known regarding the provisional matrix, which forms at biomaterial and medical device interfaces in vivo. Attractive hypotheses have been presented regarding the presumed ability of materials and protein-adsorbed materials to modulate cellular interactions through their interactions with adhesive molecules and cells. In considering the bioavailability, pharmacokinetics, and pharmacodynamics of proteins and peptides released from controlled-release systems, it must be appreciated that each of the events/reactions indicated in Table 1 may alter the efficacy and function of the protein or peptide release system.

Temporal Sequence of Inflammation and Wound Healing

Inflammation is generally defined as the reaction of vascularized living tissue to local injury. Inflammation serves to contain, neutralize, dilute, or wall off the injurious agent or process. In addition, it sets into motion a series of events that may heal and reconstitute the implant site through replacement of the injured tissue by regeneration of native parenchymal cells, formation of fibroblastic scar tissue, or a combination of these two processes (11,12).

The sequence of events following implantation of a drug delivery system, medical device, or biomaterial is illustrated in Figure 1. The size, shape, and chemical and physical properties of the biomaterial and/or the physical dimensions and properties of the prosthesis or device may be responsible for variations in the intensity and time duration of the inflammatory and wound-healing processes. Thus, intensity and/or time duration of inflammatory reaction responses may characterize the biocompatibility of a biomaterial, medical device, or drug delivery system.

In general, the biocompatibility of a material with tissue has been described in terms of the acute and chronic inflammatory responses and of the fibrous capsule formation that are seen over various time periods following implantation. Histologic evaluation of tissue adjacent to implanted materials as a function of implant time has been the most commonly used method of evaluating the biocompatibility. Classically, the biocompatibility of an implanted material has been described in terms of the morphologic appearance of the inflammatory reaction to the material. However, the inflammatory response is a series of complex reactions involving various types of cells, the densities, activities, and functions of which are controlled by various endogenous and autocoid mediators. The simplistic view of the acute inflammatory response progressing to the chronic inflammatory response may be misleading with respect to biocompatibility studies and an understanding of the inflammatory response to implants. Studies using the cage implant system show that monocytes and macrophages are present in highest concentrations when neutrophils are also at their highest concentrations, that is, the acute inflammatory response (27,28). Neutrophils have short lifetimes—hours to days—and disappear from the exudate more rapidly than macrophages, which have lifetimes of days to weeks to months. Eventually, monocytes become the predominant cell type in the exudate, resulting in a chronic inflammatory response. Monocytes rapidly differentiate into macrophages, the cells principally responsible for normal wound healing in the foreign body reaction (Fig. 2). Classically, the development of granulation tissue has been considered to be a part of chronic inflammation, but because of unique tissue-material interactions, it is preferable to differentiate the foreign body reaction—with its varying degree of granulation tissue development, including macrophages, fibroblasts, and capillary formation—from chronic inflammation.

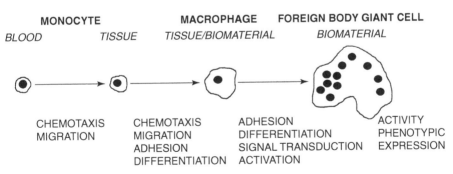

FIGURE 2 In vivo transition from blood-borne monocyte to biomaterial adherent monocyte/ macrophage to FBGC at the tissue/biomaterial interface. Little is known regarding the indicated biologic responses that are considered to play important roles in the transition to FBGC development. *Abbreviation*: FBGC, foreign body giant cell.

Acute Inflammation

Acute inflammation is of relatively short duration, lasting from minutes to days, depending on the extent of injury. The main characteristics of acute inflammation are the exudation of fluid and plasma proteins (edema) and the emigration of leukocytes (predominantly neutrophils). Neutrophils and other motile white cells emigrate or move from the blood vessels to the perivascular tissues and the injury (implant) site (Fig. 3) (29–31).

The accumulation of leukocytes, in particular neutrophils and monocytes, is the most important feature of the acute and chronic inflammatory reactions, respectively. Leukocytes accumulate through a series of processes including margination, adhesion, emigration, phagocytosis, and extracellular release of leukocyte products (32). Increased leukocytic adhesion in inflammation involves specific interactions between complementary "adhesion molecules" present on the leukocyte and endothelial surfaces (33,34). The surface expression of these adhesion molecules is modulated by inflammatory agents; mechanisms of interaction include stimulation of leukocyte adhesion molecules (C5a, LTB_4), stimulation of endothelial adhesion molecules [interleukin (IL)-1], or both effects [tumor necrosis factor (TNF)-α]. Integrins comprise a family of transmembrane glycoproteins that modulate cell-matrix and cell-cell relationships by acting as receptors to extracellular protein ligands and also as direct adhesion molecules (35). An important group of integrins (adhesion molecules) on leukocytes include the CD11/CD18 family of adhesion molecules. Leukocyte-endothelial cell interactions are also controlled by endothelium-leukocyte adhesion molecules [(ELAMs) E-selectins] or intracellular adhesion molecules [(ICAMs) ICAM-1, ICAM-2, and VCAMs] on endothelial cells (36).

White cell emigration is controlled in part by chemotaxis, which is the unidirectional migration of cells along a chemical gradient. A wide variety of exogenous and endogenous substances have been identified as chemotactic agents (13,29–41). Important to the emigration or movement of leukocytes is the presence of specific receptors for chemotactic agents on the cell membranes of leukocytes. These and other receptors may also play a role in the activation of leukocytes. Following localization of leukocytes at the injury (implant) site, phagocytosis and the release of enzymes occur following activation of

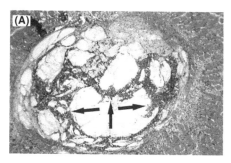 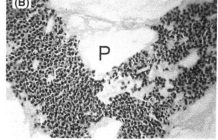

FIGURE 3 Acute inflammatory response of a swellable, bioresorbable polymer matrix at three days' implantation. (**A**) Low-magnification view showing fragmentation of the polymer matrix with polymorphonuclear leukocyte acute inflammatory response (*arrows*) predominantly to polymer particle surfaces. (**B**) High-magnification view of the polymorphonuclear leukocytes present between particles of the swellable, resorbable polymer matrix (P). Hematoxylin and eosin stain.

neutrophils and macrophages. The major role of the neutrophils in acute inflammation is to phagocytose microorganisms and foreign materials. Phagocytosis is seen as a three-step process in which the injurious agent undergoes recognition and neutrophil attachment, engulfment, and killing or degradation. With regard to biomaterials, engulfment and degradation may or may not occur depending on the size and properties of the biomaterial.

Although biomaterials are not generally phagocytosed by neutrophils or macrophages because of the size disparity (i.e., the surface of the biomaterial is greater than the size of the cell), certain events in phagocytosis may occur. The process of recognition and attachment is expedited when the injurious agent is coated by naturally occurring serum factors called opsonins. The two major opsonins are IgG and the complement-activated fragment C3b. Both of these plasma-derived proteins are known to adsorb to biomaterials, and neutrophils and macrophages have corresponding cell membrane receptors for these opsonization proteins. These receptors may also play a role in the activation of the attached neutrophil or macrophage. Because of the size disparity between the biomaterial surface and the attached cell, "frustrated phagocytosis" may occur (37,38). This process does not involve engulfment of the biomaterial but does cause the extracellular release of leukocyte products in an attempt to degrade the biomaterial. Neutrophils adherent to complement-coated and immunoglobulin-coated nonphagocytosable surfaces may release enzymes by direct extrusion or exocytosis from the cell (37–39). The amount of enzyme released during this process depends on the size of the polymer particle, with larger particles inducing greater amounts of enzyme release. This suggests that the specific mode of cell activation in the inflammatory response in tissue is dependent on the size of the implant and that a material in a phagocytosable form (e.g., powder or particulate) may provoke a degree of inflammatory response different from that of the same material in a nonphagocytosable form (e.g., film).

In considering protein or peptide delivery systems, the shape and size of the carrier may play a significant role in mechanisms by which proteins and peptides may be modified. Nanoparticles and microparticles, no larger than 5 μm in diameter, may be phagocytosed where the intracellular lysosomal and phagosomal compartments may play a direct role in protein/peptide modification. Microparticles with a diameter larger than 10 μm do not normally undergo phagocytosis, but cells at the material surface may undergo frustrated phagocytosis. Both phagocytosis and frustrated phagocytosis involve three general categories of reactive chemicals that can modify proteins and peptides, thus leading to significant changes in their bioavailability, efficacy, and pharmacokinetics. These general categories are acids, enzymes, and reactive oxygen species (ROS).

Chronic Inflammation
Chronic inflammation is histologically less uniform than acute inflammation. In general, chronic inflammation is characterized by the presence of monocytes and lymphocytes (11,12,42,43) (Fig. 4). It must be noted that many factors modify the course and histologic appearance of chronic inflammation.

Persistent inflammatory stimuli lead to chronic inflammation. Although the chemical and physical properties of the biomaterial may lead to chronic

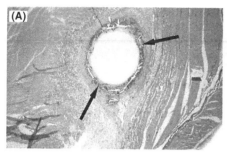 

FIGURE 4 Chronic inflammatory response of a swellable, bioresorbable polymer matrix at three days' implantation. (A) Low-magnification view showing initial swelling and fragmentation of the surface of the polymer matrix at the tissue/implant interface with focal chronic inflammation (*arrows*). (B) High-magnification view of the implant/tissue interface of the swellable bio-resorbable polymer (P) showing two interfacial responses: initial swelling and fragmentation at the surface of the polymer matrix and a focal chronic inflammatory response (*arrows*). A zone of resolving inflammation with macrophages and exudate is present between the polymer matrix (P) with a focal chronic inflammatory response (*arrows*) and SM. Hematoxylin and eosin stain. *Abbreviation*: SM, skeletal muscle.

inflammation, motion in the implant site by the biomaterial may also produce chronic inflammation. The chronic inflammatory response to biomaterials is confined to the implant site. Inflammation with the presence of mononuclear cells, including lymphocytes and plasma cells, is given the designation chronic inflammation, whereas the foreign body reaction with granulation tissue development is considered the normal wound-healing response to implanted biomaterials (i.e., the normal foreign body reaction). Chronic inflammation with biocompatible materials is usually of very short duration, that is, a few days.

The presence of chronic inflammation with lymphocytes and monocytes beyond the first two weeks following injection or implantation usually indicates an adverse reaction secondary to infection or cytotoxicity of the bioactive agent that is being released. Figure 5 demonstrates a response to cytotoxic naltrexone released from a polylactic acid–glycolic acid (PLGA) matrix. Figure 5B indicates that the placebo polymer alone does not initiate focal chronic inflammation, whereas focal chronic inflammation is seen with the naltrexone-containing PLGA matrix in Figure 5A (7,44).

Lymphocytes and plasma cells are involved principally in immune reactions and are key mediators of antibody production and delayed hypersensitivity responses. Their roles in nonimmunologic injuries and inflammation are largely unknown. Little is known regarding humoral immune responses and cell-mediated immunity to synthetic biomaterials. The role of macrophages and dendritic cells must be considered in the possible development of immune responses to synthetic biomaterials. Macrophages and dendritic cells process and present the antigen to immunocompetent cells and are thus key mediators in the development of immune reactions.

Chronic inflammation is also seen in injection sites where cell-mediated immune reactions are elicited by the bioactive agent, which functions as an antigen to initiate this response. Figure 6 demonstrates the expected response when an antigen, recombinant human growth hormone, is released into another

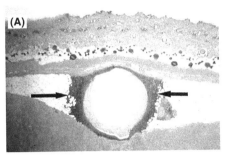
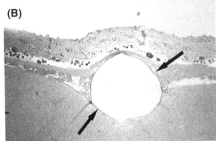

FIGURE 5 Tissue responses to a naltrexone-containing PLGA matrix and placebo matrix at 14 days' implantation. (**A**) Marked focal chronic inflammation composed of monocytes and lymphocytes (*arrows*) is present at the naltrexone-containing bead surface and surrounds the bead in a well-demarcated zone (*arrow*). (**B**) The placebo PLGA bead shows a minimal foreign body reaction at its surface and slight fibrous capsule formation (*arrows*) with no focal chronic inflammation. Hematoxylin and eosin stain. *Abbreviation*: PLGA, polylactic acid–glycolic acid.

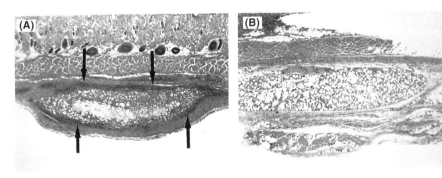

FIGURE 6 Chronic inflammatory response secondary to release of a bioactive agent from PLGA microparticles at 14 days' implantation. (**A**) Marked focal chronic inflammatory response (*arrows*) with monocytes and lymphocytes at the tissue/particle aggregate interface of recombinant human growth hormone–containing (Nutropin®) PLGA microparticles. (**B**) Focal macrophage and foreign body giant cell responses to placebo PLGA microparticles. Hematoxylin and eosin stain. *Abbreviation*: PLGA, polylactic acid–glycolic acid.

species and results in an expected immune reaction because of cross-species reactivity. Neutralizing immune responses have been identified after 14 to 16 days of infusing human growth factors into rats (45).

Therapeutic proteins and peptides released from implanted delivery systems must be considered as potential antigens capable of initiating immune reactions (46). Because of the potential antigenicity of proteins and peptides in implantable delivery systems, immunotoxicity testing is a requirement in the overall biocompatibility and tissue response evaluation of protein/peptide delivery systems. Immune responses to injected cytokines and growth factors are well documented in the literature (47–49).

The macrophage is probably the most important cell in chronic inflammation because of the great number of biologically active products it produces.

Important classes of products produced and secreted by macrophages include neutral proteases, chemotactic factors, arachidonic acid metabolites, reactive oxygen metabolites, complement components, coagulation factors, growth-promoting factors, and cytokines.

Growth factors such as PDGF, fibroblast growth factor (FGF), TGF-β, TGF-α/endothelial growth factor (EGF), and IL-1 or TNF are important for the growth of fibroblasts and blood vessels and the regeneration of epithelial cells. Growth factors, released by activated cells, stimulate production of a wide variety of cells; initiate cell migration, induce cellular differentiation, initiate tissue remodeling, and may be involved in various stages of wound healing (19,50–54). It is clear that there is a lack of information regarding interaction and synergy among various cytokines and growth factors and their abilities to exhibit chemotactic, mitogenic, and angiogenic properties.

Granulation Tissue

Within one day following implantation of a biomaterial (i.e., injury), the healing response is initiated by the action of monocytes and macrophages, followed by proliferation of fibroblasts and vascular endothelial cells at the implant site, leading to the formation of granulation tissue, the hallmark of healing inflammation. Granulation tissue derives its name from the pink, soft granular appearance on the surface of healing wounds, and its characteristic histologic features include the proliferation of new small blood vessels and fibroblasts. Depending on the extent of injury, granulation tissue may be seen as early as three to five days following implantation of a biomaterial.

The new small blood vessels are formed by budding or sprouting of preexisting vessels in a process known as neovascularization or angiogenesis (55–57). This process involves proliferation, maturation, and organization of endothelial cells into capillary tubes. Fibroblasts also proliferate in developing granulation tissue and are active in synthesizing collagen (fibrous tissue) and proteoglycans. In the early stages of granulation tissue development, proteoglycans predominate; later, however, collagen—especially type I collagen—predominates and forms the fibrous capsule.

LATE, PERSISTENT TISSUE/MATERIAL RESPONSES
Foreign Body Reaction: Macrophages and Foreign Body Giant Cells

Two factors that play a role in monocyte/macrophage adhesion and activation and FBGC formation are the surface chemistry of the substrate onto which the cells adhere and the protein adsorption that occurs before cell adhesion. These two factors have been hypothesized to play significant roles in the inflammatory and wound-healing responses to biomaterials, medical devices, and drug delivery systems.

Macrophage interactions with biomaterials are initiated when blood-borne monocytes in the early, transient responses migrate to the implant site and adhere to the blood protein–adsorbed biomaterial through monocyte-integrin interactions. Following adhesion, adherent monocytes differentiate into macrophages, which may then fuse to form FBGCs (Fig. 2).

Figure 7 shows monocyte and macrophage infiltration into a PLGA microcapsule aggregate at 14 days of implantation. Infiltration of monocytes and

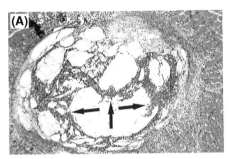
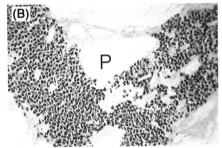

FIGURE 5.3 Acute inflammatory response of a swellable, bioresorbable polymer matrix at three days' implantation. (**A**) Low-magnification view showing fragmentation of the polymer matrix with polymorphonuclear leukocyte acute inflammatory response (*arrows*) predominantly to polymer particle surfaces. (**B**) High-magnification view of the polymorphonuclear leukocytes present between particles of the swellable, resorbable polymer matrix (P). Hematoxylin and eosin stain (*see page 112*).

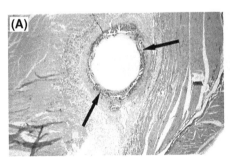

FIGURE 5.4 Chronic inflammatory response of a swellable, bioresorbable polymer matrix at three days' implantation. (**A**) Low-magnification view showing initial swelling and fragmentation of the surface of the polymer matrix at the tissue/implant interface with focal chronic inflammation (*arrows*). (**B**) High-magnification view of the implant/tissue interface of the swellable bioresorbable polymer (P) showing two interfacial responses: initial swelling and fragmentation at the surface of the polymer matrix and a focal chronic inflammatory response (*arrows*). A zone of resolving inflammation with macrophages and exudate is present between the polymer matrix (P) with a focal chronic inflammatory response (*arrows*) and SM. Hematoxylin and eosin stain. *Abbreviation*: SM, skeletal muscle (*see page 114*).

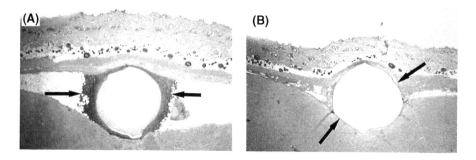

FIGURE 5.5 Tissue responses to a naltrexone-containing PLGA matrix and placebo matrix at 14 days' implantation. (**A**) Marked focal chronic inflammation composed of monocytes and lymphocytes (*arrows*) is present at the naltrexone-containing bead surface and surrounds the bead in a well-demarcated zone (*arrow*). (**B**) The placebo PLGA bead shows a minimal foreign body reaction at its surface and slight fibrous capsule formation (*arrows*) with no focal chronic inflammation. Hematoxylin and eosin stain. *Abbreviation*: PLGA, polylactic acid–glycolic acid (*see page 115*).

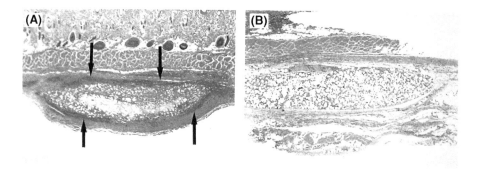

FIGURE 5.6 Chronic inflammatory response secondary to release of a bioactive agent from PLGA microparticles at 14 days' implantation. (**A**) Marked focal chronic inflammatory response (*arrows*) with monocytes and lymphocytes at the tissue/particle aggregate interface of recombinant human growth hormone–containing (Nutropin®) PLGA microparticles. (**B**) Focal macrophage and foreign body giant cell responses to placebo PLGA microparticles. Hematoxylin and eosin stain. *Abbreviation*: PLGA, polylactic acid–glycolic acid (*see page 115*).

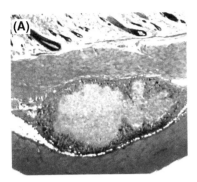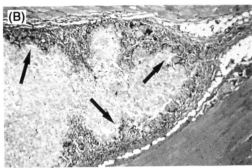

FIGURE 5.7 Monocyte, macrophage, FBGC infiltration into polylactic acid–glycolic acid micro-capsule aggregates at 14 days' implantation. (A) Low-magnification view showing cellular infiltration and the foreign body response (*arrows*) within the particle aggregate with fibrous capsule formation encapsulating the microcapsule aggregate. (B) High-magnification view showing macrophages and FBGCs at the microcapsule surfaces in a typical foreign body response within the microcapsule aggregate. Hematoxylin and eosin stain. *Abbreviation*: FBGC, foreign body giant cell (*see page 117*).

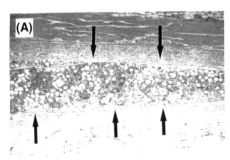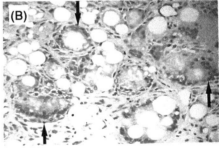

FIGURE 5.8 FBGC response and fibrous capsule formation to an injected aggregate of PLGA microcapsules at 42 days. (A) Low-magnification view. The PLGA microcapsule aggregate has been completely infiltrated by monocytes, macrophages, FBGCs, and fibroblasts. Resolution of the acute and chronic inflammatory responses has occurred. A thin fibrous capsule (*arrows*) completely encapsulates the PLGA microcapsule aggregate. (B) A high-magnification view of the core of the PLGA microcapsule aggregate showing monocytes, macrophages, and FBGCs at the surfaces of the PLGA microcapsules. FBGCs (*arrows*) are seen. The interstitial space between PLGA microcapsules is sufficient to provide for fibroblast infiltration with subsequent collagen deposition. Hematoxylin and eosin stain. *Abbreviations*: FBGC, foreign body giant cell; PLGA, polylactic acid–glycolic acid (*see page 117*).

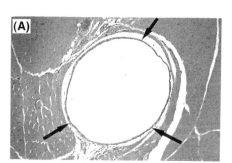 

FIGURE 5.9 Fibrous capsule formation at the end stage of the healing response to non-biodegradable or slowly biodegradable polymer matrices at one month implantation. (**A**) A thin fibrous capsule (*arrows*) is present surrounding the polymer implant within skeletal muscle. A minimal foreign body reaction with a single layer of macrophages and foreign body giant cells is present at the implant surface. (**B**) A relatively acellular, nonvascularized fibrous capsule is present at the polymer surface. No blood vessels are identified in the acellular fibrous capsule (*bar*), but they are present in the soft tissue outside of the fibrous capsule (*arrows*). Hematoxylin and eosin stain (*see page 120*).

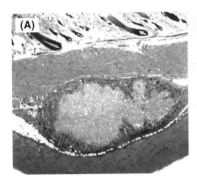
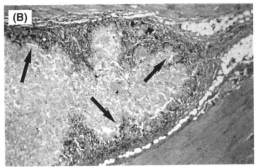

FIGURE 7 Monocyte, macrophage, FBGC infiltration into polylactic acid–glycolic acid micro-capsule aggregates at 14 days' implantation. (**A**) Low-magnification view showing cellular infiltration and the foreign body response (*arrows*) within the particle aggregate with fibrous capsule formation encapsulating the microcapsule aggregate. (**B**) High-magnification view showing macrophages and FBGCs at the microcapsule surfaces in a typical foreign body response within the microcapsule aggregate. Hematoxylin and eosin stain. *Abbreviation*: FBGC, foreign body giant cell.

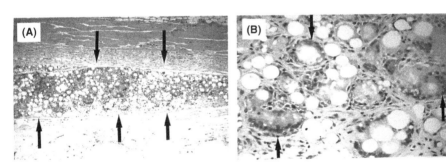

FIGURE 8 FBGC response and fibrous capsule formation to an injected aggregate of PLGA microcapsules at 42 days. (**A**) Low-magnification view. The PLGA microcapsule aggregate has been completely infiltrated by monocytes, macrophages, FBGCs, and fibroblasts. Resolution of the acute and chronic inflammatory responses has occurred. A thin fibrous capsule (*arrows*) completely encapsulates the PLGA microcapsule aggregate. (**B**) A high-magnification view of the core of the PLGA microcapsule aggregate showing monocytes, macrophages, and FBGCs at the surfaces of the PLGA microcapsules. FBGCs (*arrows*) are seen. The interstitial space between PLGA microcapsules is sufficient to provide for fibroblast infiltration with subsequent collagen deposition. Hematoxylin and eosin stain. *Abbreviations*: FBGC, foreign body giant cell; PLGA, polylactic acid–glycolic acid.

macrophages is a time-dependent process, and it is clear that days to weeks are necessary for complete cellular infiltration into the interstices of microcapsule aggregates. Figure 8 shows complete macrophage infiltration into an injected aggregate of PLGA microspheres at 42 days. FBGCs are seen on the surfaces of the microcapsules within the aggregate, and a fibrous capsule surrounds the entire aggregate.

The foreign body reaction is composed of FBGCs and the components of granulation tissue, which consist of macrophages, fibroblasts, and capillaries in varying amounts, depending on the form and topography of the implanted material. Relatively flat and smooth surfaces, such as those found on breast prostheses, have a foreign body reaction that is composed of a layer of macrophages one to two cells in thickness. Relatively rough surfaces, such as those found on the outer surfaces of expanded poly(tetrafluoroethylene) (ePTFE) vascular prostheses or polymethyl-methacrylate bone cement, have a foreign body reaction composed of several layers of macrophages and FBGCs at the surface. Fabric and porous materials generally have a surface response composed of macrophages and FBGCs with varying degrees of granulation tissue subjacent to the surface response. The foreign body reaction consisting mainly of macrophages and/or FBGCs may persist at the tissue-implant interface for the lifetime of the implant (1–10,58–61).

As previously discussed, the form and topography of the surface of the biomaterial determines the composition of the foreign body reaction. With biocompatible materials, the composition of the foreign body reaction in the implant site may be controlled by the surface properties of the biomaterial, the form of the implant, and the relationship between the surface area of the biomaterial and the volume of the implant. For example, high surface-to-volume implants such as fabrics, porous materials, or microparticle aggregates (Figs. 7 and 8) will have higher ratios of macrophages and FBGCs in the implant site than smooth-surface implants, which will have fibrosis as a significant component of the implant site.

Cytoskeletal and adhesive structure studies of in vitro macrophages and FBGCs have demonstrated that podosomal structures, and not focal contacts, are the major adhesive structures present within macrophages and FBGCs on surfaces. The podosomal structures are present at the ventral periphery of the FBGCs and contain vinculin, talin, and paxillin in a ring-like structure surrounding an F-actin core. These podosomal adhesion structures are similar to those identified for osteoclast adhesion. The podosomal structure present at the ventral and peripheral macrophage and FBGC surface implies a functional polarization and suggests the presence of frustrated phagocytosis via the formation of a closed compartment between the macrophage or FBGC and the underlying substrate where acid, degradative enzymes, reactive oxygen intermediates, and/or other products are secreted.

The adhesion of macrophages and FBGCs at the surfaces of biomaterials, prostheses, medical devices, and controlled-release systems to produce a closed compartment, that is, privileged microenvironment that exists between the cell membrane and the surface of the material, has significant implications in regard to the biodegradation of the biomaterial as well as the chemical modification and potential degradation of bioactive agents released from controlled-release systems. In a process described by Henson as frustrated phagocytosis, macrophages and FBGCs can release mediators of degradation such as acid, degradative enzymes, and ROS into this privileged zone between the cell membrane and the biomaterial surface so that immediate buffering or inhibition of these mediators is delayed or reduced. Phagolysosomes in macrophages can have acidity as low as a pH of 4, and direct microelectrode studies of this acid environment have determined pH levels as low as 3.5. Moreover, only several hours are necessary to achieve these acid levels following adhesion of macrophages (62–66). Proteins and peptides released from controlled-release systems would immediately encounter this highly acid environment. Conformational

TABLE 3 Macrophage and Neutrophil Degradation Products

Enzymes	Reactive oxygen species
Acid hydrolases	Superoxide anion
Neutral hydrolases	Hydrogen peroxide
Proteases	Hydroxyl radical
DNases	Nitric oxide
Lipases	Peroxynitrite
Cathepsins	Hypochlorous acid
Elastase	Chloramines
Phosphatases	
Lysozyme	
Collagenases	
Glucuronidases	
Glucosamidases	
Gelatinases	
Mannosidase	
Myeloperoxidase	

changes in proteins and peptides may occur in acidic environments with potential impact on efficacy and/or immunogenicity. Macrophages and FBGCs are also known to release a myriad of enzymes that could also degrade and modify proteins or peptides released from carriers. As macrophages and FBGCs also release ROS, oxidation of proteins or peptides released from carriers could be significantly modified so that their activity, bioavailability, and pharmacokinetics are markedly inhibited. Table 3 identifies some of the wide variety of enzymes and ROS that have been identified in macrophages as well as neutrophils. Unlike macrophages that have an extended lifetime, neutrophils in the acute inflammatory response have only a short lifespan, and thus, the interaction of proteins or peptides with these degradative agents would be expected to be short. In further support of the highly degradative nature of adherent macrophages and FBGCs on biomaterials, these cells have been implicated in the biodegradation of polymeric medical devices (67–69).

The foreign body reaction consisting mainly of macrophages and/or FBGCs may persist at the tissue-implant interface for the lifetime of the implant (1–3). Generally, fibrosis (i.e., fibrous encapsulation) surrounds the biomaterial or implant with its interfacial foreign body reaction, isolating the implant and foreign body reaction from the local tissue environment. Figure 9 illustrates fibrous capsule formation at the end stage of the healing response to nondegradable drug delivery systems. Early in the inflammatory and wound-healing response, the macrophages are activated upon adherence to the material surface. Although it is generally considered that the chemical and physical properties of the biomaterial are responsible for macrophage activation, the nature of the subsequent events regarding the activity of macrophages at the surface is not clear. Adherent macrophages, derived from circulating blood monocytes, may coalesce to form multinucleated FBGCs. Very large FBGCs containing large numbers of nuclei are typically present on the surface of biomaterials. Although these FBGCs may persist for the lifetime of the implant, it is not known if they remain activated, releasing their lysosomal constituents, or become quiescent. However, it can be anticipated that cytokines, growth factors, and other proteins or peptides released from controlled-release systems may not

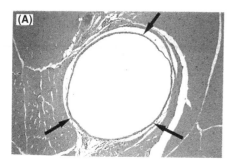

FIGURE 9 Fibrous capsule formation at the end stage of the healing response to non-biodegradable or slowly biodegradable polymer matrices at one month implantation. (**A**) A thin fibrous capsule (*arrows*) is present surrounding the polymer implant within skeletal muscle. A minimal foreign body reaction with a single layer of macrophages and foreign body giant cells is present at the implant surface. (**B**) A relatively acellular, nonvascularized fibrous capsule is present at the polymer surface. No blood vessels are identified in the acellular fibrous capsule (*bar*), but they are present in the soft tissue outside of the fibrous capsule (*arrows*). Hematoxylin and eosin stain.

only function as chemoattractants but also be activators for adherent monocytes, macrophages, and FBGCs.

Efforts in our laboratory have focused on differential lymphokine regulation of macrophage fusion that leads to morphologic variants of multinucleated giant cells, and the role played by the surface chemistry and other properties of the foreign material in facilitating monocyte adhesion, macrophage development, and giant cell formation (70–81). In our studies, human IL-4 induced the formation of FBGC from human monocyte-derived macrophages, an effect that was optimized with either granulocyte-macrophage colony-stimulating factor (GM-CSF) or IL-3, depending on the concentration of IL-4, and specifically prevented by anti-IL-4 (70,72). A distinct difference in adhesion was seen in our studies when IL-4 was added to freshly adherent (2 hours) monocytes (71). IL-4, under these conditions, resulted in a detachment of adherent cells and an inhibition of initial monocyte adhesion by IL-4. Figure 10 demonstrates the sequence of events involved in inflammation and wound healing when biomaterials, medical devices, and drug delivery systems are implanted. In general, the PMN predominant acute inflammatory response and the lymphocyte/monocyte predominant chronic inflammatory response resolve quickly, that is, within two weeks, depending on the type and location of implant. Studies utilizing IL-4 by us and others demonstrate the role for T_h2 helper lymphocytes in the development of the foreign body reaction at the tissue/material interface. T_h2 helper lymphocytes have been described as "anti-inflammatory" on the basis of their cytokine profile of which IL-4 is a significant component. T_h2 helper lymphocytes also produce IL-13, and we have utilized this to demonstrate its similar effect to IL-4 on FBGC formation. In this regard, it is noteworthy that anti-IL-4 antibody does not inhibit IL-13-induced FBGC formation, nor does anti-IL-13 antibody inhibit IL-4-induced FBGC formation. In our IL-4 and IL-13 FBGC culture systems, the macrophage mannose receptor (MMR) has been identified as critical to the fusion of macrophages in the formation of FBGC (73,74). FBGC formation can be prevented by competitive

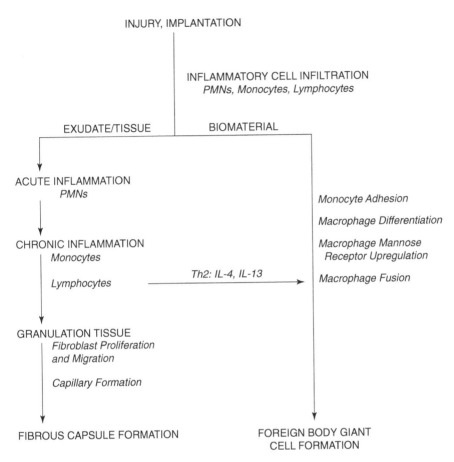

FIGURE 10 Sequence of events involved in inflammatory and wound-healing responses leading to FBGC formation. This shows the importance of T$_H$2 lymphocytes in the transient chronic inflammatory phase, with the production of IL-4 and IL-3 that can induce monocyte/macrophage fusion to form FBGCs. *Abbreviations*: IL, interleukin; FBGC, foreign body giant cell.

inhibitors of MMR activity, that is, α-mannan, or inhibitors of glycoprotein processing that restrict MMR surface expression.

Fibrous Encapsulation

The end-stage healing response to biomaterials is generally fibrosis or fibrous encapsulation. However, there may be exceptions to this general statement (e.g., porous materials inoculated with parenchymal cells or porous materials implanted into bone). Repair of implant sites involves two distinct processes: regeneration, which is the replacement of injured tissue by parenchymal cells of the same type, or replacement by connective tissue that constitutes the fibrous capsule. These processes are generally controlled by either (*i*) the proliferative capacity of the cells in the tissue receiving the implant and the extent of injury as it relates to the destruction or (*ii*) persistence of the tissue framework of the implant site. Retention of the framework may lead to restitution of the normal tissue structure, whereas destruction of the framework most commonly leads to fibrosis. It is important to consider the species-dependent nature of the regenerative capacity of cells. For example, cells from the same organ or tissue but from different species may exhibit different regenerative capacities and/or connective tissue repair.

Figure 9 provides two examples of the fibrous capsule, which surrounds drug delivery devices in the end stage of the healing response. Fibrous capsules form within the first three weeks following implantation and, like the foreign body reaction, remain for the lifetime of the device in tissue (82). Figure 9B illustrates the bioavailability barrier characteristics of the fibrous capsule as the tissue zone closest to the implant does not contain blood vessels necessary for vascular distribution of drugs.

Modulation of Tissue Responses to Implanted Drug Delivery Systems

Drug delivery systems have been incorporated in medical devices to enhance performance and improve outcome (83). In addition, drug delivery devices have also been used to modulate tissue responses in model systems to investigate mechanisms of cell/material interactions. The following are several examples of the use of drug delivery systems to achieve enhanced performance or to modulate cell/material interactions.

The use of a dexamethasone-releasing silicone rubber drug delivery system in the cage implant system has allowed elucidation of the cellular mechanism of in vivo stress cracking in polyurethane materials utilized in pacemaker leads (67). Dexamethasone released from the silicone rubber modulated and downregulated the function of monocytes and macrophages so that these cells were incapable of facilitating polyurethane degradation following adhesion to the polyurethane surfaces. These studies demonstrated that monocyte/macrophage/FBGC adhesion was necessary for polyurethane biodegradation. In addition, these studies also demonstrated that the dexamethasone markedly inhibited fibrous capsule formation around the stainless steel mesh cages.

Daugherty et al. have demonstrated that the tissue response to polylactic-co-glycolide acid microspheres can be modulated (84). Using the controlled release of bioactive agents such as cyclosporin, dexamethasone, TGF-β, or insulin-like growth factor (IGF)-I, both the pattern and time course of cellular and fibrotic events associated with the tissue response to the microspheres could be modified.

Site-specific drug delivery from coronary stents has been suggested to modulate restenosis. Anti-inflammatory agents, antiproliferative agents, and cellular migration inhibitors are currently being investigated to inhibit or control the inflammatory and wound-healing restenosis mechanisms with coronary artery stents (85–87).

Controlled delivery systems providing growth factors and cytokines to local tissue environments have facilitated significant advancement in the areas of regenerative medicine and tissue engineering (88–102). In particular, growth factors such as FGF-1, vascular endothelial growth factor (VEGF), and PDGF-BB have provided useful constructs to enhance endothelial cell migration and proliferation as well as angiogenesis. In addition to providing new blood vessels for tissue-engineered devices, these systems may also reduce the barrier function of the fibrous capsule by facilitating vascularization within the fibrous capsule. Potentially, these systems can markedly improve the bioavailability of drugs, proteins, peptides, and other therapeutic agents released from controlled drug delivery systems.

FUTURE PERSPECTIVES

While this chapter focuses on the significant potential of inflammatory responses, foreign body reaction, and fibrous encapsulation as being principal barriers to the bioavailability, pharmacokinetics, pharmacodynamics, and metabolism of proteins and peptides released from implanted controlled-release systems, it must be clearly noted that these barriers have been overcome for a few systems and clinically successful implants have been achieved for the release of proteins and peptides. The most obvious example is the clinical use of leuprolide acetate sustained-release systems for treatment of prostate cancer and endometriosis. The clinical success of these systems clearly demonstrates that efficacious bioavailability, pharmacokinetics, pharmacodynamics, and metabolism can be achieved with implantable controlled-release systems.

Even though the effect of acid, enzymes, and oxygen radicals generated in the inflammatory and foreign body reactions and the effect of the fibrous capsule on the diffusion and transport of proteins and peptides are generally unknown and considered to be specific to the chemistry of a given therapeutic agent, it is clear that these interactions may be minimal and may not significantly affect the achievement of clinically relevant therapeutic levels of proteins and peptides. Tissue responses to implantable peptide and protein delivery systems must be considered in the overall design, research, and development of these systems, especially in regard to the potential for protein and peptide biodegradation as well as modulation of mass transport and diffusion; they need not be a significant barrier to the development of clinically useful devices. Biodegradable polymer depots, that is, microcapsules and microspheres, pumps, and multireservoir microchip array devices, would all be expected to demonstrate the in vivo tissue responses described in this chapter (103–106). However, they do not necessarily lead to failure of the device. Therapeutic polypeptides with high potency and clinical efficacy can be delivered in spite of these tissue responses.

The clinical success as well as in vivo research studies of protein and peptide controlled-release systems of various types demonstrate that the inflammatory responses, foreign body reactions, and fibrous capsule formation do not present insurmountable barriers to the achievement of therapeutic levels of the respective protein or peptide. These observations indicate that the future development of these systems for clinical application is very promising. While the chemical and physical characteristics of each protein or peptide will ultimately determine its fate following release from a controlled-release system, attention can now be turned toward other variables unique to each controlled-release system that will determine the efficacy and appropriate and adequate function of each respective system. Thus, the future appears bright and positive for the development of protein and peptide controlled-release systems that can produce appropriate and adequate bioavailability, pharmacokinetics, and pharmacodynamics for therapeutic application.

REFERENCES

1. Anderson JM. Biological responses to materials. Annu Rev Mater Res 2001; 31: 81–110.
2. Anderson JM, Rodriguez A, Chang DT. Foreign body reaction to biomaterials. Semin Immunol 2008; 20:86–100.
3. Anderson JM. Mechanisms of inflammation and infection with implanted devices. Cardiovasc Pathol 1993; 2:199S–208S
4. Anderson JM. Inflammation and the foreign body response. Probl Gen Surg 1994; 11:17–160.
5. Anderson JM. Inflammatory response to implants. ASAIO J 1988; 11:101–107.
6. Anderson JM, Marchant RE. Tissue responses to drug delivery systems. In: Anderson JM, Kim SW, eds. Recent Advances in Drug-Delivery Systems. New York: Plenum Press, 1984:23–39.
7. Yamaguchi Y, Anderson JM. Biocompatibility studies of naltrexone sustained release formulations. J Control Release 1992; 19:299–314.
8. Yamaguchi K, Anderson JM. In vivo biocompatibility studies of Medisorb® 65/35 D,L-lactide/glycolide copolymer microspheres. J Control Release 1993; 24:81–93.
9. Anderson JM. In vivo biocompatibility of implantable delivery systems and biomaterials. Eur J Pharm Biopharm 1994; 40:1–8.
10. Anderson JM, Shive MS. Biodegradation and biocompatibility of PLA and PLGA microspheres. Adv Drug Deliv Rev 1997; 28:5–24.
11. Acute and chronic inflammation, tissue renewal and repair: regeneration, healing, and fibrosis. In: Kumar V, Abbas AK, Fausto N, eds. Pathologic Basis of Disease. 7th ed. Philadelphia: WB Saunders Company, 2005:47–118.
12. Gallin JI, Synderman R, eds. Inflammation: Basic Principles and Clinical Correlates. 2nd ed. New York: Raven Press, 1999.
13. Weissmann G, Smolen JE, Korchak HM. Release of inflammatory mediators from stimulated neutrophils. N Engl J Med 1980; 303:27–34.
14. Salthouse TN. Cellular enzyme activity at the polymer-tissue interface: a review. J Biomed Mater Res 1976; 10:197–229.
15. Clark RA, Lanigan JM, DellePelle P, et al. Fibronectin and fibrin provide a provisional matrix for epidermal cell migration during wound reepithelialization. J Invest Dermatol 1982; 79:264–269.
16. Tang L, Eaton JW. Fibrin(ogen) mediates acute inflammatory responses to biomaterials. J Exp Med 1993; 178:2147–2156.
17. Tang L. Mechanism of pro-inflammatory fibrinogen: biomaterial interactions. J Biomat Sci Polym Edn 1998; 9:1257–1266.
18. Riches DWF. Macrophage involvement in wound repair, remodeling, and fibrosis. In: Clark RAF, Henson PM, eds. The Molecular and Cellular Biology of Wound Repair. New York: Plenum Press, 1998:95–141.
19. Wahl SM, Wong H, McCartney FN. Role of growth factors in inflammation and repair. J Cell Biochem 1989; 40:193–199.
20. Dvorak HF, Harvey VS, Estrella P, et al. Fibrin containing gels induce angiogenesis. Implications for tumor stroma generation and wound healing. Lab Invest 1987; 57:673–686.
21. Broadley KN, Aquino AM, Woodward SC, et al. Monospecific antibodies implicate basic fibroblast growth factor in normal wound repair. Lab Invest 1989; 61:571–575.
22. Sporn MB, Roberts AB. Peptide growth factors are multifunctional. Nature 1988; 332:217–219.
23. Muller G, Behrens J, Nussbaumer U, et al. Inhibitor action of transforming growth factor beta on endothelial cells. Proc Natl Acad Sci U S A 1987; 84:5600–5604.
24. Madri JA, Pratt BM, Tucker AM. Phenotypic modulation of endothelial cells by transforming growth factor-beta depends upon the composition and organization of the extracellular matrix. J Cell Biol 1988; 106:1375–1384.
25. Wahl SM, Hunt DA, Wakefield LM, et al. Transforming growth factor-beta (TGF-β) induces monocyte chemotaxis and growth factor production. Proc Natl Acad Sci U S A 1987; 84:5788–5792.

26. Ignotz R, Endo T, Massague J. Regulation of fibronectin and type I collagen mRNA levels by transforming growth factor-beta. J Biol Chem 1987; 262:6443.
27. Marchant R, Hiltner A, Hamlin C, et al. In vivo biocompatibility studies: I. The cage implant system and a biodegradable hydrogel. J Biomed Mater Res 1983; 17:301–325.
28. Spilizewski KL, Marchant RE, Hamlin CR, et al. The effect of hydrocortisone acetate loaded poly(DL-lactide) films on the inflammatory response. J Control Release 1985; 2:197–203.
29. Ganz T. Neutrophil receptors. In: RI Lehrer, moderator. Neutrophils and Host Defense. Ann Intern Med 1988; 109:172–182.
30. Henson PM, Johnston RB Jr. Tissue injury in inflammation: oxidants, proteinases, and cationic proteins. J Clin Invest 1987; 79:669–674.
31. Malech HL, Gallin JL. Current concepts: immunology. Neutrophils in human diseases. N Engl J Med 1987; 317:687–694.
32. Jutila MA. Leukocyte traffic to sites of inflammation. APMIS 1992; 100:191–201.
33. Pober JS, Cotran RS. The role of endothelial cells in inflammation. Transplantation 1990; 50:537–544.
34. Cotran RS, Pober JS. Cytokine-endothelial interactions in inflammation, immunity, and vascular injury. J Am Soc Nephrol 1990; 1:225–235.
35. Hynes RO. Integrins: versatility, modulation, and signaling in cell adhesion. Cell 1992; 69:11–25.
36. Butcher EC. Leukocyte-endothelial cell recognition: three (or more) steps to specificity and diversity. Cell 1991; 67:1033–1036.
37. Henson PM. The immunologic release of constituents from neutrophil leukocytes. II. Mechanisms of release during phagocytosis, and adherence to nonphagocytosable surfaces. J Immunol 1971; 107:1547–1557.
38. Henson PM. Mechanisms of exocytosis in phagocytic inflammatory cells. Am J Pathol 1980; 101:494–511.
39. Henson PM. The immunologic release of constituents from neutrophil leukocytes. I. The role of antibody and complement on nonphagocytosable surfaces or phagocytosable particles. J Immunol 1971; 107(6):1535–1546.
40. Weiss SJ. Tissue destruction by neutrophils. N Engl J Med 1989; 320:365–376.
41. Paty PB, Greaff RW, Mathes SJ, et al. Superoxide production by wound neutrophils: evidence for increased activity of the NADPH oxidase. Arch Surg 1990; 125:65–69.
42. Johnston RB Jr. Monocytes and macrophages. N Engl J Med 1988; 318:747–752.
43. Williams GT, Williams WJ. Granulomatous inflammation—a review. J Clin Pathol 1983; 36:723–733.
44. Hulse GK, Stalenberg V, McCallum D, et al. Histological changes over time around the site of sustained release naltrexone-poly(DL-lactide) implants in humans. J Control Release 2005; 108:43–55.
45. Fielder PJ, Mortensen DL, Mallet P, et al. Differential long-term effects of insulin-like growth factor-I (IGF-I), growth hormone (GH), and IGF-I plus GH on body growth and IGF binding proteins in hypophysectomized rats. Endocrinology 1996; 137:1913–1920.
46. Anderson JM, Langone JJ. Issues and perspectives on the biocompatibility and immunotoxicity evaluation of implanted controlled release systems. J Control Release 1999; 57:107–113.
47. Rice GP, Paszner B, Oger J, et al. The evolution of neutralizing antibodies in multiple sclerosis patients treated with interferon beta-1b. Neurology 1999; 52(6):1277–1279.
48. Meager A, Wadhwa M, Bird C, et al. Spontaneously occurring neutralizing antibodies against granulocyte-macrophage colony-stimulating factor in patients with autoimmune disease. Immunology 1999; 97(3):526–532.
49. Wadhwa M, Skog AL, Bird C, et al. Immunogenicity of granulocyte-macrophage colony-stimulating factor (GM-CSF) products in patients undergoing combination therapy with GM-CSF. Clin Cancer Res 1999; 5(6):1353–1361.
50. Sporn MB, Roberts AB, eds. Peptide Growth Factors and Their Receptors I. New York: Springer-Verlag, 1990.
51. Fong Y, Moldawer LL, Shires GT, et al. The biologic characteristics of cytokines and their implication in surgical injury. Surg Gynecol Obstet 1990; 170:363–378.

52. Kovacs EJ. Fibrogenic cytokines: the role of immune mediators in the development of scar tissue. Immunol Today 1991; 12:17–23.
53. Golden MA, Au YP, Kirkman TR, et al. Platelet-derived growth factor activity and RNA expression in healing vascular grafts in baboons. J Clin Invest 1991; 87:406–414.
54. Mustoe TA, Pierce GF, Thomason A, et al. Accelerated healing of incisional wounds in rats induced by transforming growth factor. Science 1987; 237:1333–1336.
55. Maciag T. Molecular and cellular mechanisms of angiogenesis. In: DeVita VT, Hellman S, Rosenberg S, eds. Important Advances in Oncology. Philadelphia: JB Lippincott, 1990:85–98.
56. Thompson JA, Anderson KD, DiPetro JM, et al. Site-directed neovessel formation in vivo. Science 1988; 241:1349–1352.
57. Ziats NP, Miller KM, Anderson JM. In vitro and in vivo interactions of cells with biomaterials. Biomaterials 1988; 9:5–13.
58. Rae T. The macrophage response to implant materials. Crit Rev Biocompat 1986; 2:97–126.
59. Greisler H. Macrophage-biomaterial interactions with bioresorbable vascular prostheses. ASAIO Trans 1988; 34:1051–1057.
60. Chambers TJ, Spector WG. Inflammatory giant cells. Immunobiology 1982; 161:283–289.
61. Anderson JM. Mutinucleated giant cells. Curr Opin Hematol 2000; 7:40–47.
62. Haas A. The phagosome: compartment with a license to kill. Traffic 2007; 8:311–330.
63. Klebanoff SJ. Myeloperoxidase: friend and foe. J Leukoc Biol 2005; 77:598–625.
64. Segal AW. How neutrophils kill microbes. Annu Rev Immunol 2005; 23:197–223.
65. Jankowski A, Scott CC, Grinstein S. Determinants of the phagosomal pH in neutrophils. J Biol Chem 2002; 277:6059–6066.
66. Silver IA, Murrills R, Etherington DJ. Microelectrode studies on the acid environment beneath adherent macrophages and osteoclasts. Exp Cell Res 1988; 175:266–276.
67. Zhao Q, Agger MP, Fitzpatrick M, et al. Cellular interactions with biomaterials: in vivo cracking of pre-stressed Pellethane 2363-80A. J Biomed Mater Res 1990; 24:621–637.
68. Zhao Q, Topham N, Anderson JM, et al. Foreign-body giant cells and polyurethane biostability: in vivo correlation of cell adhesion and surface cracking. J Biomed Mater Res 1991; 25:177–183.
69. Wiggins MJ, Wilkoff B, Anderson JM, et al. Biodegradation of polyether polyurethane inner insulation in bipolar pacemaker leads. J Biomed Mater Res 2001; 58:302–307.
70. McNally AK, Anderson JM. Complement C3 participation in monocyte adhesion to different surfaces. Proc Natl Acad Sci U S A 1994; 91:10119–10123.
71. McNally AK, Anderson JM. Interleukin-4 induces foreign body giant cells from human monocytes/macrophages. Differential lymphokine regulation of macrophage fusion leads to morphological variants of multinucleated giant cells. Am J Pathol 1995; 147:1487–1499.
72. Kao WJ, McNally AK, Hiltner A, et al. Role for interleukin-4 in foreign-body giant cell formation on a poly(etherurethane urea) in vivo. J Biomed Mater Res 1995; 29:1267–1276.
73. DeFife KM, McNally AK, Colton E, et al. Interleukin-13 induces human monocyte/macrophage fusion and macrophage mannose receptor expression. J Immunol 1997; 158:319–328.
74. McNally AK, DeFife KM, Anderson JM. Interleukin-4-induced macrophage fusion is prevented by inhibitors of mannose receptor activity. Am J Pathol 1996; 149:975–985.
75. Jenney CR, DeFife KM, Colton E, et al. Human monocyte/macrophage adhesion, macrophage motility, and IL-4-induced foreign body giant cell formation on silane-modified surfaces in vitro. J Biomed Mater Res 1998; 41:171–184.
76. Jenney CR, Anderson JM. Effects of surface-coupled polyethylene oxide on human macrophage adhesion and foreign body giant cell formation in vitro. J Biomed Mater Res 1998; 44:206–216.

77. Jenney CR, Anderson JM. Alkylsilane-modified surfaces: inhibition of human macrophage adhesion and foreign body giant cell formation. J Biomed Mater Res 1999; 46:11–21.
78. Jones JA, Chang DT, Meyerson H, et al. Proteomic analysis and quantification of cytokines and chemokines from biomaterial surface-adherent macrophages and foreign body giant cells. J Biomed Mater Res 2007; 83A:585–596.
79. Jones JA, McNally AK, Chang DT, et al. Matrix metalloproteinases and their inhibitors in the foreign body reaction on biomaterials. J Biomed Mater Res 2008; 84A:158–166.
80. Anderson JM, Jones JA. Phenotypic dichotomies in the foreign body reaction. Biomaterials 2007; 28:5114–5120.
81. Chang DT, Jones JA, Meyerson H, et al. Lymphocyte/macrophage interactions: biomaterial surface-dependent cytokine, chemokine, and matrix protein production. J Biomed Mater Res A 2008; 87:676–687.
82. Anderson JM, Niven H, Pelagalli J, et al. The role of the fibrous capsule in the function of implanted drug-polymer sustained release systems. J Biomed Mater Res 1981; 15:889–902.
83. Schierholz JM. Drug delivery devices to enhance performance and improve outcome. Drug Deliv Syst Sci 2001; 1(2):52–56.
84. Daugherty AL, Cleland JL, Duenas EM, et al. Pharmacological modulation of the tissue response to implanted polylactic-co-glycolic acid microspheres. Eur J Pharm Biopharm 1997; 44(1637):89–102.
85. Lewis AL, Vick TA. Site-specific drug delivery from coronary stents. Drug Deliv Syst Sci 2001; 1(3):65–71.
86. Marx SO, Marks AR. The development of Rapamycin and its application to stent restenosis. Circulation 2001; 104:852–855.
87. Suzuki T, Kopia G, Hayashi S-I, et al. Stent-based delivery of sirolimus reduces neointimal formation in a porcine coronary model. Circulation 2001; 104:1188–1193.
88. Richardson TP, Peters MC, Ennett AB, et al. Polymeric system for dual growth factor delivery. Nat Biotech 2001; 19:1029–1034.
89. Cleland JL, Daugherty A, Mrsny R. Emerging protein delivery methods. Curr Opin Biotechnol 2001; 12:212–219.
90. Babensee JE, McIntire LV, Mikos AG. Growth factor delivery for tissue engineering. Pharm Res 2000; 17(5):497–504.
91. Cleland JL, Duenas ET, Park A, et al. Development of poly-(D,L-lactid-coglycolide) microsphere formulations containing recombinant human vascular endothelial growth factor to promote local angiogenesis. J Control Release 2001; 72:13–24.
92. Fischbach C, Mooney DJ. Polymers for pro- and anti-angiogenic therapy. Biomaterials 2007; 28:2069–2076.
93. Sands RW, Mooney DJ. Polymers to direct cell fate by controlling the microenvironment. Curr Opin Biotechnol 2007; 18:448–458.
94. Sun Q, Chen RR, Shen Y, et al. Sustained vascular endothelial growth factor delivery enhances angiogenesis and perfusion in ischemic hind limb. Pharm Res 2005; 22 (7):1110–1116.
95. Silva EA, Mooney DJ. Spatiotemporal control of vascular endothelial growth factory delivery from injectable hydrogels enhances angiogenesis. J Thromb Haemost 2007; 5:590–598.
96. Holland TA, Bodde EWH, Cuijpers V, et al. Degradable hydrogel scaffolds for in vivo delivery of single and dual growth factors in cartilage repair. Osteoarthritis Cartilage 2007; 15:187–197.
97. Hao X, Silva E, Månsson-Broberg A, et al. Angiogeneic effects of sequential release of VEGF-A(165) and PDGF-BB with alginate hydrogels after myocardial infarction. Cardiovasc Res 2007; 75:178–185.
98. Ennett AB, Kaigler D, Mooney DJ. Temporally regulated delivery of VEGF in vitro and in vivo. J Biomed Mater Res A 2006; 79A:176–184.
99. Daugherty AL, Rangell LK, Eckert R, et al. Sustained release formulations of rhVEGF165 produce a durable response in a non-compromised murine model of peripheral angiogenesis. (Personal communication.)

100. Holland TA, Tabata Y, Mikos AG. Dual growth factor delivery from degradable oligo(poly(ethylene glycol) fumarate) hydrogel scaffolds for cartilage tissue engineering. J Control Release 2005; 101(1–3):111–125.
101. Kim HK, Park TG. Comparative study on sustained release of human growth hormone from semi-crystalline poly(L-lactic acid) and amorphous poly(D,L-lactic-co-glycolic acid) microspheres: morphological effect on protein release. J Control Release 2004; 98:115–125.
102. Street J, Bao M, deGuzman L, et al. Vascular endothelial growth factor stimulates bone repair by promoting angiogenesis and bone turnover. Proc Natl Acad U S A 2002; 99(15):9656–9661.
103. Maloney JM, Uhland SA, Polito BF, et al. Electrothermally activated microchips for implantable drug delivery and biosensing. J Control Release 2005; 109:244–255.
104. Prescott JH, Lipka S, Baldwin S, et al. Chronic, programmed polypeptide delivery from an implanted, multireservoir microchip device. Nat Biotech 2006; 24(4): 437–438.
105. Prescott JH, Krieger TJ, Lipka S, et al. Dosage form development, in vitro release kinetics, and in vitro-in vivo correlation for leuprolide released from an implantable multi-reservoir array. Pharm Res 2007; 24(7):1252–1261.
106. Proos ER, Prescott JH, Staples MA. Long-term stability and in vitro release of hPTH (1-34) from a multi-reservoir array. Pharm Res 2008; 25(6):1387–1395.

6 Engineering as a Means to Improve the Pharmacokinetics of Injected Therapeutic Proteins

Mark S. Dennis, Yik A. Yeung, and Henry B. Lowman
Department of Antibody Engineering, Genentech, Inc., South San Francisco, California, U.S.A.

Robert F. Kelley
Departments of Antibody Engineering and Protein Engineering, Genentech, Inc., South San Francisco, California, U.S.A.

INTRODUCTION

A central paradigm of the biopharmaceutical industry is to identify a natural protein having a desired biochemical activity and then use recombinant methods to produce that protein in sufficient quantities for biological testing and ultimately, for therapeutic use, usually as an injected drug. For proteins intended for use as human pharmaceuticals, a key experiment will be to test the protein for efficacy in an animal model of the human disease. Testing of these proteins, however, may be limited by factors other than their intrinsic biochemical activity. Poor solubility, stability, and pharmacokinetic half-life are common issues hampering preclinical and clinical testing of protein pharmaceuticals. In particular, low-molecular-weight proteins usually show rapid clearance in vivo, making it difficult to deliver a high-enough dose to achieve the desired biological effect. Although smaller molecular size may have advantages, for example, in penetration into solid tumors, for many injected therapeutics, prolonged half-life and infrequent dosing are often preferred.

Protein engineering can alter many functional properties of peptides and proteins and, combined with high-throughput screening or molecular diversity approaches, such as phage display, may often lead to dramatic alterations in binding affinity toward a target receptor or antigen. In recent years, protein engineering efforts have also turned to improvement of the pharmacokinetic properties of first- or second-generation polypeptide drugs, which are often limited in their pharmacokinetic half-life by glomerular filtration in the kidney (1–3). Monoclonal antibodies (MAbs) and other protein drugs may be modified with substitutions of methionine or asparagine residues that may be otherwise subject to oxidation or deamidation, respectively, in vitro or in vivo, to yield reduced functional activity over time. In addition, proteolytic sites, such as dibasic sequences (consecutive arginine or lysine residues), have been recognized as potential sources of in vivo instability and may be altered through protein engineering (4,5). In this chapter, we omit discussion of such protein-specific issues, which differ greatly even among therapeutic antibodies according to their diverse complimentarity-determining regions (CDRs), and focus instead on three general strategies for improving the pharmacokinetic half-lives of injected peptides and proteins (Fig. 1).

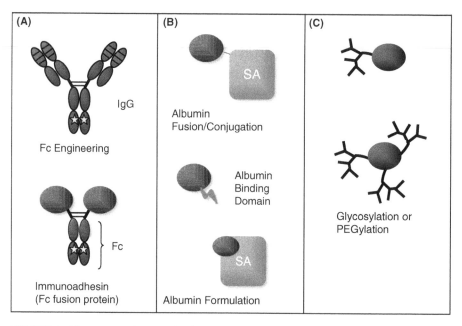

FIGURE 1 Three general strategies for enhancing the pharmacokinetic half-life of injected peptides and proteins: (**A**) use of an antibody (IgG) platform or Fc fusion (immunoadhesin), which can be further engineered for enhanced FcRn binding; (**B**) covalent or noncovalent association with serum albumin; and (**C**) chemical modification to increase hydrodynamic volume.

Key considerations in the application of protein engineering methods for therapeutic polypeptides are that the final molecule should have appropriate specificity, target-binding affinity, and half-life to achieve the intended therapeutic benefit while retaining manufacturing, formulation, and delivery feasibility. The modified or mutated protein must also evade the human immune system, particularly if it is to be used over an extended period of time. The consequences of an immunogenic response to a drug may include severe adverse events such as anaphylactic (hyperimmune) antibody responses, or simply faster pharmacokinetic clearance as a result of complexes formed with antidrug antibodies. We have also omitted discussion of approaches to reduce immunogenicity, which have been reviewed elsewhere for antibodies and other proteins (6).

MAbs represent a particularly attractive class of injected protein therapeutics, in part because of the highly developed technologies available for the discovery, humanization, and affinity maturation of highly specific, targeted agents that in full-length immunoglobulin G (IgG) form typically have long pharmacokinetic half-lives on the order of three weeks in humans, as reviewed in the previous chapter of this volume. However, even these relatively long-lived proteins may benefit from protein engineering efforts. Here we review recent studies that have shown significant increases in circulating half-life in nonhuman primates for antibodies that have been mutated at a few sites in the Fc region of the molecule. These mutations serve to enhance binding to the neonatal Fc receptor (FcRn), an abundant cell surface receptor that is found throughout the reticuloendothelial system and that serves to scavenge IgG from the circulation, thus protecting it from metabolic degradation.

Serum albumin, another long-lived serum protein, also benefits from interaction with FcRn, persisting in the circulation with a half-life of about three weeks. This observation has given rise to a variety of strategies to use albumin as a "carrier protein" for peptides and small proteins that can be fused to—or otherwise engineered to associate with—serum albumin.

Finally, the pharmacokinetic half-life of small proteins and peptides can often be greatly increased by simply increasing the hydrodynamic radius through chemical modification, for example, by conjugation to polyethylene glycol (PEG) or similar polymers. An established strategy has been to modify the protein with one or more PEG groups, with the goal of achieving a larger effective molecular weight (EMW), and this strategy has proved successful in several marketed therapeutics, including pegvisomant (Somavert®), which was engineered for higher binding affinity to its target [human growth hormone (hGH)] receptor, for antagonist activity, and for longer half-life through PEGylation (5). Yet even larger, particulate formulations are also being investigated using conjugates that form micelles or nanoparticles (7). These approaches create additional opportunities for delivering otherwise short-lived drugs in Trojan horse fashion, either specifically or nonspecifically, to target tissues.

It is worth noting at this point that not all therapeutic polypeptides will benefit from engineering efforts to maximize half-life. This is illustrated by insulin and other family members, which can have unexpected or undesired consequences if dosed on a schedule incompatible with their metabolic role (8). Among antibodies, bevacizumab (Avastin®) anti-VEGF benefits from long half-life allowing for dosing by intravenous infusion every three or four weeks (9). However, a different anti-VEGF, ranibizumab (Lucentis®), a Fab fragment engineered for high antigen-binding affinity and tissue penetration to the retina, is delivered intravitreally (10). In this case, prolonged residence in the eye may be beneficial, whereas increased systemic half-life may raise safety issues.

The central role of FcRn in mediating the half-life of antibodies and the fact that FcRn effects are mediated through interaction with a specific region of the antibody—the Fc—translates to a simple approach to improve the half-life of many small proteins, including receptor extracellular domains, namely, by fusing the protein to the Fc region of an IgG. Such an approach has additional advantages of (*i*) providing a bivalent molecule, which may benefit in potency from avidity effects, (*ii*) minimizing immunogenicity risks because both protein (e.g., an extracellular domain (ECD)) and antibody Fc may be of human origin (though the junction site is typically a nonnative sequence), and (*iii*) facilitating manufacturing and clinical assay development activities that are analogous to those for therapeutic antibodies. Etanercept (Enbrel®) is a marketed anti-TNF-α (tumor necrosis factor α) agent in this category and is discussed below in comparison with anti-TNF antibodies. Both types of molecules may benefit from Fc engineering that enhances binding interactions with FcRn.

ENGINEERING BINDING TO FcRn
The Neonatal Fc Receptor: Function and Expression
Typical serum proteins have half-lives of less than one week, for example, fibrinogen (1–3 days), IgD (2–5 days), IgM (4–6 days), IgA (3–7 days), and haptoglobin (~5 days) (11–14). However, serum IgG and albumin have half-lives of approximately three weeks. The prolonged half-lives of IgGs and

albumins are mainly due to the protective action of the FcRn (15,16). FcRn is also known as Fc receptor Brambell (17) (FcRB) and Fc receptor protection (FcRp), named after Professor Brambell who first described it and its protective function, respectively. In the late 1950s, a saturable receptor system was proposed by Professor Brambell for mediating the transport of IgG from mothers to infants through the yolk sacs and intestines (17,18). Observing the similarity between passive transmission and catabolism of IgG, he later postulated that a similar or identical receptor system was responsible for the protection of IgG from catabolism (19). It was not until thirty years later that this IgG receptor, FcRn, was cloned (18) and confirmed to carry out both the important functions of transcytosis (18) and homeostasis of IgG (20–23). FcRn of multiple mammalian species have been cloned (National Center for Biotechnology Information), and functional expression of FcRn has been reported in mammals like rat, mouse, rabbit, sheep, bovine, nonhuman primate, and human (17,24–26).

FcRn, which is structurally homologous to major histocompatibility complex (MHC) class I molecules, is a heterodimer consisting of a transmembrane α-chain and β_2-microglobulin (β_2m) (27,28). However, unlike MHC molecules, FcRn is incapable of binding peptides because the counterpart of the MHC peptide-binding groove in FcRn is occupied by its own residues (27,28). FcRn can simultaneously bind both IgG and albumin, but the binding stoichiometries are different, with an FcRn/IgG ratio of 2:1 and a FcRn/albumin ratio of 1:1 (29–32). Although a crystal complex structure of human Fc-FcRn is still not available, the major contact residues in the human complex can be deduced from the crystal structure of a rat Fc:FcRn complex (Fig. 2A) (29,33), and site-directed mutagenesis studies (35–38). FcRn binds the Fc portion of IgG at the CH2-CH3 interface (28,29,36). Major contact residues of the human FcRn are Glu115, Asp130, Trp131, Glu133, and Leu135 on the α-chain and Ile1 on the β_2m (Fig. 2B). On the Fc side, residues Ile253, Ser254, His310, His435, and Tyr436 are important for the interaction (Fig. 2C), as alanine substitution at these positions results in a greater than 10-fold reduction in binding to FcRn (36). Meanwhile, the contact residues for FcRn and albumin are not as clearly defined as those for the FcRn and IgG interaction. FcRn was found to bind albumin around His166 (Fig. 2B), opposite from the Fc binding region (39); this may explain why FcRn can simultaneously bind both IgGs and albumins. It is not clear whether β_2m is involved in binding to the albumin, but β_2m is required for the expression of FcRn (40).

FcRn protects IgGs and albumins from catabolism in a pH-dependent manner. IgGs and albumins bind FcRn with high affinity at acidic pH; as the pH is raised to neutral, the binding affinity drops considerably. The pH-dependent interaction is mainly attributed to the titration of histidine residues (41). Specifically, pinocytosed IgGs or albumin are captured by FcRn in the acidic endosome (42), recycled back to the cell surface (43) and then released back into the circulation at physiological pH of 7.4 (44). IgGs or albumins that are not bound by FcRn are targeted to the lysosome and degraded (42). Previous studies in knockout mice illustrated that the serum half-lives of IgGs and albumin in FcRn- or β_2m-deficient mice were greatly reduced (16,20–22,30). It was also observed in familial hypercatabolic hypoproteinemia patients that their low levels of serum IgG and albumin were caused by the reduction of FcRn expression, resulting from β_2m deficiency (23). Functional FcRn expression has been reported in a variety of tissues and cells such as vascular endothelium (45), monocytes (46), macrophages (46), dendritic cells (46), intestinal epithelium (47),

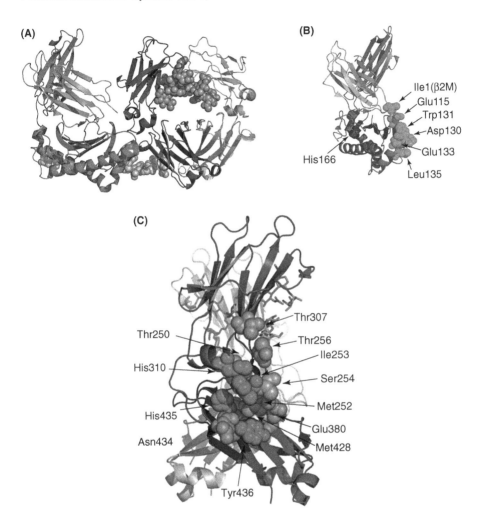

FIGURE 2 The crystal structures of a rat Fc:FcRn complex (**A**), human FcRn (**B**), and human IgG1 Fc (**C**). (**A**) A rat Fc:FcRn complex structure (PDB accession number 1I1A) shows that CH2-CH3 interface of the IgG Fc region (*right side ribbons*) interacts with both the α-chain and the β₂-microglobulin of FcRn left side ribbons. Carbohydrates are shown as spheres. (**B**) Critical binding residues on human FcRn (PDB 1EXU) for IgG and albumin interactions are shown as labeled spheres. Human IgGs and albumin appear to bind at the opposite sides of a human FcRn. The α-chain and β₂-microglobulin of human FcRn are shown as ribbons. (**C**) Residues where alanine substitutions result in greater than 10-fold reduction in FcRn affinity are shown as Ile253, Ser254, His310, His435, Tyr436 spheres on a human Fc structure (PDB 1DN2). These critical residues are located at the CH2-CH3 interface of a human IgG1 Fc. Mutations of the Fc residues shown as Thr250, Met252, Thr256, Thr307, Glu380, Met428, Asn434 spheres have been shown to improve human FcRn affinity. These residues are surrounding the critical binding residues. Molecular models in this figure were created using PyMol. *Abbreviation*: PDB, Protein Data Bank. *Source*: From Ref. 34.

brain and choroids plexus endothelium (48,49), podocytes (50), placental endo-thelium (51,52), and lung epithelium (53). FcRn has been shown in these studies to either recycle or transport IgG across the cellular barriers. Overall, FcRn has been shown to extend the half-lives of IgGs and albumins in circulation.

Immunoadhesins: Fc Fusions

Different engineering strategies have been employed to harness the protective action of FcRn to prolong the half-life of injected therapeutics. One of these strategies is to genetically or chemically fuse the protein or peptide of interest (referred to as the ligand in this text) to either the whole IgG, the Fc portion of an IgG, or an albumin, all of which interact with FcRn. The premise is that instead of rapid clearance by the kidney, or catabolism, IgG-, Fc-, or albumin-fused ligand can be recycled and protected by FcRn. We will first focus our discussion on Fc fusions. Albumin fusions will be discussed later in this chapter.

CD4-Fc: a Prototype for Immunoadhesins

CD4 is a cell surface glycoprotein present on a subset of peripheral T cells that recognize antigens presented by class II MHC molecules (54). It consists of four immunoglobulin-like domains followed by a transmembrane segment and a C-terminal cytoplasmic tail (55–57). CD4 biology became a hot research topic in the 1980s as CD4 was shown to mediate high-affinity ($K_D \sim$ nM) interaction with human immunodeficiency virus (HIV) envelop glycoprotein gp120 (58). Soluble recombinant CD4 ectodomain (rCD4) was created and successfully shown to block HIV infectivity, syncytium formation, and cell killing by gp120 (58). It was later shown that truncated rCD4 containing the first two N-terminal domains (trCD4) was also as effective as the full-length rCD4 (59,60). However, one major initial shortcoming of developing rCD4 as a drug was its short serum half-life ($t_{1/2} \sim 4$ hours in rabbits and $t_{1/2} \sim 10$ hours in human patients) (61,62).

Although FcRn function was not clearly elucidated at that time, Capon et al. observed that the Fc portion of an IgG, but not the Fab, had a long serum half-life like the whole IgG (61). They then constructed the first ligand-IgG fusion (named immunoadhesins) by genetically fusing the rCD4 or trCD4 to the constant region (CH1-CH2-CH3) of the human IgG1 heavy chain (61). Both the immunoadhesins showed similar binding to gp120 as rCD4 but exhibited more than 25-fold improved plasma half-lives in rabbits compared with rCD4 (61). An improved version of CD4 immunoadhesin was later generated by genetically fusing trCD4 to the Fc (CH2-CH3; residues 216 to 441, Kabat numbering) of a human IgG1 (63,64). The half-life of this trCD4-Fc variant was approximately two days in patients, a significant improvement over the $t_{1/2}$ of approximately 10 hours observed for rCD4 following the same dosing regiment (62,65,66). However, the $t_{1/2}$ of this CD4 immunoadhesin was still not as long as the $t_{1/2}$ of the anti-gp120 antibody 2G12, which was determined to be approximately 15 to 20 days in patients (67).

Fc Fusions in Research and in the Clinic

In general, proteins and peptides of interest can be constructed as N- or C-terminal fusions to an IgG Fc. Linkers with or without protease sites can be inserted between the ligand and Fc region to increase the flexibility of the two domains, and the presence of a protease site allows release of the ligand from the fusion protein. Numerous studies have shown that fusing an Fc to a ligand can improve the ligand's half-life by more than 100-fold. Fc fusions also provide other benefits such as ease of purification using protein A or G, stability improvement achieved by the fusion, recruitment of Fc effector functions, and ability for placental transport. Currently, there are four marketed Fc fusion therapeutics; they are all constructed as C-terminal fusions to the human IgG1 Fc (Table 1). Multiple other Fc fusions are under development for various disease indications (68).

Given the proven utility of Fc fusions as therapeutic agents and the relative ease of their construction, one may ask whether an Fc fusion, instead of an antibody, should be developed for a particular target. Unfortunately, there is no straightforward answer, but several factors should be considered in choosing either format. Parts of the following discussion are based on current anti-TNF therapies, where both Fc fusions and antibodies were developed to block TNF: Etanercept (TNF-R-p75-IgG1-Fc fusion, Amgen, Inc., California, U.S., and Wyeth, Pennsylvania, U.S.), Lenercept (TNF-R-p55-IgG1-Fc fusion, Roche, New Jersey, U.S.), Adalimumab (fully human anti-TNF IgG1, Abbott, Illinois, U.S.), and Infliximab (chimeric anti-TNF IgG1, Centocor, Malvern, PA, U.S.).

Mechanisms of Action

Fc fusion proteins are an important class of protein therapeutics that may have activities not readily achieved with MAbs. Antibodies are usually developed to elicit antagonistic activities by blocking the antigens' interactions with their endogenous binding partners. In some cases, agonistic antibodies can be generated, often functioning though cross-linking the target receptors (69–71). Therefore, if a certain agonistic activity is required for the therapeutic action such as activating a cytokine receptor to trigger downstream signaling cascade, developing an Fc fusion, with the endogenous ligand being part of the fusion, should be more straightforward. Conversely, both Fc fusions and antibodies can be readily developed for antagonistic activities, as in the case of blocking TNF.

Binding Properties of an Antibody Vs. an Fc Fusion

Binding specificity is of concern when comparing the development of an antibody versus that of an Fc fusion. Antibodies typically mediate highly specific interactions with their antigens, so cross-reactions with other non–target proteins are very rare. In contrast, Fc fusions are sometimes made up of endogenous ligands or receptors, which can have broad specificities in vivo. For example, Etanercept binds both TNF and lymphotoxin, whereas Adalimumab and Infliximab bind only TNF. There may be unexpected activities as well as toxicities associated with the bindings of these secondary targets. Therefore, it is important to investigate if an endogenous ligand–Fc fusion has any undesirable off-target binding partner in vivo.

Despite binding the same target, an antibody and an Fc fusion can have different binding properties such as affinity, binding kinetics, binding epitopes, and stoichiometry, all of which can impact the pharmacokinetics and pharmacodynamics of a drug. For example, Etanercept has a faster association rate than Adalimumab and Infliximab in binding soluble TNF, hence, it is more effective at neutralizing low concentrations of TNF (72,73). Moreover, because of the difference in binding stoichiometry (TNF/Etanercept 1:1 and TNF/Infliximab 1:3),

TABLE 1 Currently Marketed Immunoadhesin (Fc Fusion) Drugs

Drug name	Product	Target	MW (kDa)	Brand name	Company	Indications	Half-life (d)	Reference[a]
Etanercept	TNF-R2-IgG1 Fc	TNF, LT	~150	Enbrel®	Amgen and Wyeth	RA, PA	~4	77
Alefacept	LFA3-IgG1 Fc	CD2	~90	Amevive®	Astellas/Biogen Idec	Psoriasis	~12	188
Abatacept	CTLA-4-IgG1 Fc	B7	~90	Orencia®	Bristol-Myers Squibb	RA, PA	~15	See below
Rilonacept	IL-1RI-IL-1RAcP-IgG1 Fc	IL-1	~250	Arcalyst®	Regeneron	CAPS	~7	189

[a]Information for these products was obtained from the manufacturer's prescribing information: Etanercept (Immunex Corp., California, U.S., 2007; http://www.enbrel.com/pdf/enbrel_pi.pdf); Alefacept (Astellas Pharmaceuticals, Inc., Illinois, U.S., 2006; http://www.astellas.us/docs/amevive.xml); Abatacept (Bristol-Myers Squibb Corp., New Jersey, U.S., 2005; http://packageinserts.bms.com/pi/pi_orencia.pdf); Rilonacept (Regeneron Pharmaceuticals, Inc., New Jersey, U.S., 2008; http://www.regeneron.com/ARCALYST-fpi.pdf).
Abbreviations: TNF, tumor necrosis factor; LT, lymphotoxin; LFA3, lymphocyte function associated antigen 3; CTLA-4, cytotoxic T-lymphocyte antigen 4; RA, rheumatoid arthritis; PA, psoriatic arthritis; CAPS, cryopyrin-associated periodic syndromes.

more Infliximab molecules are bound to the transmembrane TNF than Etanercept molecules, potentially triggering a higher level of cross-linking and hence a stronger reverse signaling through transmembrane TNF (71,74). Therefore, certain binding properties that are crucial for the therapeutic outcome may dictate the development of either format. Currently, proteins can be readily engineered for desired properties such as high affinity and rapid association and dissociation rates using the current protein engineering tools. However, engineering the ligand part of an endogenous protein–Fc fusion may incur the risk of immunogenicity, which can unfavorably affect the PK and the PD of the fusion.

Molecular Modification of an Fc Fusion
The molecular size of a therapeutic agent can impact its diffusivity and hence targeting ability. The smaller the protein, the faster its diffusion and hence the better its targeting ability. Depending on the size of the ligand and the amount of posttranslational modification on the ligand, an antibody and an Fc fusion can have dramatic differences in size. Fc fusions like peptide fusions (75), which are smaller in sizes than full-length antibodies, have an advantage over antibodies in terms of targeting ability. Alternatively, Fc fusions involving large, complex, multidomain ligands are not only unfavorable in terms of targeting ability but can also be difficult to manufacture because of their complexities. Extra efforts may be required to improve such an Fc fusion; possible methods include truncating the nonbinding domains of the ligand and fusing only one ligand per Fc fusion (76). Most importantly, any molecular modification is to be done under the condition that it does not significantly sacrifice efficacy, in vivo half-life, and stability.

In Vivo Half-Life Differences Between an Antibody and an Fc Fusion
Since FcRn protects both antibodies and Fc fusions, typical Fc fusions are expected to have half-lives close to those of antibodies. Interestingly, most Fc fusions have a half-life of about a week or less; this is much shorter than that of a typical antibody. For example, Etanercept and Lenercept have half-lives of approximately four and seven days, respectively (77,78), whereas Adalimumab has a half-life of about two to three weeks (79,80). The shorter half-lives of Fc fusions could be due to the fact that steric hindrance from the ligand domain of the Fc fusion prevents optimal binding of the fusion to the membrane-bound FcRn. In vitro FcRn-binding assays can shed light on the relative binding of an Fc fusion to FcRn compared with an antibody. Another reason could be a stability difference between an antibody and an Fc fusion. Antibodies, in general, are very stable, whereas stabilities of Fc fusions vary greatly depending on the ligand and linker parts of the Fc fusion. Stability of an Fc fusion needs to be closely examined to ensure that the fusion does not degrade rapidly in serum. Other possibilities that may result in different clearances between Fc fusions and antibodies include differences in ability to withstand protease activities, antigen-dependent clearance, and posttranslational modifications.

Immunogenicity
Antidrug antibodies can significantly reduce drug exposure and efficacy in vivo. It is difficult to predict the immunogenicity of a recombinant therapeutic protein in humans (6). In principle, proteins normally expressed in humans and those

sharing greater sequence identity with them are expected to be less immunogenic as therapeutics than those having more sequence differences. For example,
Infliximab, a chimeric antibody, has greater immunogenicity in patients than
Adalimumab, a fully human antibody (81,82). An Fc fusion involves fusing two
human proteins together, the ligand of interest and the IgG Fc region, and is
therefore expected to have low immunogenicity. However, the fusion junctions
between the two domains (the ligand and the Fc) and between the linker and
either domain of the fusion may represent novel immunogenic epitopes.
Another more worrisome immunogenicity issue is the antibody responses to the
recombinant ligand domain of the Fc fusion; these neutralizing antibodies may
cross-deplete endogenous ligand and block its biological activity. For example,
neutralizing antibodies to recombinant thrombopoietin developed in a small
portion of patients cross-react with the endogenous thrombopoietin, leading to
thrombocytopenia (83) and pancytopenia (84). Overall, antibodies may have
fewer causes for concern than Fc fusions in terms of immunogenicity.

Manufacturability: Glycosylation Variations

The ligand domain of an Fc fusion protein can be glycosylated and, in some
cases, may be heavily glycosylated. Careful monitoring of the glycan structures
attached on the protein is required during the manufacturing process, as glycan
variations can dramatically impact the PK of the fusion. Glycoproteins with
certain oligosaccharide structures are recognized and cleared rapidly by specific
receptors like asialoglycoprotein receptor and mannose/N-acetylglucosamine
receptor (85–90). Jones et al. have also observed that the terminal N-acetylglu-
cosamine level on Lenercept from different batches correlated with the observed
30% to 40% differences in area under the curve (AUC) (78,91). These variations
in carbohydrate can therefore present challenges in the development of a
glycoprotein-Fc fusion.

Fc Engineering to Modulate Pharmacokinetics

As FcRn recycles IgGs or Fc fusions through its interaction with the Fc region,
engineering the FcRn:Fc interaction is one of the methods for modifying the PK
and PD of an IgG or Fc fusion (15,92). A number of studies have demonstrated
that improving the affinity of an Fc variant against FcRn at acidic pH can prolong its serum half-life. Ghetie et al. first showed that an engineered murine Fc
variant (T252L/T254S/T254F), which had threefold increase in murine FcRn
affinity at pH 6, exhibited approximately 30% to 60% extension in serum half-life
in mice (93). Subsequent studies have identified and validated various favorable
mutations on human IgG1 Fc residues, namely Thr250, Met252, Thr254, Ser256,
Thr307, Glu380, Met428, and Asn434 (Fig. 1C) (36,94–96). These mutations can
be combined synergistically to give Fc variants of different FcRn affinities. For
example, one human IgG1 Fc variant carrying the double mutations T250Q/
M428L exhibited a half-life of approximately 35 days in Rhesus monkeys
(Rhesus macaques), a significant improvement over its wild-type counterpart
(~ 14 days). However, it was observed that affinity increase of an Fc variant at
acidic pH was usually coupled with an undesirable affinity increase at neutral
pH (94,97,98). High levels of binding at neutral pH hinder the release of an
FcRn-bound IgG variant back into the circulation and increase the binding of
circulating IgGs to the cell surface–expressing FcRn, effectively accelerating the

clearance of IgG and canceling out the benefit of increased affinity at acidic pH (94). Therefore, FcRn-binding affinities at both acidic and physiological pH are important determinants to balance for the PK engineering of an IgG or Fc fusion.

One important aspect of engineering the half-lives of IgGs is the use of suitable preclinical animal models to evaluate the variant's half-life in vivo and predict the variant's PK parameters in humans. Earlier studies involved testing human IgG1 Fc variants in mouse model; however, because of the recently discovered differences in human IgG's affinity and specificity to murine and human FcRn (99), PK results of the variants in mice were not expected to truly reflect the variants' PK in humans (94,100). Currently, the closest model system to humans is nonhuman primates, as nonhuman primates have similar levels of endogenous IgG and human IgG1 binds nonhuman primate FcRn with similar affinity and specificity as human FcRn (96,100,101). A promising alternative model to nonhuman primates is the human FcRn transgenic mouse, which lacks endogenous murine FcRn but expresses human FcRn (95). These mice allow the in vivo study of human IgG Fc variants' interactions with human FcRn in a higher throughput and more cost-effective manner than nonhuman primates. However, further developments like control of FcRn expression levels and patterns and endogenous IgG levels are needed to render these transgenic mice more humanlike.

An IgG can also be engineered to have a shorter in vivo half-life by reducing its FcRn affinity. It was shown that Fc variants' half-lives reduced in correspondence with the variants' decreasing FcRn affinities (92). These short-lived IgG Fc variants can be useful in situations where shorter IgG exposure is desired (e.g., imaging and toxin-conjugated antibody therapy).

PK-modifying IgG Fc variants are still in the research and early development phases; none of the variants has been evaluated in humans. Fc engineering does appear to be a promising way to modulate the PK of an antibody or Fc fusion, but it remains to be seen how much PK improvement and immunogenicity these Fc variants will have in humans.

ENGINEERING BINDING TO CARRIER PROTEINS
Properties of Serum Albumin

Serum protein binding is a general property frequently responsible for extending the pharmacokinetic properties of small-molecule drugs. This property has been exploited as a strategy to increase the serum half-life of protein therapeutics. This approach takes advantage of the long half-life inherent in many highly abundant and long-lived serum proteins to enhance the properties of many therapeutic molecules. One of the most commonly exploited proteins is serum albumin.

At a concentration of 35 to 50 mg/mL (~ 600 μM), albumin is the most highly abundant protein in serum, comprising roughly 60% of the total serum protein (102). Albumin is broadly distributed throughout the body, serving as an important vascular and extravascular carrier of a wide variety of endogenous molecules such as bilirubin, thyroxine, fatty acids, and metals. Albumin has also been recognized as a major carrier of several small-molecule therapeutics such as ibuprofen, warfarin, and diazepam. Albumin plays an important role in maintaining the solubility of these compounds and protecting them from rapid

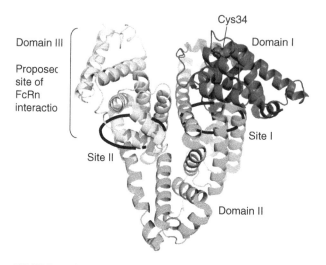

FIGURE 3 Structure of albumin. A ribbon diagram is shown on the basis of the X-ray crystal structure (Protein Data Bank accession number 2BXH), rendered using PyMol. The proposed site of FcRn interaction is indicated at left, and the region of small-molecule binding sites I and II are circled. *Source*: From Refs. 103 and 34.

renal clearance. Many functional and structural studies of albumin have revealed multiple binding sites that enable albumin to accommodate the binding of this diverse set of ligands. Binding specificity and potential drug-drug interactions resulting from multiple highly adaptable binding sites on albumin have been investigated in a comprehensive study by Ghuman et al. (103).

Albumin is a 66-kDa globular protein consisting of three homologous domains oriented in a heart shape (Fig. 3). Although numerous drug interaction sites are distributed throughout the protein, there are two major drug-binding sites, sites 1 and 2, located in domains II and III, respectively, in addition to the metal-binding site near Cys_{34} in domain I. Albumin contains a total of 17 disulfides and one free thiol at Cys_{34} (102). Importantly, the very high serum concentration of albumin makes Cys_{34} the most abundant free thiol in serum.

In human, albumin has a half-life of 19 days, similar to that of IgG1. This long serum persistence can be partially explained by albumin's size that helps it to avoid renal clearance. On the other hand, like IgG, the fractional catabolic rate of albumin is directly related to its serum concentration. For example, in analbuminemic patients, the half-life of albumin is as long as 115 days (104). This observation led Schultze and Heremans to propose, in a similar fashion to Brambell's hypothesis concerning the protection of IgG by FcRn, the existence of a set of albumin receptors that protect albumin from degradation (105). Only recently and quite accidentally, it was discovered that both albumin and IgG are protected in a noncompetitive manner by the same receptor, FcRn (30,31). FcRn is proposed to bind both albumin and IgG at low pH ($\sim$pH 6) during cellular pinocytosis, protecting them from lysosomal degradation. Upon recycling of the endocytic vessel back to the cell surface, both IgG and albumin are released back into serum at neutral pH. Molecules associated with albumin or IgG that do not interfere with FcRn binding and are able to maintain their association during this pH change would be expected to survive this pinocytosis process as well.

The abundance, long half-life in serum, and broad biodistribution of albumin make it an important target for enhancing the pharmacokinetic and pharmacodynamic properties of many therapeutic small molecules and protein therapeutics. General features and examples of the many approaches to capture the properties of albumin are described below and are outlined in Table 2. Although this review focuses on proteins and peptides, some of these approaches originate from attempts to extend the pharmacokinetic properties of small-molecule drugs; these pioneering examples have been included where appropriate.

Strategies to Reduce Clearance Through Association with Albumin
Conjugation to Albumin

Early work by Poznansky and colleagues demonstrated many of the benefits of associating proteins with albumin. While searching for an approach to enhance enzyme therapy, glutaraldehyde was used as a cross-linking agent to form soluble conjugates of enzymes such as superoxide dismutase (SOD) with an excess of albumin (106). While unconjugated SOD was rapidly cleared with a half-life of six minutes, the SOD-albumin conjugates, fractionated by molecular weight and injected into rats, were found to clear through the liver with half-lives of up to 15 hours. Conjugates of increasing molecular weight were cleared progressively faster, suggesting an optimum conjugation ratio. Interestingly, conjugation to albumin also reduced the immunogenicity and antigenicity of SOD in rabbits. Similar findings of reduced immunogenicity were found in an earlier study linking uricase to albumin (107). These early studies demonstrated the potential of albumin conjugates as a means to prevent rapid clearance and reduce immunogenicity of heterologous proteins (108).

While encouraging in principal, these studies suffered from the heterogeneous preparations that were generated. Glutaraldehyde conjugation to albumin was random, and stoichiometry was difficult to control. Further, bioactivity of the conjugated protein was often compromised. For example, in a study conjugating porcine growth hormone (GH) to albumin, the average complex consisted of two molecules of albumin and six molecules of GH and although the conjugate increased the half-life of GH from five minutes to two to three hours, it did not stimulate growth in hypophysectomized rats (109).

An effective use of this approach has been to deliver cytostatic agents like methotrexate (MTX) to tumors. In a cancer therapy setting, albumin can provide additional benefits besides enhanced pharmacokinetics. Albumin accumulates in and is a major energy and nitrogen source for tumors, making it an ideal delivery vehicle for anticancer agents (110). In addition, association with albumin can alter the metabolism of associated agents, as clearance is shifted from the kidney to the liver. As a result of longer serum persistence, the slower metabolic release of free cytotoxic compounds can also lead to a reduction in toxicity (111). Finally, because of its smaller size, albumin is thought to diffuse through tissues ten times faster than IgG. In a study examining the loading rate of MTX on albumin, high MTX to albumin ratios were associated with more rapid clearance and decreased tumor uptake in rats, while a 1:1 ratio

TABLE 2 Characteristics of Albumin-Based Strategies

	Characteristics	Effectiveness	Features
Covalent attachment to albumin			
Random conjugation in vitro	Heterogeneous product; aggregates; poor reproducibility; very complex manufacturing process anticipated	Can achieve modest PK improvement, up to the full half-life of albumin	Early demonstration of the use and benefits of albumin for prolonging half-life; difficult to control or optimize linkage
Specific conjugation via C_{34} in vitro	Defined product; requires albumin for in vitro conjugation; relatively complex manufacturing		Optimization required to identify linkage that maintains functionality of therapeutic molecule; possible to engineer time- or condition-dependent reversibility into linkage
Specific conjugation via C_{34} in vivo	Simple defined product; conjugation to endogenous albumin becomes dominant species in vivo; relatively simple to manufacture		
Albumin fusions	Simple defined product that can be made recombinantly; manufacture requires expression in yeast or mammalian cell culture		Application limited to proteins and peptides; linkage limited to N- or C-terminal fusions that often impair functionality of therapeutic molecule; similar to Fc fusions but lacks $Fc\gamma$ receptor functions
Noncovalent association to albumin			
serum albumin ligands	Design derived from known albumin ligands; typically low affinity for albumin	Modest half-life extension; affinity for albumin affects half-life and can vary across species	Requires numerous analogues to identify a variant with appropriate albumin-binding affinity while maintaining functionality of therapeutic molecule FDA-approved example (Levemir)
Bacterial domains	Bacterially derived surface antigens; modest to very high affinity for albumin; affinity relatively easy to modify	Can achieve modest PK improvement, up to the full half-life of albumin; affinity for albumin affects half-life and can vary across species	Several examples have been shown to enhance immunogenicity of fused molecule; may be ideal for vaccine development
Engineered Binding Domains	Derived from engineered peptides or immunglobulin domains; modest to very high affinity for albumin; affinity relatively easy to modify		Optimization required to identify linkage that maintains functionality of therapeutic molecule; possible to engineer time- or condition-dependent reversibility into linkage

demonstrated enhanced pharmacokinetics and tumor accumulation (112). Thus, an optimum conjugation ratio must be identified, which does not interfere with the circulating properties of albumin. MTX-albumin is currently undergoing multiple clinical trials.

To make conjugates with a defined stoichiometry, the free thiol present in albumin (Cys_{34}) has been conjugated via a 1,6-bismaleimidohexane linker to the free thiol on a Fab' to achieve a 1:1 molar ratio with a defined linkage (113). This coupling chemistry increased the half-life of an anti-TNF-α Fab' from 0.7 hours to around 20 hours in rats without any loss in TNF-α binding. A similar approach has been taken for a series of biologically active peptides [e.g., atrial nutriuretic peptide (114), dynorphine (115), the kringle 5 region of plasminogen (116)] with the aim of increasing half-life.

In a novel twist, Shechter et al. used a similar approach to conjugate an amino group of insulin to Cys_{34} of albumin through an Fmoc linkage. The conjugate retained only 12% of the biological activity of insulin. Over a 24-hour period, however, the linkage slowly and spontaneously hydrolyzed to release unmodified insulin, resulting in a prolonged glucose-lowering pattern in mice and streptozotocin-induced diabetic rats following a single subcutaneous or intraperitoneal dose (117).

Genetic Fusion to Albumin

Another simple strategy to capitalize on the properties of albumin is to genetically fuse a protein of interest to the N- or C-terminal of albumin. This approach eliminates potential heterogeneity introduced by nonspecific cross-linking technologies. The first albumin fusion consisted of the extracellular 179 residues of CD4 to the C-terminus of human serum albumin (HSA) (118). The resulting HSA-CD4 fusion had a half-life of 34 hours in rabbits compared with an initial half-life of 15 minutes for CD4 and 47 hours for HSA. Unlike the CD4 immunoadhesin example described above, the HSA-CD4 fusion lacked all Fcγ receptor–binding functions, making this a favored approach under conditions where immune cell recruitment would be undesirable.

Several other fusions linking biologically active molecules with albumin are summarized in Table 3; however, pharmacokinetic comparisons across species are difficult. Although the half-life of albumin is reported to be 19 days in human, it is considerably shorter in other species: mouse (1 day), rat (1.9 days), and rabbit (5.6 days) (131–134). Albumin fusions include hormones (insulin, hGH, granulocyte colony-stimulating factor (G-CSF), portions of angiostatin, basic natriuretic peptide, glucagon-like peptide (GLP)-1 (123,125,126,128,129), cytokines (interferon-α, interferon-β, IL-2) (122,124,127), and other biologically active peptides (hirudin, extendin-4, and barbourin) (119,120,129). Improved pharmacokinetic properties have also been described for immunoglobulin-derived fragments fused to HSA (113,130). Albumin fusions offer the potential to reduce immunogenicity and extend pharmacokinetic properties and thus improve therapeutic efficacy of these molecules. Several of these albumin fusions, developed by Human Genome Sciences, are currently undergoing clinical trials (135).

The complexity of albumin (35 cysteines) necessitates that fusions be generated through recombinant expression in yeast or mammalian cell lines; often the yields are fairly low. Further, while both N- and C-terminal fusions to albumin are possible, fusion to albumin can often hinder activity of the fused

TABLE 3 Range of Pharmacokinetic Effects for Albumin-Based Drugs

Molecule	$T_{1/2}\beta$ (hr)	Fusion	$T_{1/2}\beta$ (hr)	Improvement (fold)	Species	Comments	References
CD4	0.25	HSA-CD4	34	136	Rabbit	C-terminal fusion	118
Hirudin	0.7	RSA-hirudin	110	158	Rabbit	N-terminal fusion active/C-terminal fusion inactive production problems	119
Fab'	0.7	scFv-HSA	16		Rat	N-terminal fusion active	113
Barbourin	3.6	Barbourin-RSA	82	23	Rabbit	N-terminal fusion active	120
G-CSF	2.5	G-CSF-HSA	5.4–8.3	3	Mouse	Possible receptor-mediated clearance	121
INF-α	5[a]	HSA-INF-α	68	14	Cynomolgus monkey	C-terminal fusion/activity reduced 20×	122
INF-α	2–3[a]	HSA-INF-α	159[a]		Human	Phase I/II/no significant anti-drug response	190
hGH		hGH-HSA		4[a]	Rat		123
hGH		hGH-HSA		6[a]	Monkey		123
INF-β	8[a]	INF-β-HSA	24	3	Rhesus	N-terminal fusion/activity reduced 5–10×	124
Adk3		Adk3-HSA		*prolonged*	Mouse	Prolonged pharmacodynamics/increased survival	125
BNP	0.05	BNP-HSA	11.2	140	Mouse	N-terminal fusion/only 1.5% active	75
GLP-1	*rapid*	GLP-1-HSA	*prolonged*		Mouse	Prolonged pharmacodynamics/N-terminal fusion only 1% active	126
IL-2	0.3	IL-2-HSA	7.8	24	Mouse	N-terminal fusion active/(Yao'04) notes different biodistribution	127, 191
Insulin	0.17	insulin-HSA	6.7	39	Mouse	N-terminal fusion has 3–6× reduced binding	128
Ex-4	2.4	Ex-4-HSA	56.7	24	Cynomolgus monkey	N-terminal fusion only 5% active	129
scFv	4	scFv-HSA-scFv	47	12	Mouse		130
scDb	16	scDb-HSA	43	3	Mouse		130
tandemFv	26	tandemFv-HSA	40	2	Mouse		130

[a]These studies utilized subcutaneous rather than intravenous administration.

Abbreviations: HSA, human serum albumin; G-CSF, granulocyte colony-stimulating factor; hGH, human growth hormone; BNP, basic natriuretic peptide; GLP, glucagon-like peptide.

protein (119,129,136). In addition, fusions do not always attain the half-life of albumin and may be cleared quickly because of receptor-mediated clearance (121), proteolytic instability, or the interference of potential recycling by FcRn.

In Vivo Conjugation: Prodrug Strategies That Target Albumin

Specific in vivo conjugation of drugs to endogenous albumin is an approach that overcomes both the drawbacks of random conjugation to albumin and the challenge of expressing albumin fusions. A prodrug strategy developed by Kratz and coworkers takes advantage of the most abundant free thiol in serum, namely Cys_{34} on albumin, to conjugate a maleimide derivative of doxorubicin in situ, thus extending its pharmacokinetic and pharmacodynamic properties (137,138).

In vivo bioconjugation to Cys_{34} on serum albumin has also been extended to maleimido derivatives strategically placed on bioactive peptides including GLP-1, insulin, and hGH-releasing factor (139–141). Unlike fusions that are limited to N- or C-terminal attachment to albumin, the most appropriate site on the protein of interest for conjugation can be assessed ex vivo to identify attachments with minimal loss of bioactivity (141).

Noncovalent Association with Albumin

The incorporation of known albumin ligands, or bacterially derived or de novo engineered albumin-binding domains (ABDs) into the therapeutic molecule of interest can enable noncovalent association with albumin and, as a result, enhance pharmacokinetic and pharmacodynamic properties. An early and successful example is Insulin detemir (Levemir®), an FDA-approved insulin analogue conjugated to myristic acid. This fatty acid binds weakly to multiple sites on albumin (142) and, as a result, extends the circulating half-life of insulin by more than five hours in human (143–145). Like myristic acid, ibuprofen and other small molecules that bind albumin have also been used as conjugates to enhance the pharmacokinetic and tissue distribution properties of peptides and antisense oligonucleotides (146,147). The very high concentration of albumin in vivo assures that even weakly bound ligands will be associated with albumin in serum the majority of the time and thus spared from renal clearance.

Many gram-positive bacteria associate with serum proteins, and a number of homologous ABDs have been identified on their cell surface (148). Improved pharmacokinetics resulting from a fusion to an ABD was first demonstrated for a protein consisting of two fragments of streptococcal protein G fused N- and C-terminally to CD4. The ABD-CD4-ABD fusion had a half-life in mice comparable to a CD4-Fc fusion protein; however, a 90% reduction in CD4 binding to gp120 was also noted (149). Likewise, an ABD fusion to a soluble form of complement receptor type I (CR1) increased the half-life of CR1 from 1.6 to 5 hours in rats (150), and a diabody-ABD fusion with low nanomolar affinity for mouse albumin extended the half-life of the diabody to 28 hours similar to a diabody-HSA fusion (130,151). Similar results obtained using a bivalent scFv-ABD fusion also revealed a 25-fold reduction in renal uptake in comparison with the nonfused molecule (152). In addition to the variable half-life of albumin across different species, the binding affinity between different ABDs and these serum albumins can differ widely (153). Both of these parameters are important considerations in assessing pharmacokinetics for ABD fusions (154).

A complex between an ABD and HSA has revealed ABD helices 2 and 3 of the three-helix bundle directly involved in the binding to domain II of HSA (155), while the FcRn-binding site has been proposed to reside in domain III of albumin (16,31). Thus, most likely, the ABD- and FcRn-binding sites are located on separate regions of albumin.

Foreign sequences present in these constructs raise concerns regarding immunogenicity. In fact, an ABD from streptococcus has been used to increase the half-lives of small peptide and polysaccharide antigens, resulting in an enhanced immune response due to preexisting T- and B-cell epitopes generated by the ABD (156–158).

Peptides or immunoglobulin fragments, engineered to bind albumin, have also been shown to be effective at prolonging the half-life of therapeutic molecules. Using a peptide library displayed on phage, we have identified a peptide that binds to a conserved epitope on albumin across multiple species (159). When fused to a Fab to form an albumin-binding Fab (AB.Fab), the half-life of the Fab was increased by 15-fold to 20 hours in mice, while in rats and rabbits, the half-life was extended to 27 and 69 hours, respectively (154). Since pharmacokinetic comparisons in different species are difficult because of the changing half-life of albumin and the changing affinity of the albumin-binding peptide for the different albumins, we created a series of albumin-binding peptides with a range of affinities for albumin within a single species. This enabled us to demonstrate a clear correlation between albumin affinity and serum half-life (Fig. 4) and to estimate a half-life for an AB.Fab in human of approximately four days (154).

Bispecific immunoglobulin-based fragments can be engineered to target both a serum protein such as IgG or albumin and another antigen of choice. Initially, Hollinger et al. engineered one arm of a bispecific diabody to target IgG to recruit Fcγ receptor functionality and found that binding to IgG increased the half-life of the diabody by fivefold (160). Similarly, bispecific molecules such as a bispecific Fab'$_2$, and single-domain antibodies (dAbs) have also been engineered

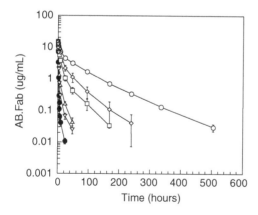

Fab Variant	Rabbit Albumin Binding (Kd, nM)
AB.Fab4D5-H	36 ± 2
AB.Fab4D5-H8	247 ± 36
AB.Fab4D5-H4	444 ± 25
AB.Fab4D5-H10	1065 ± 87
AB.Fab4D5-H11	1110 ± 32
Fab4D5	nd

FIGURE 4 Pharmacokinetics of AB.Fab variants in rabbit. Fab4D5, AB.Fab4D5-H, AB.Fab4D5-H4, AB.Fab4D5-H8, AB.Fab4D5-H10, and AB.Fab4D5-H11 were dosed at 0.5 mg/kg intravenous bolus in New Zealand White rabbits (three rabbits/group). Samples taken at the indicated times were assayed using a HER2 binding ELISA.

to target albumin to extend plasma half-life (113,161). Bifunctional dAbs that bind albumin (AlbudAb) were shown to match the half-life of albumin in mice (161).

Although each of the above examples can be engineered to increase the noncovalent association with albumin through higher affinity, potential benefits of very high affinity have not been demonstrated (153). Once an affinity is reached that ensures all noncovalently associated molecules are essentially bound to albumin, the half-life of albumin should be attained (barring other clearance mechanisms) and further pharmacokinetic enhancements are unlikely. The reports described above suggest that this affinity is in the single-digit nanomolar range owing to the high concentration of albumin in vivo.

Capturing Albumin Properties Through Formulation
Albumin has also been used very successfully as a formulating agent. A key advantage of Abraxane® is the use of albumin to formulate paclitaxel-albumin nanoparticles that are free of the toxic side effects due to cremophor, a solvent normally used to formulate paclitaxel. This formulation, however, also results in a pharmacokinetic profile that differs from the commercial formulation (Taxol). Abraxane was shown to present lower plasma and higher tissue levels of paclitaxel with wider, more rapid distribution and slower metabolism. As a result, Abraxane is 59-fold less toxic than Taxol and 29-fold less toxic than its excipients (162). From the examples above, it is likely that many of these benefits can be attributed to the properties of albumin, a carrier that can extend the half-life of therapeutic molecules, alter their metabolism, and enhance their biodistribution.

ENGINEERING INCREASED HYDRODYNAMIC VOLUME
Rationale for Engineering Protein Hydrodynamic Volume
A general strategy for engineering improved half-life is through modification of the protein with hydrophilic polymers. Among polymers that could be attached to proteins, PEG has been the predominant choice for half-life extension. The focus of this section is on strategies to engineer improved half-life of these proteins through covalent modification with polyethelene glycol. Modulation of in vivo half-life may be required to realize the therapeutic potential of both non-antibody proteins and antibody fragments lacking FcRn-binding capacity.

Kidney Clearance of Low-Molecular-Weight Proteins
Glomerular filtration in the kidneys appears to be the primary route of elimination for proteins of molecular weight less than about 40 kDa. Shape and molecule charge also contribute to the rate of clearance via the kidneys. Serum albumin, with a molecular weight of about 67 kDa, is not cleared through kidney filtration. Kidney clearance of low-molecular-weight proteins has also been demonstrated through in vivo experiments in which the kidneys were bypassed. For example, the half-life of interleukin-2 (IL-2; MW = 19.5 kDa) in the mouse was significantly increased upon ligation of the renal pedicles (163). Since clearance of IL-2 was not completely blocked by this procedure, other clearance mechanisms may become evident upon removal of this rate-determining process.

Antibody fragments can have half-lives considerably shorter than intact IgG. In experiments with anti-IL-8 antibody fragments, Koumenis et al. (164)

found that the clearance of Fab (MW = 49 kDa) and Fab$'_2$ (MW = 98 kDa) in rabbits was 220-fold and 28-fold faster than that of IgG (MW = 150 kDa), respectively. Although not directly compared by Koumenis et al., other studies (165) indicate that clearance of scFv (MW ~ 24 kDa) is several-fold faster than that measured for Fab. Since the molecular size of Fab$'_2$ is larger than the kidney filtration limit and is two-thirds of the molecular weight of IgG, it might have been expected to have a clearance rate closer to that of IgG. However, the interaction of IgG, but not fragments lacking Fc, with FcRn further prolongs the half-life of IgG. As mentioned above, in addition to molecular size, shape and charge are factors contributing to kidney clearance and may have an impact on the clearance of antibody fragments.

Increasing Hydrodynamic Volume Through Pegylation

Molecular size can be readily increased through modification of the protein with the hydrophilic polymer PEG. Linear PEG chains have a varied number (n) of ethylene glycol units, are usually capped at one end with a methoxy group to eliminate cross-linking reactions, and have a functional group (X) at the other end for covalent reaction with the target protein. Thus, the general formula is $CH_3O(CH_2CH_2O)_n$-X. PEG chains ranging in molecular weight from 2000 to about 60,000 are commercially available. The chemistry employed for synthesis of PEG chains does not generate a homogenous polymer, instead, there is a distribution of chain lengths around an average mass. Nonetheless, attachment of longer chain length PEGs produces a larger increase in protein molecular size.

PEGs are usually attached to proteins through reaction with protein free amino or sulfhydryl groups. Protein modification can also be accomplished using aldehyde groups introduced via periodate oxidation of carbohydrate. The free amino groups on the protein are supplied by lysine side chains or the protein N-terminus, whereas only cysteine residues have a free sulfhydryl. For reaction with free amino groups, PEGs functionalized with amine-reactive moieties such as N-hydroxysuccinimide (NHS)-activated esters, or aldehydes, are employed. Maleimide-derivatized PEG is used for modifying cysteine residues. Amino-PEG is used for modification of introduced aldehyde groups.

Attachment of PEG to a protein (PEGylation) can have a large effect on apparent hydrodynamic volume. This is because the PEG chain has an extended conformation and is heavily hydrated, thus sweeping out a large hydrodynamic volume. A measure of hydrodynamic volume, EMW, is conveniently measured from the elution position on a size-exclusion chromatography (SEC) column calibrated with globular proteins of known molecular weight. By using an in-line static light-scattering detector, the molar mass and polydispersity of the peaks eluting from the SEC column can be calculated. Alternatively, dynamic light scattering can be used to calculate the hydrodynamic volume of PEGylated proteins in solution without need for a SEC column. Experiments of this type generally show that the change in hydrodynamic volume from PEGylation is greater than expected from the change in molar mass. For example, modification of a Fab$'_2$ with a single 20-kDa PEG chain results in a sevenfold increase in EMW (166).

Increased EMW is correlated with decreased pharmacokinetic clearance. Variable EMW is easily attained by changing the chain length or number of PEG attachment sites. As shown in Figure 5, increasing the EMW has the largest

(A)

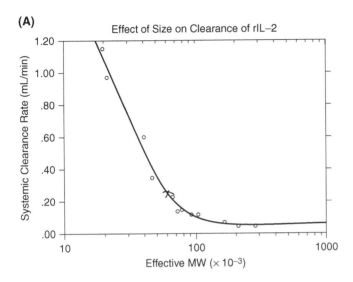

(B)

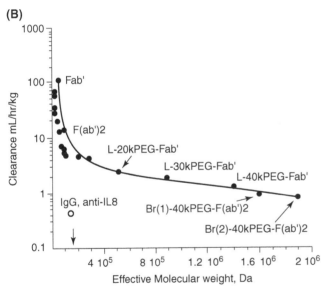

FIGURE 5 Effect of molecular size on pharmacokinetic clearance. (**A**) rIL-2 PEGylated forms. (**B**) PEGylated Fab fragments.

effect on clearance when the EMW is below the kidney filtration limit of 70 kDa. Further increase in EMW by addition of more or larger PEG chains results in smaller increments of decrease in clearance. This trend is observed for both IL-2 (164) (Fig. 5A) and anti-IL-8 Fab$'_2$ (Fig. 5B).

Multiple sites of PEG attachment on a protein can increase the chance of loss of bioactivity upon PEGylation. If the coupling site is within or near the binding site for receptor, then the PEG chain can sterically interfere with receptor binding. Therefore, use of larger PEGs with single or few points of attachment is preferable to many coupling sites with smaller PEG chains. Since linear PEG chains can result in larger volumes of distribution and may deposit in kidney vacuoles (167), branched PEGs (168,169) have been developed. Branched PEGs, having two or more PEG chains built off a single functional group, produce a larger change in mass with fewer sites of modification.

Site-specific PEGylation is strongly preferred to random modification. Since proteins generally have several accessible lysine residues that can react with amine-reactive PEG, a distribution of products will be obtained with this chemistry. Reaction at some sites will produce losses in activity, and it is difficult to drive the reaction to completion at one site without modification at other sites. The result is a heterogeneous product, with variability in specific activity between lots that may be difficult to characterize. Site-specific PEGylation may be obtained by directing chemistry toward the N-terminus (170,171) or by engineering Cys residues into the protein for reaction with PEG-maleimide (172). The approach of using engineered Cys residues has a distinct advantage in the specific and facile reaction of maleimide with sulfhydryl groups. Knowledge of the protein structure is required so that surface sites, spatially distinct from known binding sites, are chosen for Cys substitution. This approach may not be applicable to proteins having multiple disulfide bonds since the extra Cys residue may initiate disulfide shuffling and misfolding of the protein. Care must also be taken to avoid formation of disulfide-linked oligomers through the engineered Cys residue. This usually involves formulating the protein in reducing agent until it is ready for PEG modification and performing the maleimide reaction at lower protein concentration. There is potential for increased immunogenicity due to the amino acid change; however, the site will be masked by covalent attachment of the PEG chain. PEG chains have very low antigenicity (173), and PEGylation has been shown to reduce the immunogenicity of foreign proteins in humans (174).

Examples of PEGylated Proteins Approved for Therapeutic Use in Humans

Three examples of PEGylated proteins currently in clinical use are described below. Two of these, PEG-asparaginase and PEG-interferon, are produced using random PEG modification of the protein. The third, PEG-G-CSF, is produced via site-specific modification of the protein N-terminus.

PEG-Asparaginase

L-Asparaginase is used in the treatment of acute lymphoblastic leukemia. Leukemic cells require L-asparagine for their rapid and malignant growth. L-Asparaginase reduces L-asparagine levels, thus starving the leukemia cells. The enzyme derived from *Escherichia coli* has been used in humans; however, patients often develop hypersensitivity to this foreign protein. A PEGylated version of *E. coli*, L-asparaginase (Oncaspar®, Enzon, Inc., Piscataway, NJ, U.S.), was developed and shown to have reduced incidence of severe hypersensitivity reactions (175). Oncaspar is produced via random PEGylation on lysine ε-amino groups and has a significantly increased half-life relative to the unmodified enzyme. As a consequence, asparagine levels are kept low for longer periods when the PEGylated enzyme is administered.

PEGylated Interferon

Interferon-α plus ribavirin has been the standard of care for patients chronically infected with hepatitis C virus. However, since interferon-α has a short half-life, it must be administered three times per week, and the response rate to this combination therapy is only 40% to 50%. Two PEGylated interferons have been developed and approved for clinical use in hepatitis C–infected patients. PegIntron® (Schering Plough, New Jersey, U.S.) is interferon-α-2b modified randomly with a single, 12-kDa linear chain of PEG. PEGASYS® (Roche) is interferon-α-2a mono-PEGylated on a free amino group with a 40-kDa branched PEG chain. Both PEGylated interferons given once per week are more efficacious than unmodified interferon given three times per week (175,176).

PEGASYS is prepared through random modification of interferon-α-2a on lysine residues by reaction with 40K PEG_2-NHS (177). This reaction produces a mixture of mono-PEGylated products in which the PEG chain is found on one of four Lys residues. PEG_2-interferon-α-2a retains 7% of the in vitro antiviral activity of unmodified interferon-α-2a. Nonetheless, PEG_2-interferon-α-2a shows superior antiviral activity in vivo most likely due to the 70-fold increase in serum half-life relative to unmodified interferon-α-2a. It would be interesting to test a site-specific PEGylated version of interferon-α-2a, which might retain a higher level of antiviral activity than the randomly PEGylated molecule, for efficacy in hepatitis C–infected patients.

PEGylated Granulocyte Colony-Stimulating Factor

Recombinant human G-CSF; marketed as Neupogen® by Amgen, Inc.) is in clinical use for neutropenia resulting from chemotherapy. G-CSF is given as daily subcutaneous injections during a standard 14-day cycle of chemotherapy treatment. A PEGylated version (Neulasta®, Amgen, Inc.) was developed with the goal of reducing the frequency of injections while maintaining efficacy for neutrophil recovery. G-CSF is mono-PEGylated on the N-terminus with a 20-kDa PEG chain by reductive alkylation to generate Neulasta. This material retains full biological activity and has a serum half-life of 42 hours, as compared with the 4-hour half-life observed for G-CSF (178). The half-life of G-CSF is determined both by renal clearance and by receptor-mediated clearance by neutrophils. PEGylation appears to reduce the contribution from kidney filtration. In clinical experiments, a single dose of PEGylated G-CSF has been shown to be as effective in raising neutrophil counts as daily dosing of G-CSF (179). This is the first example of a site-specific PEGylated protein that has been approved by the FDA for clinical use.

Future Directions in Polymer Conjugation

Although two of the approved PEGylated proteins described above resulted from random modification, it seems unlikely that this approach will yield viable biopharmaceuticals in many cases. Instead, site-specific PEGylation will be required for many candidate therapeutics. Assuming that structural information can be obtained for the target protein, engineering of surface-accessible cysteine residues is an attractive approach. For antibody fragments retaining the hinge cysteine residue, it has been demonstrated that modification of this residue with PEG-maleimide generates a product that retains high-affinity antigen binding and has prolonged half-life in vivo (180). PEGylation of Fab or scFv antibody fragments can exploit engineered cysteine mutants that have been selected for high expression and reactivity (181). It is expected that further refinements in PEGylation technology will yield additional options for cysteine modification. This will include other reactive functional groups on the PEG chain, greater variety in the length and branching of PEG chains, and less polydispersity in the average chain length. Such developments will enable fine-tuning of the pharmacokinetic/pharmacodynamic properties while ensuring a reliable product.

For proteins where site-specific modification is not possible or losses in activity result with this approach, releasable PEGs (182) may be a viable option. Such PEGs either use chemistry where the PEG is released slowly by hydrolysis under physiological conditions or employ a cleavable linker that is recognized by extracellular or intracellular enzymes. Releasable PEGs are particularly important for small-molecule drugs. Analogous to a prodrug strategy, the PEG is used for extending half-life, but the polymer interferes with activity. With the appropriate cleavable linker and upon uptake into cells, the PEG chain is removed by lysosomal enzymes and the small-molecule drug is rendered active. This approach has been used to increase the antitumor activity of doxorubicin (183).

The goal of improved drug exposure may be attained through slow-release formulations rather than covalent modification of the protein. Much work has been done to design biodegradable microspheres for slow release of therapeutic proteins. For example, Nutropin Depot® consists of hGH contained in microspheres formed from poly-(D,L-lactide-coglycolide) copolymer. When implanted subcutaneously, hGH is slowly released, initially by diffusion followed by both diffusion and polymer degradation. Efficacious doses of hGH could be delivered via this method, which are comparable to what can be achieved with more frequent IV injections (184).

Alternative Polymers

An alternative posttranslational modification to increase half-life is to engineer additional glycosylation sites into a protein that is produced in mammalian cells (185). The serum half-life of human erythropoietin (EPO), which has 3 N-linked and 1 O-linked glycosylation sites, has been found to be proportional to the sialic acid–containing carbohydrate content of the molecule. Addition of 2 N-linked glycosylation sites to EPO resulted in a molecule (Aranesp®, Amgen, Inc.) with threefold longer half-life than EPO. Less frequent dosing of Aranesp is required to achieve efficacy equal to EPO in anemic patients. The mechanism explaining the longer half-life of Aranesp is unclear; however, it is unlikely to result simply from a change in kidney clearance since the addition of two carbohydrate moieties does not have a significant effect on hydrodynamic volume.

Polymers of *N*-acetylneuraminic acid, polysialic acids, have shown potential for increasing the half-life of proteins (186). Like PEG, polysialic acids are built from repeating units, are hydrophilic, can be produced in varied chain lengths, and can be activated at one end for coupling to proteins. Since polysialic acids are built from natural modules, they appear to be nonimmunogenic. Unlike PEG, polysialic acids are biodegradable, so conjugates may have improved safety profiles. Polysialylated proteins have been shown to have improved serum half-life; however, the flexibility in chain length and chemistry of attachment is not yet on par with PEGylation. In addition, since sialic acid has a negative charge, polysialylation will add negative charge to the protein. For each target protein, the consequence of increased negative charge will need to be examined.

Results from modification of EPO and G-CSF with hydroxyethyl starch (HESylation) have been recently reported (PEGS meeting, 2008). Hydroxyethyl starch is a blood plasma substitute, is well tolerated in humans, has low immunogenicity, and is biodegradable. HESylated EPO and G-CSF were shown to have pharmacodynamic properties similar to the PEGylated versions of these molecules. Utility of these HESylated molecules and potential for additional applications of HESylation await further research.

SUMMARY AND FUTURE DIRECTIONS
In this chapter, we have focused on approaches for improving the half-life of therapeutic peptides and proteins through direct modulation of their interaction with FcRn, through engineered association with serum albumin (whose clearance properties are also modulated through binding to FcRn), and through chemical manipulation that serves to increase the hydrodynamic radius of the drug through PEGylation or other chemical modification. These approaches may significantly increase the circulating half-life of an injected drug, thereby providing improved drug exposure and patient compliance with the treatment regimen as a result of the increased convenience of fewer office visits to physicians or fewer self-administered injections. Both components have the potential to improve treatment outcomes and patient quality of life.

Looking ahead, it seems likely that the modular elements of these approaches will be combined, for example, by engineering immunoadhesins with even longer half-life through engineering of higher-affinity FcRn binding. In addition, new conjugation chemistry (187) may provide further opportunities to enhance chemical modification techniques for larger hydrodynamic volume and nanoparticle approaches.

REFERENCES
1. Brown LR. Commercial challenges of protein drug delivery. Expert Opin Drug Deliv 2005; 2(1):29–42.
2. Renkin EM, Robinson RR. Glomerular filtration. N Engl J Med 1974; 290(14):785–789.
3. Felgenhauser K. Protein filtration and secretion at human body fluid barriers. Pflugers Arch 1980; 384:9–17.
4. Daugherty AL, Mrsny RJ. Formulation and delivery issues for monoclonal antibody therapeutics. Adv Drug Deliv Rev 2006; 58:686–706.
5. Frokjaer S, Otzen DE. Protein drug stability: a formulation challenge. Nat Rev Drug Discov 2005; 4:298–306.
6. Barbosa MDFS, Celis E. Immunogenicity of protein therapeutics and the interplay between tolerance and antibody responses. Drug Discov Today 2007; 12:674–681.
7. Labhasetwar V, Song C, Levy RJ. Nanoparticle drug delivery systems for restenosis. Adv Drug Deliv Rev 1997; 24(1):63–85.
8. Saukkonen T, Amin R, Williams RM, et al. Dose-dependent effects of recombinant human insulin-like growth factor (IGF)-I/IGF binding protein-3 complex on overnight growth hormone secretion and insulin sensitivity in type 1 diabetes. J Biol Chem 2004; 89(9):4634–4641.
9. Ferrara N, Hillan KJ, Gerber HP, et al. Discovery and development of bevacizumab, an anti-VEGF antibody for treating cancer. Nat Rev Drug Discov 2004; 3:391–400.
10. Ferrara N, Damico L, Shams N, et al. Development of ranibizumab, an anti-vascular endothelial growth factor antigen binding fragment, as therapy for neovascular age-related macular degeneration. Retina 2006; 26:859–870.

11. Stein TP, Leskiw MJ, Wallace HW. Measurement of half-life human plasma fibrinogen. Am J Physiol 1978; 234(5):D504–D510.
12. Rogentine GN Jr., Rowe DS, Bradley J, et al. Metabolism of human immunoglobulin D (IgD). J Clin Invest 1966; 45(9):1467–1478.
13. Barth WF, Wochner RD, Waldmann TA, et al. Metabolism of human gamma macroglobulins. J Clin Invest 1964; 43:1036–1048.
14. Strober W, Wochner RD, Barlow MH, et al. Immunoglobulin metabolism in ataxia telangiectasia. J Clin Invest 1968; 47(8):1905–1915.
15. Roopenian DC, Akilesh S. FcRn: the neonatal Fc receptor comes of age. Nat Rev Immunol 2007; 7(9):715–725.
16. Anderson CL, Chaudhury C, Kim J, et al. Perspective—FcRn transports albumin: relevance to immunology and medicine. Trends Immunol 2006; 27(7):343–348.
17. Brambell FW, Halliday R, Morris IG. Interference by human and bovine serum and serum protein fractions with the absorption of antibodies by suckling rats and mice. Proc R Soc Lond B Biol Sci 1958; 149(934):1–11.
18. Simister NE, Mostov KE. An Fc receptor structurally related to MHC class I antigens. Nature 1989; 337(6203):184–187.
19. Brambell FW, Hemmings WA, Morris IG. A theoretical model of gamma-globulin catabolism. Nature 1964; 203:1352–1354.
20. Ghetie V, Hubbard JG, Kim JK, et al. Abnormally short serum half-lives of IgG in beta 2-microglobulin-deficient mice. Eur J Immunol 1996; 26(3):690–696.
21. Israel EJ, Wilsker DF, Hayes KC, et al. Increased clearance of IgG in mice that lack beta 2-microglobulin: possible protective role of FcRn. Immunology 1996; 89(4): 573–578.
22. Junghans RP, Anderson CL. The protection receptor for IgG catabolism is the beta2-microglobulin-containing neonatal intestinal transport receptor. Proc Natl Acad Sci U S A 1996; 93(11):5512–5516.
23. Wani MA, Haynes LD, Kim J, et al. Familial hypercatabolic hypoproteinemia caused by deficiency of the neonatal Fc receptor, FcRn, due to a mutant beta2-microglobulin gene. Proc Natl Acad Sci U S A 2006; 103(13):5084–5089.
24. Mayer B, Zolnai A, Frenyo LV, et al. Localization of the sheep FcRn in the mammary gland. Vet Immunol Immunopathol 2002; 87(3–4):327–330.
25. Leach JL, Sedmak DD, Osborne JM, et al. Isolation from human placenta of the IgG transporter, FcRn, and localization to the syncytiotrophoblast: implications for maternal-fetal antibody transport. J Immunol 1996; 157(8):3317–3322.
26. Kacskovics I, Kis Z, Mayer B, et al. FcRn mediates elongated serum half-life of human IgG in cattle. Int Immunol 2006; 18(4):525–536.
27. Burmeister WP, Gastinel LN, Simister NE, et al. Crystal structure at 2.2 A resolution of the MHC-related neonatal Fc receptor. Nature 1994; 372(6504):336–343.
28. West AP Jr., Bjorkman PJ. Crystal structure and immunoglobulin G binding properties of the human major histocompatibility complex-related Fc receptor. Biochemistry 2000; 39(32):9698–9708.
29. Martin WL, West AP Jr., Gan L, et al. Crystal structure at 2.8 A of an FcRn/heterodimeric Fc complex: mechanism of pH-dependent binding. Mol Cell 2001; 7(4):867–877.
30. Chaudhury C, Mehnaz S, Robinson JM, et al. The major histocompatibility complex-related Fc receptor for IgG (FcRn) binds albumin and prolongs its lifespan. J Exp Med 2003; 197(3):315–322.
31. Chaudhury C, Brooks CL, Carter DC, et al. Albumin binding to FcRn: distinct from the FcRn-IgG interaction. Biochemistry 2006; 45(15):4983–4990.
32. Sanchez LM, Penny DM, Bjorkman PJ. Stoichiometry of the interaction between the major histocompatibility complex-related Fc receptor and its Fc ligand. Biochemistry 1999; 38(29):9471–9476.
33. Burmeister WP, Huber AH, Bjorkman PJ. Crystal structure of the complex of rat neonatal Fc receptor with Fc. Nature 1994; 372(6504):379–383.
34. The PyMOL Molecular Graphics System. DeLano Scientific, 2002. Available at: http://www.pymol.org.

35. Kim JK, Firan M, Radu CG, et al. Mapping the site on human IgG for binding of the MHC class I-related receptor, FcRn. Eur J Immunol 1999; 29(9):2819–2825.
36. Shields RL, Namenuk AK, Hong K, et al. High resolution mapping of the binding site on human IgG1 for Fc gamma RI, Fc gamma RII, Fc gamma RIII, and FcRn and design of IgG1 variants with improved binding to the Fc gamma R. J Biol Chem 2001; 276(9):6591–6604.
37. Zhou J, Johnson JE, Ghetie V, et al. Generation of mutated variants of the human form of the MHC class I-related receptor, FcRn, with increased affinity for mouse immunoglobulin G. J Mol Biol 2003; 332(4):901–913.
38. Zhou J, Mateos F, Ober RJ, et al. Conferring the binding properties of the mouse MHC class I-related receptor, FcRn, onto the human ortholog by sequential rounds of site-directed mutagenesis. J Mol Biol 2005; 345(5):1071–1081.
39. Andersen JT, Dee Qian J, Sandlie I. The conserved histidine 166 residue of the human neonatal Fc receptor heavy chain is critical for the pH-dependent binding to albumin. Eur J Immunol 2006; 36(11):3044–3051.
40. Praetor A, Hunziker W. Beta(2)-Microglobulin is important for cell surface expression and pH-dependent IgG binding of human FcRn. J Cell Sci 2002; 115(pt 11):2389–2397.
41. Raghavan M, Bonagura VR, Morrison SL, et al. Analysis of the pH dependence of the neonatal Fc receptor/immunoglobulin G interaction using antibody and receptor variants. Biochemistry 1995; 34(45):14649–14657.
42. Ober RJ, Martinez C, Vaccaro C, et al. Visualizing the site and dynamics of IgG salvage by the MHC class I-related receptor, FcRn. J Immunol 2004; 172(4):2021–2029.
43. Prabhat P, Gan Z, Chao J, et al. Elucidation of intracellular recycling pathways leading to exocytosis of the Fc receptor, FcRn, by using multifocal plane microscopy. Proc Natl Acad Sci U S A 2007; 104(14):5889–5894.
44. Ober RJ, Martinez C, Lai X, et al. Exocytosis of IgG as mediated by the receptor, FcRn: an analysis at the single-molecule level. Proc Natl Acad Sci U S A 2004; 101(30):11076–11081.
45. Borvak J, Richardson J, Medesan C, et al. Functional expression of the MHC class I-related receptor, FcRn, in endothelial cells of mice. Int Immunol 1998; 10(9): 1289–1298.
46. Zhu X, Meng G, Dickinson BL, et al. MHC class I-related neonatal Fc receptor for IgG is functionally expressed in monocytes, intestinal macrophages, and dendritic cells. J Immunol 2001; 166(5):3266–3276.
47. Dickinson BL, Badizadegan K, Wu Z, et al. Bidirectional FcRn-dependent IgG transport in a polarized human intestinal epithelial cell line. J Clin Invest 1999; 104(7):903–911.
48. Schlachetzki F, Zhu C, Pardridge WM. Expression of the neonatal Fc receptor (FcRn) at the blood-brain barrier. J Neurochem 2002; 81(1):203–206.
49. Deane R, Sagare A, Hamm K, et al. IgG-assisted age-dependent clearance of Alzheimer's amyloid beta peptide by the blood-brain barrier neonatal Fc receptor. J Neurosci 2005; 25(50):11495–11503.
50. Akilesh S, Huber TB, Wu H, et al. Podocytes use FcRn to clear IgG from the glomerular basement membrane. Proc Natl Acad Sci U S A 2008; 105(3):967–972.
51. Antohe F, Radulescu L, Gafencu A, et al. Expression of functionally active FcRn and the differentiated bidirectional transport of IgG in human placental endothelial cells. Hum Immunol 2001; 62(2):93–105.
52. Simister NE. Placental transport of immunoglobulin G. Vaccine 2003; 21(24):3365–3369.
53. Spiekermann GM, Finn PW, Ward ES, et al. Receptor-mediated immunoglobulin G transport across mucosal barriers in adult life: functional expression of FcRn in the mammalian lung. J Exp Med 2002; 196(3):303–310.
54. Maddon PJ, Dalgleish AG, McDougal JS, et al. The T4 gene encodes the AIDS virus receptor and is expressed in the immune system and the brain. Cell 1986; 47(3):333–348.
55. Clark SJ, Jefferies WA, Barclay AN, et al. Peptide and nucleotide sequences of rat CD4 (W3/25) antigen: evidence for derivation from a structure with four immunoglobulin-related domains. Proc Natl Acad Sci U S A 1987; 84(6):1649–1653.

56. Chamow SM, Peers DH, Byrn RA, et al. Enzymatic cleavage of a CD4 immu-noadhesin generates crystallizable, biologically active Fd-like fragments. Biochem-istry 1990; 29(42):9885–9891.
57. Garrett TP, Wang J, Yan Y, et al. Refinement and analysis of the structure of the first two domains of human CD4. J Mol Biol 1993; 234(3):763–778.
58. Smith DH, Byrn RA, Marsters SA, et al. Blocking of HIV-1 infectivity by a soluble, secreted form of the CD4 antigen. Science 1987; 238(4834):1704–1707.
59. Traunecker A, Luke W, Karjalainen K. Soluble CD4 molecules neutralize human immunodeficiency virus type 1. Nature 1988; 331(6151):84–86.
60. Berger EA, Fuerst TR, Moss B. A soluble recombinant polypeptide comprising the amino-terminal half of the extracellular region of the CD4 molecule contains an active binding site for human immunodeficiency virus. Proc Natl Acad Sci U S A 1988; 85(7):2357–2361.
61. Capon DJ, Chamow SM, Mordenti J, et al. Designing CD4 immunoadhesins for AIDS therapy. Nature 1989; 337(6207):525–531.
62. Hodges TL, Kahn JO, Kaplan LD, et al. Phase 1 study of recombinant human CD4-immunoglobulin G therapy of patients with AIDS and AIDS-related complex. Antimicrob Agents Chemother 1991; 35(12):2580–2586.
63. Byrn RA, Mordenti J, Lucas C, et al. Biological properties of a CD4 immunoadhesin. Nature 1990; 344(6267):667–670.
64. Traunecker A, Schneider J, Kiefer H, et al. Highly efficient neutralization of HIV with recombinant CD4-immunoglobulin molecules. Nature 1989; 339(6219):68–70.
65. Kahn JO, Allan JD, Hodges TL, et al. The safety and pharmacokinetics of recombinant soluble CD4 (rCD4) in subjects with the acquired immunodeficiency syndrome (AIDS) and AIDS-related complex. A phase 1 study. Ann Intern Med 1990; 112(4):254–261.
66. Mordenti J, Chen SA, Moore JA, et al. Interspecies scaling of clearance and volume of distribution data for five therapeutic proteins. Pharm Res 1991; 8(11):1351–1359.
67. Joos B, Trkola A, Kuster H, et al. Long-term multiple-dose pharmacokinetics of human monoclonal antibodies (MAbs) against human immunodeficiency virus type 1 envelope gp120 (MAb 2G12) and gp41 (MAbs 4E10 and 2F5). Antimicrob Agents Chemother 2006; 50(5):1773–1779.
68. Jazayeri JA, Carroll GJ. Fc-based cytokines: prospects for engineering superior therapeutics. BioDrugs 2008; 22(1):11–26.
69. Deng B, Banu N, Malloy B, et al. An agonist murine monoclonal antibody to the human c-Mpl receptor stimulates megakaryocytopoiesis. Blood 1998; 92(6):1981–1988.
70. Zhang L, Zhang X, Barrisford GW, et al. Lexatumumab (TRAIL-receptor 2 mAb) induces expression of DR5 and promotes apoptosis in primary and metastatic renal cell carcinoma in a mouse orthotopic model. Cancer Lett 2007; 251(1):146–157.
71. Mitoma H, Horiuchi T, Hatta N, et al. Infliximab induces potent anti-inflammatory responses by outside-to-inside signals through transmembrane TNF-alpha. Gastroenterology 2005; 128(2):376–392.
72. Kaymakcalan Z, Kalghatgi L, Xiong L. Differential TNF-neutralizing potencies of adalimumab, etanercept and infliximab. Clin Immunol 2006; 119(suppl 1):S77–S78.
73. Kaymakcalan Z, Sakorafas P, Bose S. Etanercept, infliximab and adalimumb bind to soluble and transmembrane TNF with similar affinities. Clin Immunol 2006; 119(S1):S75.
74. Scallon B, Cai A, Solowski N, et al. Binding and functional comparisons of two types of tumor necrosis factor antagonists. J Pharmacol Exp Ther 2002; 301(2):418–426.
75. Wang B, Nichol JL, Sullivan JT. Pharmacodynamics and pharmacokinetics of AMG 531, a novel thrombopoietin receptor ligand. Clin Pharmacol Ther 2004; 76(6):628–638.
76. Bitonti AJ, Dumont JA, Low SC, et al. Pulmonary delivery of an erythropoietin Fc fusion protein in non-human primates through an immunoglobulin transport pathway. Proc Natl Acad Sci U S A 2004; 101(26):9763–9768.
77. Korth-Bradley JM, Rubin AS, Hanna RK, et al. The pharmacokinetics of etanercept in healthy volunteers. Ann Pharmacother 2000; 34(2):161–164.

78. Jones AJ, Papac DI, Chin EH, et al. Selective clearance of glycoforms of a complex glycoprotein pharmaceutical caused by terminal N-acetylglucosamine is similar in humans and cynomolgus monkeys. Glycobiology 2007; 17(5):529–540.
79. Weisman MH, Moreland LW, Furst DE, et al. Efficacy, pharmacokinetic, and safety assessment of adalimumab, a fully human anti-tumor necrosis factor-alpha monoclonal antibody, in adults with rheumatoid arthritis receiving concomitant methotrexate: a pilot study. Clin Ther 2003; 25(6):1700–1721.
80. den Broeder A, van de Putte L, Rau R, et al. A single dose, placebo controlled study of the fully human anti-tumor necrosis factor-alpha antibody adalimumab (D2E7) in patients with rheumatoid arthritis. J Rheumatol 2002 29(11):2288–2298.
81. Tracey D, Klareskog L, Sasso EH, et al. Tumor necrosis factor antagonist mechanisms of action: a comprehensive review. Pharmacol Ther 2008; 117(2):244–279.
82. Anderson PJ. Tumor necrosis factor inhibitors: clinical implications of their different immunogenicity profiles. Semin Arthritis Rheum 2005; 34(5 suppl 1):19–22.
83. Li J, Yang C, Xia Y, et al. Thrombocytopenia caused by the development of antibodies to thrombopoietin. Blood 2001; 98(12):3241–3248.
84. Basser RL, O'Flaherty E, Green M, et al. Development of pancytopenia with neutralizing antibodies to thrombopoietin after multicycle chemotherapy supported by megakaryocyte growth and development factor. Blood 2002; 99(7):2599–2602.
85. Morell AG, Gregoriadis G, Scheinberg IH, et al. The role of sialic acid in determining the survival of glycoproteins in the circulation. J Biol Chem 1971; 246(5):1461–1467.
86. Stockert RJ. The asialoglycoprotein receptor: relationships between structure, function, and expression. Physiol Rev 1995; 75(3):591–609.
87. Stahl PD. The mannose receptor and other macrophage lectins. Curr Opin Immunol 1992; 4(1):49–52.
88. Park EI, Mi Y, Unverzagt C, et al. The asialoglycoprotein receptor clears glycoconjugates terminating with sialic acid alpha 2,6GalNAc. Proc Natl Acad Sci U S A 2005; 102(47):17125–17129.
89. Lee SJ, Evers S, Roeder D, et al. Mannose receptor-mediated regulation of serum glycoprotein homeostasis. Science 2002; 295(5561):1898–1901.
90. Park EI, Manzella SM, Baenziger JU. Rapid clearance of sialylated glycoproteins by the asialoglycoprotein receptor. J Biol Chem 2003; 278(7):4597–4602.
91. Keck R, Nayak N, Lerner L, et al. Characterization of a complex glycoprotein whose variable metabolic clearance in humans is dependent on terminal N-acetylglucosamine content. Biologicals 2008; 36(1):49–60.
92. Olafsen T, Kenanova VE, Wu AM. Tunable pharmacokinetics: modifying the in vivo half-life of antibodies by directed mutagenesis of the Fc fragment. Nat Protoc 2006; 1(4):2048–2060.
93. Ghetie V, Popov S, Borvak J, et al. Increasing the serum persistence of an IgG fragment by random mutagenesis. Nat Biotechnol 1997; 15(7):637–640.
94. Dall'Acqua WF, Woods RM, Ward ES, et al. Increasing the affinity of a human IgG1 for the neonatal Fc receptor: biological consequences. J Immunol 2002; 169(9): 5171–5180.
95. Petkova SB, Akilesh S, Sproule TJ, et al. Enhanced half-life of genetically engineered human IgG1 antibodies in a humanized FcRn mouse model: potential application in humorally mediated autoimmune disease. Int Immunol 2006; 18(12):1759–1769.
96. Hinton PR, Xiong JM, Johlfs MG, et al. An engineered human IgG1 antibody with longer serum half-life. J Immunol 2006; 176(1):346–356.
97. Datta-Mannan A, Witcher DR, Tang Y, et al. Monoclonal antibody clearance. Impact of modulating the interaction of IgG with the neonatal Fc receptor. J Biol Chem 2007; 282(3):1709–1717.
98. Datta-Mannan A, Witcher DR, Tang Y, et al. Humanized IgG1 variants with differential binding properties to the neonatal Fc receptor: relationship to pharmacokinetics in mice and primates. Drug Metab Dispos 2007; 35(1):86–94.
99. Vaccaro C, Bawdon R, Wanjie S, et al. Divergent activities of an engineered antibody in murine and human systems have implications for therapeutic antibodies. Proc Natl Acad Sci U S A 2006; 103(49):18709–18714.

100. Dall'acqua WF, Kiener PA, Wu H. Properties of human IgG1s engineered for enhanced binding to the neonatal Fc receptor (FcRn). J Biol Chem 2006; 281(33): 23514–23524.
101. Hinton PR, Johlfs MG, Xiong JM, et al. Engineered human IgG antibodies with longer serum half-lives in primates. J Biol Chem 2004; 279(8):6213–6216.
102. Peters T Jr. All About Albumin. San Diego: Academic Press, Inc, 1996:432.
103. Ghuman J, Zunszain PA, Petitpas I, et al. Structural basis of the drug-binding specificity of human serum albumin. J Mol Biol 2005; 353:38–52.
104. Cormode EJ, Lyster DM, Israels S. Analbuminemia in a neonate. J Pediatr 1975; 86:862–867.
105. Schultze HE, Heremans JF. Molecular biology of human proteins: with special reference to plasma proteins. Vol 1. Nature and Metabolism of Extracellular Proteins. Amsterdam, London, New York: Elsevier, 1966.
106. Wong K, Cleland LG, Poznanski MJ. Enhanced anti-inflammatory effects and reduce immunogenicity of bovine liver superoxide dismutase by conjugation with homologous albumin. Agent Action 1980; 10:231–239.
107. Remy MH, Poznansky MJ. Immunogenicity and antigenicity of soluble cross-linked enzyme/albumin polymers: advantages for enzyme therapy. Lancet 1978; 2(8080): 68–70.
108. Poznansky MJ, Soluble enzyme-albumin conjugates: new possibilities for enzyme replacement therapy. Methods Enzymol 1988; 137:566–574.
109. Poznansky MJ, Halford J, Taylor D. Growth hormone-albumin conjugates, reduced renal toxicity and altered plasma clearance. FEBS Lett 1988; 239(1):18–22.
110. Stehle G, Sinn H, Wunder A, et al. Plasma protein (albumin) catabolism by the tumor itself—implicaions for tumor metabolism and genesis of cachexia. Crit Rev Oncol Hematol 1997; 26:77–100.
111. Fiume L, Bolondi L, Busi C, et al. Doxorubicin coupled to lactosaminated albumin inhibits the growth of hepatocellular carcinomas induced in rats by diethylnitrosamine J Hepatology 2005; 43:645–652.
112. Stehle G, Sinn H, Wunder A, et al. The loading rate determines tumor targeting properties of methotrexate-albumin conjugates in rats. Anticancer Drugs 1997; 8:677–685.
113. Smith BJ, Popplewell A, Athwal D, et al. Prolonged in vivo residence times of antibody fragments associated with albumin. Bioconjug Chem 2001; 12:750–756.
114. Leger R, Robitaille M, Quraishi O, et al. Synthesis and in vitro analysis of atrial natriuretic peptide–albumin conjugates. Bioorg Med Chem Lett 2003; 13:3571–3575.
115. Holmes DL, Thibaudeau K, L'Archeveque B, et al. Site specific 1:1 opioid:albumin conjugate with in vitro activity and long in vivo duration. Bioconjugate Chem 2000; 11(4):439–444.
116. Leger R, Benquet C, Huang X, et al. Kringle 5 peptide–albumin conjugates with antimigratory activity. Bioorg Med Chem Lett 2004; 14:841–845.
117. Shechter Y, Mironchik M, Rubinraut S, et al. Albumin-insulin conjugate releasing insulin slowly under physiological conditions: a new concept for long-acting insulin. Bioconjugate Chem 2005; 16:913–920.
118. Yeh P, Landais D, Lemaitre M, et al. Design of yeast-secreted albumin derivatives for human therapy: biological and antiviral properties of a serum albumin-CD4 genetic conjugate. Proc Natl Acad Sci U S A 1992; 89:1904–1908.
119. Syed S, Schuyler P, Kulczycky M, et al. Potent antithrombin activity and delayed clearance from the circulation characterize recombinant hirudin genetically fused to albumin. Blood 1997; 89(9):3243–3252.
120. Marques JA, George JK, Smith IJ, et al. A barbourin-albumin fusion protein that is slowly cleared in vivo retains the ability to inhibit platelet aggregation in vitro. Thromb Haemost 2001; 86:902–908.
121. Halpern W, Riccobene TA, Agostini H, et al. Albugranin, a recombinant human granulocyte colony stimulating factor (G-CSF) genetically fused to recombinant human albumin induces prolonged myelopoietic effects in mice and monkeys. Pharm Res 2002; 19(11):1720–1729.

122. Osborn BL, Olsen HS, Nardellli B, et al. Pharmacokinetic and pharmacodynamic studies of a human serum albumin-interferon-alpha fusion protein in cynomolgus monkeys. JPET 2002; 303:540–548.
123. Osborn BL, Sekut L, Corcoran M, et al. Albutropin: a growth hormone–albumin fusion with improved pharmacokinetics and pharmacodynamics in rats and monkeys. Eur J Pharm 2002; 456(1–3):149–158.
124. Sung C, Nardellli B, Lafleur DW, et al. An IFN-b-albumin fusion protein that displays improved pharmacokinetic and pharmacodynamic properties in nonhuman primates. J Interferon Cytokine Res 2003; 23:25–36.
125. Bouquet C, Frau E, Opolon P, et al. Systemic administration of a recombinant adenovirus encoding a HSA–angiostatin kringle 1–3 conjugate inhibits MDA-MB-231 tumor growth and metastasis in a transgenic model of spontaneous eye cancer. Mol Ther 2003; 7(2):174–184.
126. Wang W, Ou Y, Shi Y. Albubnp, a recombinant B-type natriuretic peptide and human serum albumin fusion hormone, as a long-term therapy of congestive heart failure. Pharm Res 2004; 21(11):2105–2111.
127. Melder RJ, Osborn BL, Riccobene T, et al. Pharmacokinetics and in vitro and in vivo anti-tumor response of an interleukin-2-human serum albumin fusion protein in mice. Cancer Immunol Immunother 2005; 54:535–547.
128. Duttaroy A, Kanakaraj P, Osborn B, et al. Development of a long-acting insulin analog using albumin fusion technology. Diabetes 2005; 54(1):251–258.
129. Huang Y-S, Chen Z, Chen Y-Q, et al. Preparation and characterization of a novel exendin-4 human serum albumin fusion protein expressed in Pichia pastoris. J Peptide Sci 2008; 14:588–595.
130. Muller D, Karle A, Meißburger B, et al. Improved pharmacokinetics of recombinant bispecific antibody molecules by fusion to human serum albumin. J Biol Chem 2007; 282:12650–12660.
131. Sterling K. The turnover rate of serum albumin in man as measured by I-131-tagged albumin. J Clin Invest 1957; 30:1228–1237.
132. Reed RG, Peters T Jr. Turnover of serum albumin-palmitate complexes. Fed Proc 1984; 43:1858.
133. Stevens DK, Eyre RJ, Bull RJ. Adduction of hemoglobin and albumin in vivo by metabolites of trichloroethylene, trichloroacetate, and dichloroacetate in rats and mice. Fundam Appl Toxicol 1992; 19:336–342.
134. Hatton MWC, Richardson M, Winocour PD. On glucose transport and non-enzymic glycation of proteins in vivo. J Theor Biol 1993; 161:481–490.
135. Subramanian GM, Fiscella M, Lamousé-Smith A, et al. Albinterferon α-2b: a genetic fusion protein for the treatment of chronic hepatitis C. Nat Biotech 2007; 25(12):1411–1419.
136. Baggio LL, Huang Q, Brown TJ, et al. A recombinant human glucagon-like peptide (GLP)-1–albumin protein (Albugon) mimics peptidergic activation of GLP-1 receptor–dependent pathways coupled with satiety, gastrointestinal motility, and glucose homeostasis. Diabetes 2004; 53:2492–2500.
137. Kratz F, Muller-Driver R, Hofmann I, et al. A novel macromolecular prodrug concept exploiting endogenous serum albumin as a drug carrier for cancer chemotherapy. J Med Chem 2000; 43:1253–1256.
138. Mansour AM, Drevs J, Esser N, et al. A new approach for the treatment of malignant melanoma: enhanced antitumor efficacy of an albumin-binding doxorubicin prodrug that is cleaved by matrix metalloproteinase 2. Cancer Res 2003; 63:4062–4066.
139. Kim JG, Baggio LL, Bridon DP, et al. Development and characterization of a glucagon-like peptide 1-albumin conjugate: the ability to activate the glucagon-like peptide 1 receptor in vivo. Diabetes 2003; 52(3):751–759.
140. Jette L, Leger R, Thibaudeau K, et al. Human growth hormone-releasing factor (hGRF)1–29-albumin bioconjugates activate the GRF receptor on the anterior pituitary in rats: identification of CJC-1295 as a long-lasting GRF analog. Endocrinology 2005; 146(7):3052–3058.

141. Thibaudeau K, Leger R, Huang X, et al. Synthesis and evaluation of insulin-human serum albumin conjugates. Bioconjugate Chem 2005; 16:1000–1008.
142. Curry S, Mandelkow H, Brick P, et al. Crystal structure of human serum albumin complexed with fatty acid reveals an asymmetric distribution of binding sites. Nat Struct Biol 1998; 5(9):827–835.
143. Kurtzhals P, Havelund S, Jonassen I, et al. Albumin binding insulins acylated with fatty acids: characterization of the ligand-protein interaction and correlation between binding affinity and timing of the insulin effect in vivo. Biochem J 1995; 312:725–731.
144. Markussen J, Havelund S, Kurtzhals P, et al. Soluble, fatty acid acylated insulins bind albumin and show protracted action in pigs. Diabetologia 1996; 39:281–288.
145. Danne T, Lupke K, Walte K, et al. Insulin detemir is characterized by a consistent pharmacokinetic profile across age-groups in children, adolescents, and adults with type 1 diabetes. Diabetes 2003; 26(11):3087–3092.
146. Koehler MFT, Zobel K, Beresini MH, et al. Albumin affinity tags increase peptide half-life in vivo. Bioorg Med Chem Lett 2002; 12(20):2883–2886.
147. Manoharan M, Inamati GB, Lesnik EA, et al. Improving antisense oligonucleotide binding to human serum albumin: dramatic effect of ibuprofen conjugation. Chembiochem 2002; 12:1257–1260.
148. Johansson MU, Frick I-M, Nilsson H, et al. Structure, specificity, and mode of interaction for bacterial albumin-binding modules. J Biol Chem 2002; 277(10): 8114–8120.
149. Nygren P-A, Uhlen M. In vivo stabilization of a human recombinant CD4 derivative by fusion to a serum-albumin-binding receptor. Vaccine 1991; 91:363–368.
150. Makrides SC, Nygren P-A, Andrews B, et al. Extended in vivo half-life of human soluble complement receptor type I fused to a serum albumin-binding receptor. J Pharmacol Exp Ther 1996; 277(1):534–542.
151. Stork R, Muller D, Kontermann RE. A novel tri-functional antibody fusion protein with improved pharmacokinetic properties generated by fusing a bispecific single-chain diabody with an albumin-binding domain from streptococcal protein G. Protein Eng Des Sel 2007; 20(11):569–576.
152. Tolmachev V, Orlova A, Pehrson R, et al. Radionuclide therapy of HER2-positive microxenografts using a 177Lu-labeled HER2-specific affibody molecule. Cancer Res 2007; 67(6):2773–2782.
153. Jonsson A, Dogan J, Herne N, et al. Engineering of a femtomolar affinity binding protein to human serum albumin. Protein Eng Des Sel 2008; 21(8):515–527.
154. Nguyen A, Reyes AE II, Zhang M, et al. The pharmacokinetics of an albumin binding Fab (AB.Fab) can be modulated as a function of affinity for albumin. Protein Eng Des Sel 2006; 19(7):291–297.
155. Lejon S, Frick I-M, Bjorck L, et al. Crystal structure and biological implications of a bacterial albumin binding module in complex with human serum albumin. J Biol Chem 2004; 279(41):42924–42928.
156. Sjolander A, Nygren P-A, Stahl S, et al. The serum albumin-binding region of streptococcal protein G: a bacterial fusion partner with carrier-related properties. J Immunol Methods 1997; 201:115–123.
157. Libona C, Corvaïaa N, Haeuwa J-F, et al. The serum albumin-binding region of streptococcal protein G (BB) potentiates the immunogenicity of the G130–230 RSV-A protein. Vaccine 1999; 17(5):406–414.
158. Goetsch L, Haeuw JF, Champion T, et al. Identification of B- and T-cell epitopes of BB, a carrier protein derived from the G protein of Streptococcus Strain G148. Clin Diagn Lab Immunol 2003; 10(1):125–132.
159. Dennis MS, Zhang M, Meng YG, et al. Albumin binding as a general stategy for improving the pharmacokinetics of proteins. J Biol Chem 2002; 277(38):35035–35043.
160. Hollinger P, Wing M, Pound JD, et al. Retargeting serum immunoglobulins with bispecific diabodies. Nat Biotec 1997; 15:632–636.
161. Holt LJ, Basran A, Jones K, et al. Anti-serum albumin domain antibodies for extending the half-lives of short lived drugs. Protein Eng Des Sel 2008; 21(5):283–288.

162. Damascelli B, Cantu G, Mattavelli F, et al. Intraarterial chemotherapy with polyoxyethylated castor oil free paclitaxel, incorporated in albumin nanoparticles (ABI-007). Cancer 2001; 92(10):2592–2602.
163. Donohue JH, Rosenberg SA. The fate of interleukin-2 after in vivo administration. J Immunol 1983; 130(5):2203–2208.
164. Koumenis IL, Shahrokh Z, Leong S, et al. Modulating pharmacokinetics of an anti-interleukin-8 F(ab')(2) by amine-specific PEGylation with preserved bioactivity. Int J Pharm 2000; 198(1):83–95.
165. Lee LS, Conover C, Shi C, et al. Prolonged circulating lives of single-chain Fv proteins conjugated with polyethylene glycol: a comparison of conjugation chemistries and compounds. Bioconjug Chem 1999; 10(6):973–981.
166. Knauf MJ, Bell DP, Hirtzer P, et al. Relationship of effective molecular size to systemic clearance in rats of recombinant interleukin-2 chemically modified with water-soluble polymers. J Biol Chem 1988; 263(29):15064–15070.
167. Bendele A, Seely J, Richey C, et al. Short communication: renal tubular vacuolation in animals treated with polyethylene-glycol-conjugated proteins. Toxicol Sci 1998; 42(2):152–157.
168. Caliceti P, Veronese FM, Jonak Z. Immunogenic and tolerogenic properties of mono-methoxypoly(ethylene glycol) conjugated proteins. Farmaco 1999; 54(7):430–437.
169. Monfardini C, Schiavon O, Caliceti P, et al. A branched monomethoxypoly(ethylene glycol) for protein modification. Bioconjug Chem 1995; 6(1):62–69.
170. Kinstler OB, Brems DN, Lauren SL, et al. Characterization and stability of N-terminally PEGylated rhG-CSF. Pharm Res 1996; 13(7):996–1002.
171. Gaertner HF, Offord RE. Site-specific attachment of functionalized poly(ethylene glycol) to the amino terminus of proteins. Bioconjug Chem 1996; 7(1):38–44.
172. Goodson RJ, Katre NV. Site-directed pegylation of recombinant interleukin-2 at its glycosylation site. Biotechnology 1990; 8(4):343–346.
173. Richter AW, Akerblom E. Antibodies against polyethylene glycol produced in animals by immunization with monomethoxy polyethylene glycol modified proteins. Int Arch Allergy Appl Immunol 1983; 70(2):124–131.
174. Ettinger LJ, Kurtzberg J, Voute PA, et al. An open-label, multicenter study of polyethylene glycol-L-asparaginase for the treatment of acute lymphoblastic leukemia. Cancer 1995; 75(5):1176–1181.
175. Zeuzem S, Feinman SV, Rasenack J, et al. Peginterferon alfa-2a in patients with chronic hepatitis C. N Engl J Med 2000; 343(23):1666–1672.
176. Manns MP, McHutchison JG, Gordon SC, et al. Peginterferon alfa-2b plus ribavirin compared with interferon alfa-2b plus ribavirin for initial treatment of chronic hepatitis C: a randomised trial. Lancet 2001; 358(9286):958–965.
177. Bailon P, Palleroni A, Schaffer CA, et al. Rational design of a potent, long-lasting form of interferon: a 40 kDa branched polyethylene glycol-conjugated interferon alpha-2a for the treatment of hepatitis C. Bioconjug Chem 2001; 12(2):195–202.
178. Molineux G. Pegylation: engineering improved pharmaceuticals for enhanced therapy. Cancer Treat Rev 2002; 28(suppl A):13–16.
179. Holmes FA, O'Shaughnessy JA, Vukelja S, et al. Blinded, randomized, multicenter study to evaluate single administration pegfilgrastim once per cycle versus daily filgrastim as an adjunct to chemotherapy in patients with high-risk stage II or stage III/IV breast cancer. J Clin Oncol 2002; 20(3):727–731.
180. Chapman AP, Antoniw P, Spitali M, et al. Therapeutic antibody fragments with prolonged in vivo half-lives. Nat Biotechnol 1999; 17(8):780–783.
181. Junutula JR, Bhakta S, Raab H, et al. Rapid identification of reactive cysteine residues for site-specific labeling of antibody-Fabs. J Immunol Methods 2008; 332(1–2):41–52.
182. Filpula D, Zhao H. Releasable PEGylation of proteins with customized linkers. Adv Drug Deliv Rev 2008; 60(1):29–49.
183. Veronese FM, Schiavon O, Pasut G, et al. PEG-doxorubicin conjugates: influence of polymer structure on drug release, in vitro cytotoxicity, biodistribution, and anti-tumor activity. Bioconjug Chem 2005; 16(4):775–784.

184. Kemp SF, Fielder PJ, Attie KM, et al. Pharmacokinetic and pharmacodynamic characteristics of a long-acting growth hormone (GH) preparation (nutropin depot) in GH-deficient children. J Clin Endocrinol Metab 2004; 89(7):3234–3240.
185. Sinclair AM, Elliott S. Glycoengineering: the effect of glycosylation on the properties of therapeutic proteins. J Pharm Sci 2005; 94(8):1626–1635.
186. Gregoriadis G, Fernandes A, Mital M, et al. Polysialic acids: potential in improving the stability and pharmacokinetics of proteins and other therapeutics. Cell Mol Life Sci 2000; 57(13–14):1964–1969.
187. Alley SC, Benjamin DR, Jeffrey SC, et al. Contribution of linker stability to the activities of anticancer immunoconjugates. Bioconjugate Chem 2008; 19:759–765.
188. Vaishnaw AK, TenHoor CN, Pharmacokinetics, biologic activity, and tolerability of alefacept by intravenous and intramuscular administration. J Pharmacokinet Pharmacodyn 2002; 29(5–6):415–426.
189. Dianello CA, Setting the cytokine trap for autoimmunity. Nat Med 2003; 9:20–22.
190. Balan V, Nelson DR, Sulkowski MS, et al. A Phase I/II study evaluating escalating doses of recombinant human albumin-interferon-alpha fusion protein in chronic hepatitis C patients who have failed previous interferon-alpha-based therapy. Antiviral Ther 2006; 11(1):35–45.
191. Yao Z, Dai W, Perry J, et al. Effect of albumin fusion on the biodistribution of interleukin-2. Cancer Immunol Immunother 2004; 53(5):404–410.

7 Ophthalmic Delivery of Protein and Peptide Therapeutics; Pharmacokinetics, Pharmacodynamics, and Metabolism Considerations

Arto Urtti

Centre for Drug Research, University of Helsinki, Helsinki, Finland

INTRODUCTION

Clinical use of peptide and protein-based drugs is currently increasing in ophthalmology. This is due to the launching of the monoclonal antibodies for the treatment of age-related macular degeneration. These drugs have superior clinical efficacy to the prior available therapies, with improved outcomes in the treatment of age-related macular degeneration (1). Antibodies directed against vascular endothelial growth factor (VEGF) alleviate problems associated with neovascularization of the posterior segment of the eye. These antibodies are injected directly into the vitreous humor of the patients. Because of the invasive nature of this mode of drug delivery, improved methods of ocular protein and peptide delivery are being investigated.

Most ophthalmic drugs are administered as eyedrops, but this approach is not suitable for peptide and protein drug delivery. This is due to the barrier constraints set by the corneal epithelium (2). This barrier severely limits the ocular absorption of large, hydrophilic drugs, for example, peptides and proteins. Likewise, the blood-eye barriers restrict the distribution of drugs from blood circulation into ocular targets (2).

Understanding ocular pharmacokinetics is essential in the design of improved delivery of ophthalmic peptide and protein drugs. Comprehensive review articles have been published on several aspects of ocular drug delivery: general ocular pharmacokinetics (2), systemic drug absorption (3), periocular drug delivery (4), and ocular dosage forms (5,6).

This chapter is limited to discussing the delivery of peptides and proteins into the eye and describes how ocular barriers impact the ocular pharmacokinetics and pharmacodynamics of protein- and peptide-based compounds. The current state and future prospects of ophthalmic drug delivery are explored.

OPHTHALMIC PHARMACOKINETIC BARRIERS: GENERAL FEATURES

The eye is protected from xenobiotics by barriers associated with the anterior chamber (e.g., tear flow, corneal and conjunctival epithelia), the blood–aqueous humor barrier, and the blood-retina barrier (inner and outer parts). Each barrier includes both physical and metabolic components. Anatomical barriers of the eye are summarized in Figure 1. While eyedrops or other topical dosage forms can be used to treat the anterior segment (cornea, conjunctiva, iris, ciliary body, trabecular meshwork, lens), the posterior segment treatment requires other means of drug administration.

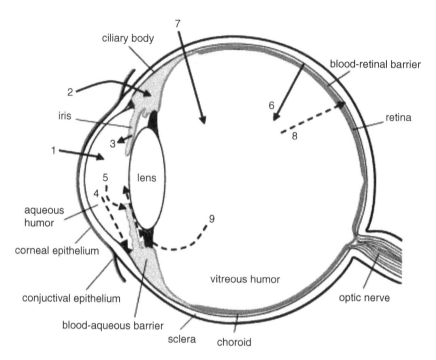

FIGURE 1 Ocular barriers and routes of drug administration. (1) Transcorneal drug permeation to the aqueous humor after topical administration. (2) Drug delivery across the conjunctiva and sclera to the anterior uvea. (3) Drug enters the anterior chamber from the systemic circulation through blood-aqueous barrier. (4) Drug elimination from the aqueous humor via aqueous turnover. (5) Drug elimination from the aqueous humor via venous circulation in the anterior uvea. (6) Drug enters the vitreal cavity from the systemic circulation via blood-retina barrier. (7) Drug delivery to the posterior segment from subconjunctival or periocular depot through sclera and chorioidea. (8) Drug elimination from the vitreous through the blood-retina barrier to the circulation of the chorioidea. (9) Drug elimination from the vitreous via anterior route (into the anterior chamber).

Most ocular drugs are administered as topical eyedrops. The extent of absorption by this route of administration is usually less than 2% of the dose (1). This is due to the following reasons. First, the applied eyedrop is drained from the ocular surface within a few minutes. Second, drugs absorbed via the conjunctival surface can enter into the blood stream. This is particularly relevant for the small-molecular-weight drugs with high conjunctival permeability. Third, the tight barrier of the corneal epithelium prevents ocular drug absorption of large, hydrophilic molecules such as therapeutic proteins and peptides.

The corneal epithelium is a stratified squamous epithelium with intercellular tight junctions only in the most apical cell layer. After its permeation across the epithelium, the drug diffuses across the loose gel-like structure of the stroma and endothelial monolayer. These layers are relevant barriers only for highly lipophilic small molecules that partition readily to the epithelium but only slowly partition further to the stroma. Otherwise, the corneal epithelium is the important barrier in the corneal drug absorption (7). Drugs can be eliminated from the anterior chamber via aqueous humor turnover and venous blood flow from the anterior uvea (1,8). Clearance from the aqueous humor is equal to the aqueous fluid turnover rate (about 3 μL/min). If the drug permeates into the venous circulation of the anterior uvea, the clearance can increase up to 20 to 30 μL/min (8). The ocular volume of distribution for topically applied drugs is difficult to determine because of the slow distribution into the posterior parts (lens, vitreous); the estimates vary from 0.3 to 2 mL (9,10).

Some portion of topically applied drugs are delivered to the eye by conjunctival permeation and subsequent transscleral diffusion (11). This route of permeation is a relatively minor contributor to the overall ocular absorption of small lipophilic compounds, but the relative importance of this route is increased at higher molecular weights and hydrophilicity (e.g., inulin). This route does not allow access of the drug to the aqueous humor, but rather the drug distributes to the iris and ciliary body (Fig. 1). This route of delivery can be improved by direct administration to the subconjunctival space where the drug is retained for longer periods and delivered across the sclera. The sclera is a hydrophilic tissue that acts as a sieve, allowing drug permeation through aqueous channels (12). In contrast to the conjunctiva and cornea, lipid solubility of a drug does not influence its permeation in the sclera. Thus, there is an increasing interest in this route of administration as a possible way of administering drug to the posterior parts of the eye, for example, chorioidea, retina, and vitreous. This approach is highly relevant for the case of peptides and proteins because of the fairly permeable nature of the sclera even for the larger molecules and will be discussed in more detail later.

The intravitreal route is used increasingly to deliver drugs into the posterior segment of the eye. Direct injection or delivery of implants guarantees access to the retina and vitreous, and chorioidea (13). Drug absorption is skipped in the case of intravitreal injection, and only distribution and elimination phases are present. Drug distribution in the vitreous is not generally a problem because of the lack of cellular barriers. The rate of elimination depends on the permeation across the blood-retina barrier [retinal pigment epithelium (RPE), retinal capillaries]. All drugs will have some elimination via the anterior route, but only some drugs are able to escape from the vitreal cavity through the blood-retina barrier (1).

TOPICAL PROTEIN ADMINISTRATION TO THE EYE

There are several reasons why proteins are not administered as eyedrops: administered liquid is rapidly cleared from the ocular surface, resulting in a contact time that is limited to a few minutes or even less. Still, ocular drop instillation is widely used in the treatment of glaucoma and other diseases with low-molecular-weight drugs (e.g., prostaglandins, β-blockers, α-2 agonists). These small, hydrophobic drugs are able to partition into the cornea epithelium at adequate rates to allow pharmacologically relevant levels of ocular drug absorption. However, larger molecules like proteins and larger peptides are (>10 kDa) are typically quite hydrophilic and do not readily partition into the corneal epithelium (14). Paracellular permeation of therapeutic proteins and peptides across the corneal epithelium is very slow because of the tight junctions. Hämäläinen et al. estimated the paracellular space of the corneal epithelium with polyethylene glycol permeation studies and subsequent effusion-based analysis of the pores (15). They concluded that the limiting paracellular pore size is about 2 nm in diameter, with a surface porosity of only 10^{-7}. Thus, these narrow paracellular pores and the lipoidal cell membranes can dramatically block the permeation of proteins with diameters greater than 2 nm.

A metabolic barrier at the surface further limits the possibilities for significant protein drug absorption from tear fluid into the anterior segment of the eye. There are both endopeptidases and aminopeptidases in the corneal epithelium that limit the absorption of peptides and proteins (16). Overall, it is clear that transcorneal delivery of proteins does not work and other approaches are needed.

More than 20 years ago, Ahmed and Patton demonstrated that large molecules can permeate from the tear fluid into the eye via conjunctiva and sclera rather than through the cornea (11). The conjunctiva is more permeable than the cornea, particularly for hydrophilic and large molecules. Also, the sclera allows permeation of large molecules, and permeability across this membrane is not dependent on lipophilicity of the drug. Why is the cornea then considered to be the main absorption route for lipophilic drugs? Permeation of lipophilic drugs is similar in these two routes, and some drug is lost to the blood stream via the circulation in the conjunctiva. For large, protein-sized compounds, the relative permeability of the cornea is much less than that in the conjunctiva and sclera (16–18). Therefore, the noncorneal route of permeation may be advantageous for proteins (18). Accordingly, also the metabolic barrier (peptidases and proteases) is less prevalent in the conjunctiva than in the cornea (16).

Still today, the potential of the noncorneal route of protein absorption from the ocular surface to the inner eye has not been fully explored. Utilization of this route would require improved formulations that prolong the retention and improve permeation into the conjunctiva. Currently, cyclosporin is the only peptide or protein drug that is administered by topical ocular route. This hydrophobic peptide is given as an eyedrop emulsion to treat the condition known as dry eyes (19).

INTRAVITREAL PROTEIN ADMINISTRATION

Intravitreal protein injection is used increasingly because of the recent approval of monoclonal antibodies for the treatment of choroidal neovascularization associated with age-related macular degeneration. The vitreous is a viscous liquid that contains mostly hyaluronan but also some collagen and glycosaminoglycans. Basically, the vitreous humor is a hydrogel with negative net charge. Because of eye movements, drug distribution in the vitreous occurs through both diffusion and convection. Drug diffusivity depends on charge and molecular weight. Mass transfer in the vitreous is driven primarily by diffusion, with convection being a minor contributor (20). Proteins are eliminated from the vitreous via either anterior or posterior route. The monoclonal antibodies, bevacizumab and ranibizumab, are eliminated from the vitreous at monoexponentially declining rate. Protein half-lives in the vitreous humor are 2.9 to 4.3 days, but even after one month, there are residual levels of an injected antibody that are pharmacologically significant (21).

Let us consider the posterior barriers of vitreal drug elimination: the retina, retinal capillaries, and RPE. Neural retina (particularly the limiting membrane) has some barrier properties that are relevant to the macromolecules (22). Although permeability coefficients of proteins in this membrane have not been estimated, retinal capillaries and RPE are considered to be the main factors that determine the elimination rate from the vitreous. Because of the limited information available, it is very difficult to estimate the clearance via retinal capillaries. Clearance via the RPE, however, is easier to estimate. Blood flow in the chorioidea results in clearance values of 30 to 300 μL/min, but the low permeability of RPE limits the clearance of large hydrophilic molecules. Based on the permeability in RPE the protein clearance would be about 0.02 μl/min, less than the turnover rate of aqueous humor (23). If we assume that the volume distribution in the vitreous is 2 mL, the vitreous half-life resulting from such RPE permeability would be around one month. This is slower than the three- to four-day half-lives measured for antibodies. Thus, it seems that the antibody elimination from the vitreous takes place also via other routes (uptake by RPE, anterior chamber elimination).

The same barriers also influence the tissue distribution and access of protein and peptides to the target site. Figure 2 illustrates this point. If the target site is in the *neural retina* (e.g., neurotrophic growth factor receptors), poor permeation across the blood-retina barrier is advantageous because of longer maintenance of therapeutic protein concentrations in the retina. Also, a therapeutic protein or peptide may be targeted to the RPE or other cells distal to the inner limiting membrane. In that case, the limiting membrane must be passed before the drug can be internalized to the RPE. Similarly, the target cells for angiogenesis inhibitors can be found in the chorioidea. For this purpose, the drug should permeate across the neural retina and RPE; high permeation rates across both layers should improve the access of the drug to the chorioidea.

Because of the invasive nature of the intravitreous drug administration via injection, it would be beneficial to increase the duration of drug action in the posterior segment. Different formulation approaches can be used for this

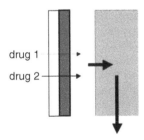

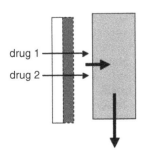

FIGURE 2 Schematic presentation of drug elimination from the vitreous. Yellow part is the neural retina, the gray part is RPE, and the pink one is chorioidea. After drug administration in the vitreous, it can get across the RPE poorly (drug 1) or at faster rate (drug 2) (*upper picture*). Poor permeation prolongs the retention in the vitreous. In disease state, the RPE barrier may not be intact (*lower picture*), and this results in faster drug elimination. The access of drug to blood circulation, even for proteins, is not limited by the vessel walls in the chorioidea. *Abbreviation*: RPE, retinal pigment epithelium.

purpose. First, controlled-release implants or microspheres can release the drug continuously into the vitreous (24). Second, elimination of a delivered protein or peptide therapeutic from a delivery system is slowed down, because particulates do not permeate across the blood-retina barrier. Third, nanoparticles (or small microparticles) can be phagocytosed by the RPE cells, and this has resulted in particle retention of four months in target cells (25). Unfortunately, these approaches have not been widely studied for the intraocular administration of protein drugs.

The effect of a disease state on posterior segment pharmacokinetics is also poorly understood. It is known that blood-retina barrier is not intact in some diseases (e.g., age-related macular degeneration, infections) (26). In the case of protein drugs, this would increase the permeation of the drug from vitreous to chorioidea and increase the drug elimination from vitreous and retina via posterior route. The net impact on drug delivery would again be dependent on the site; overall exposure time should be decreased in the retina, but access to the choroidal target cells is improved.

TRANSSCLERAL PROTEIN DELIVERY

Because of the invasiveness of intravitreous injections, subconjunctival administration of proteins has gained increasing attention. Subconjunctival injections can be given in outpatient clinics, but intravitreal injection requires operation room setup, because there is the risk of vision-threatening endophthalmitis that must be minimized. After subconjunctival administration, the drug must diffuse through sclera, chorioidea, and RPE to reach a retinal target. Some fraction of the drug will likely be eliminated from the subconjunctival space to the blood circulation before entering the sclera (1).

The sclera is a membrane about 0.5 mm thick that is composed of interspersed collagen fibrils and water channels. Drug permeability across the sclera is not dependent on the lipophilicity (27) but is dependent on molecular weight. Macromolecules have reasonable permeability in the sclera (range: $0.2–1.0 \times 10^{-7}$ cm/sec) (23). Permeation across the sclera has been widely investigated, but, in fact, the sclera is not the rate-limiting barrier in many cases. Recent investigations show that the sclera and RPE are approximately equal barriers for lipophilic small compounds, but for macromolecules and hydrophilic solutes, the RPE is tighter than the sclera (23). Although the difference appears to be approximately an order of magnitude, there is surprisingly sparse data in the literature on the permeability of RPE. In addition to the sclera and RPE, the drug should pass the chorioidea. This layer has rich blood flow, with vessels fenestrated to allow rapid movement between the blood and choroidal tissue. It has been known for a long time that the blood vessel capillaries allow take-up of plasma proteins from tissues that have escaped from the vasculature. Therefore, it is expected that protein and peptides delivered to the choroidal tissue can transfer to blood vessels, although there is very limited data about this process. Bill determined the systemic absorption of carboxyfluorescein, albumin, and IgG from the chorioidea (28). All compounds were absorbed, but the higher-molecular-weight molecules showed decreased rates of absorption. The influence of choroidal clearance on drug permeation to the retina is not known. This is determined by the blood flow and the extraction ratio. Even at low extraction ratio values, the rate of elimination from the choroid to the blood flow is very rapid. Further studies will clarify the roles of each layer.

Subconjunctival and sub-Tenon administration of pigment epithelium–derived factor (PEDF) was investigated recently. The protein was incorporated into the polymer matrix of poly (lactide-glycolide) (PLGA) that released the drug in controlled manner, with adequate delivery into the retina being achieved over a prolonged period (29). Other investigators have also demonstrated significant drug distribution from the subconjunctival site to the retina, although the clinical utility of this method of administration remains to be seen. In addition, iontophoresis has been used to deliver drugs via transscleral route into the retina (30). The iontophoresis is used to facilitate the drug permeation into the sclera and further to the deeper tissues. This approach has been used mostly for small molecules, but recent report shows that even nanoparticles can be delivered by iontophoresis across the sclera (31). Therefore, it may be possible to use this method also for protein delivery.

PROTEIN DELIVERY BY GENE TRANSFER

The goal of gene delivery typically involves the transfer of specific genetic material into the target cell populations to trigger a desired outcome. Most ocular gene therapy research has been directed against retinal diseases. Gene transfer can be accomplished with viral or nonviral vectors.

Early gene transfer research in the eye concentrated mostly on the adenovirus systems. Adenovirus vectors caused some inflammatory side effects, and therefore, the main emphasis in the field has shifted to the adeno-associated virus (AAV). Very recently, there have been breakthroughs in this field. Some retinitis patients who suffered from an inherited deficiency of RPE-65 protein were treated with AAV with RPE-65 gene (32). Viral particles were administered to the subretinal site under microscopic observation, and future studies will evaluate the clinical utility of this approach. Viral gene therapy has been reviewed elsewhere, and it is not directly in the scope of this article.

Nonviral gene therapy of the eye has the same basic difficulty as elsewhere: delivery of genes into the nuclei of the ocular target cells is not sufficiently efficient. Gene delivery can be accomplished by using the same routes of administration as described earlier in this text. Toropainen et al. (2007) showed recently that the tight junction barrier of the corneal epithelial surface was overcome by transfecting the surface cells with lipid-DNA complexes (33), with the cells subsequently secreting the marker protein [secreted alkaline phosphatase (SEAP)] into the anterior chamber. This outcome is possible because of the leakiness of the deeper layers of the cornea that allow diffusion of a secreted protein. The proteins, as such, do not permeate across the corneal epithelium. Therefore, protein eyedrops would not be effective in the treatment of intraocular targets.

Since some of the SEAP protein was secreted to the tear fluid (33), it is possible to use the corneal epithelial surface cells as a platform to secrete therapeutic proteins over prolonged times either to the tear fluid or to the anterior chamber. This approach could potentially be used for prolonged delivery of protein therapeutics to the tear fluid (e.g., to treat the dry eye syndrome). The contact period of instilled protein solution would be very short, requiring frequent instillation of expensive protein drug. It remains to be seen if the "gene eyedrops" will be an effective way in the delivery of therapeutic proteins.

PROTEIN DELIVERY FROM ENCAPSULATED CELLS
Cell encapsulation is another indirect method of protein delivery. In this case, the cells are transduced to produce therapeutic protein permanently. Thereafter, the cells are encapsulated with a polymer that allows influx of nutrients and outflow of therapeutic protein but prevents the access of defense mechanisms to violate the encapsulated cells.

Cell encapsulation has been investigated recently for the treatment of retinal degeneration (34,35). The cells were encapsulated with a growth-enabling matrix into a cylinder of semipermeable membranes. The human immortal retinal pigment epithelial cell line (ARPE-19) was stably transfected to secrete ciliary neurotrophic factor (CNTF). The device is placed to the vitreous of the eyes, and thereafter the system secretes CNTF over prolonged periods of more than six months. In principle, this system could provide life-long treatment. Ongoing clinical trials have provided promising results, and the device is being developed for the treatment of dry age-related macular degeneration and retinitis pigmentosa. The pharmacokinetics of CNTF after its release from the cell capsule obeys the principles described earlier. In this case, steady-state concentration will be reached and maintained.

RPE has advantages as the protein-secreting cell line, because these cells can maintain their viability for prolonged times. An RPE cell line has been successfully encapsulated and maintained also into alginate-based microcapsules for cell therapy (36,37).

CONCLUSIONS
Protein delivery from the ocular surface to the retina and vitreous is the most appealing delivery approach. This is not effective yet, but this would be the optimal method of protein delivery in the clinical practice. The ease of topical drug administration would replace the intravitreal injections. Currently, intravitreal injections have gained wide clinical acceptance, and it is likely that only other more convenient or more long-acting modes of protein drug delivery can improve the current situation. Systemic delivery is not practical unless prolonged action and sufficient drug targeting to the posterior eye can be provided.

Several possibilities exist for protein delivery to the ocular targets. New delivery methods are needed, and they should be integrated to the biological constraints that are set by the pathophysiology of the eye.

REFERENCES
1. Jager RD, Mieler WF, Miller JW. Age-related macular degeneration. N Engl J Med 2008; 358:2606–2617.
2. Maurice DM, Mishima S. Ocular pharmacokinetics. In: Sears ML, ed. Handbook of Experimental Pharmacology. Vol 69. Berlin: Springer Verlag, 1984:16–119.
3. Urtti A, Salminen L. Minimizing systemic absorption of topically administered ophthalmic drugs. Surv Ophthalmol 1993; 37:435–456.
4. Ranta VP, Urtti A. Transscleral drug delivery to the posterior eye: prospects of pharmacokinetic modeling. Adv Drug Deliv Rev 2006; 58:1164–1181.
5. Del Amo E, Urtti A. Current and future ophthalmic drug delivery systems. A shift to the posterior segment. Drug Discov Today 2008; 13:135–143.

6. Järvinen K, Järvinen T, Urtti A. Ocular absorption following topical delivery. Adv Drug Deliv Rev 1995; 16:3–19.
7. Huang HS, Schoenwald RD, Lach JL. Corneal penetration behavior of beta blockers. J Pharm Sci 1983; 72:1272–1279.
8. Urtti A. Challenges and obstacles of ocular pharmacokinetics and drug delivery. Adv Drug Deliv Rev 2006; 58:1131–1135.
9. Conrad JM, Robinson JR. Aqueous chamber drug distribution volume measurement in rabbits. J Pharm Sci 1977; 66:219–224.
10. Miller SC, Gokhale RD, Patton TF, et al. Pilocarpine ocular distribution volume. J Pharm Sci 1980; 69:615–616.
11. Ahmed I, Patton TF. Importance of the noncorneal absorption route in topical ophthalmic drug delivery. Invest Ophthalmol Vis Sci 1985; 26:584–587.
12. Lee SB, Geroski DH, Prausnitz MR, et al. Drug delivery through the sclera: effects of thickness, hydration, and sustained release systems. Exp Eye Res 2004; 78:599–607.
13. Charles NC, Steiner GC. Ganciclovir intraocular implant. A clinicopathologic study. Ophthalmology 1996; 103:416–421.
14. Lee VH, Carson LW, Kashi SD, et al. Metabolic and permeation barriers to the ocular absorption of topically applied enkephalins in albino rabbits. J Ocul Pharmacol 1986; 2:345–352.
15. Hämäläinen KM, Kananen K, Auriola S, et al. Characterization of paracellular and aqueous penetration routes in cornea, conjunctiva, and sclera. Invest Ophthalmol Vis Sci 1997; 38:627–634.
16. Hämäläinen KM, Ranta VP, Auriola S, et al. Enzymatic and permeation barrier of [D-Ala(2)]-Met-enkephalinamide in the anterior membranes of the albino rabbit eye. Eur J Pharm Sci 2000; 9(3):265–270.
17. Olsen TW, Edelhauser HF, Lim JI, Geroski DH. Human scleral permeability. Effects of age, cryotherapy, transscleral diode laser, and surgical thinning. Invest Ophthalmol Vis Sci 1995; 36:1893–1903.
18. Ambati J, Canakis CS, Miller JW, et al. Diffusion of high molecular weight compounds through sclera. Invest Ophthalmol Vis Sci 2000; 41:1181–1185.
19. Chiang TH, Walt JG, McMahon JP Jr., et al. Real-world utilization patterns of cyclosporine ophthalmic emulsion 0.05% within managed care. Can J Clin Pharmacol 2007; 14:e240–e245.
20. Kim H, Lizak MJ, Tansey G, et al. Study of ocular transport of drugs released from an intravitreal implant using magnetic resonance imaging. Ann Biomed Eng 2005; 33:150–164.
21. Bakri SJ, Snyder MR, Reid JM, et al. Pharmacokinetics of intravitreal ranibizumab (Lucentis). Ophthalmology 2007; 112:2179–2182.
22. Pitkänen L, Pelkonen J, Ruponen M, et al. Neural retina limits the non-viral gene transfer to RPE in an in vitro bovine eye model. AAPS J 2004; 6(3):e25.
23. Pitkänen L, Ranta VP, Moilanen H, et al. Permeability of retinal pigment epithelium: effects of permeant molecular weight and lipophilicity. Invest Ophthalmol Vis Sci 2005; 46:641–646.
24. Bourges JL, Bloquel C, Thomas A, et al. Intraocular implants for extended drug delivery: therapeutic applications, Adv Drug Deliv Rev 2006; 58:1182–1202.
25. Bourges JL, Gautier SE, Delie F, et al. Ocular drug delivery targeting the retina and retinal pigment epithelium using polylactide nanoparticles. Invest Ophthalmol Vis Sci 2003; 44:3562–3569.
26. Peng S, Rahner C, Rizzolo LJ. Apical and basal regulation of the permeability of the retinal pigment epithelium. Invest Ophthalmol Vis Sci 2003; 44:808–817.
27. Prausnitz MR, Noonan JS. Permeability of cornea, sclera, and conjunctiva: a literature analysis for drug delivery to the eye. J Pharm Sci 1998; 87:1479–1488.
28. Bill A. Capillary permeability to and extravascular dynamics of myoglobin, albumin and gammaglobulin in the uvea. Acta Physiol Scand 1968; 73:204–219.
29. Li H, Tran VV, Hu Y, et al. A PEDF N-terminal peptide protects the retina from ischemic injury when delivered in PLGA nanospheres. Exp Eye Res 2006; 83:824–833.

30. Halhal M, Renard G, Courtois Y, et al. Iontophoresis: from the lab to the bed side. Exp Eye Res 2004; 78:751–757.
31. Eljarrat-Binstock E, Orucov F, Aldouby Y, et al. Charged nanoparticles delivery to the eye using hydrogel iontophoresis. J Control Release 2008; 126:156–161.
32. Bainbridge JW, Smith AJ, Barker SS, et al. Effect of gene therapy on visual function in Leber's congenital amaurosis. N Engl J Med 2008; 358:2231–2239.
33. Toropainen E, Hornof M, Kaarniranta K, et al. Corneal epithelium as a platform for secretion of transgene products after transfection with liposomal gene eyedrops. J Gene Med 2007; 9(3):208–216.
34. Bush RA, Lei B, Tao W, et al. Encapsulated cell-based intraocular delivery of ciliary neurotrophic factor in normal rabbit: dose-dependent effects on ERG and retinal histology. Invest Ophthalmol Vis Sci 2004; 45:2420–2430.
35. Tao W. Application of encapsulated cell technology for retinal degenerative diseases. Expert Opin Biol Ther 2006; 6:717–726.
36. Wikström J, Syväjärvi H, Urtti A, et al. Kinetic simulation model of protein secretion and accumulation in the cell microcapsules. J Gene Med 2008; 10:575–582.
37. Wikström J, Elomaa M, Syväjärvi H, et al. Alginate based microencapsulation of retinal pigment epithelial cell line for cell Therapy. Biomaterials 2008; 29:869–876.

Pharmacokinetic and Pharmacodynamic Aspects of Intranasal Delivery

Henry R. Costantino, Anthony P. Sileno, Mike Templin,
Chingyuan Li, Diane Frank, Gordon C. Brandt, and Steven C. Quay
MDRNA, Inc., Bothell, Washington, U.S.A.

INTRODUCTION

The choice of administration route is critical for successful drug development. A judicious selection should take into consideration the properties of the drug as well as its intended use and the targeted patient population. For example, high-molecular-weight drugs such as peptides and proteins are typically developed as injection products since their relative fragility makes oral administration problematic. Even so, injection may not be preferred when considering other important factors such as patient comfort, convenience, and compliance. An alternative in this regard is noninvasive drug delivery via the intranasal (IN) route, the topic of the current chapter. Although the nasal mucosa provides a barrier to permeation of high-molecular-weight therapeutics such as peptides and proteins, the tight junctions within this tissue can be reversibly and safely opened; thus, IN delivery of peptides and proteins provides a feasible and advantageous option to injections. The rapid absorption achieved by the IN route may provide pharmacodynamic and safety advantages over the subcutaneous route of administration. Applications for IN delivery are presented, followed by discussion of considerations related to pharmacokinetic and pharmacodynamic performance, with examples demonstrating these principles.

CONSIDERATIONS FOR INTRANASAL DELIVERY

Intranasal administration can be employed either acutely or chronically and in the context of either local or systemic drug delivery. The sections below highlight such considerations for IN delivery, including discussion of nasal physiology, effect of nasal inflammation, and impact of dosing technique and dosage form on pharmacokinetics (PK) and pharmacodynamics (PD). Also relevant is the concept of tight junction (TJ) modulation (1). TJs are the cell-to-cell connections in epithelial and endothelial layers regulating transport across the natural boundaries that comprise nasal and other mucosal tissues. As part of normal physiological function, TJs selectively open and close in response to various stimuli, allowing the passage of large molecules or even cells across the TJ barrier. The physicochemical properties of the drug are also highly germane, for instance high-molecular-weight drugs typically exhibit low permeation across the nasal mucosa. In this case, judicious choice of TJ modulating excipients can be employed to improve PK and PD performance.

LOCAL DELIVERY

For the situation where the nasal cavity itself is the target organ for therapeutic effect, the IN route is a logical choice for treatment. Some prominent examples of drugs administered through the IN route for local treatment include antihistamines and corticosteroids for allergic rhinitis (2), as well as decongestants for nasal cold symptoms. A few examples of low-molecular-weight drugs that have extensive use in this area include the anticholinergic agent ipratropium bromide (3), and steroidal anti-inflammatory agents such as mometasone furoate (4), budesonide (5), beclomethasone (6), and triamcinolone (6).

IN antihistamines and corticosteroids have a minimal potential for systemic adverse effects when compared with oral therapy, primarily due to the effectiveness of relatively low doses that are topically administered (7). For example, the recommended therapeutic dose for IN antihistamines does not cause impairment of psychomotor function or significant sedation, whereas the larger dose that is needed for oral administration may trigger these effects. Such factors make IN delivery of corticosteroids and antihistamines an attractive and preferable route of administration, particularly if rapid, local relief of symptoms is required.

Besides low-molecular-weight drugs, there are also a number of peptide and protein drugs that may be suitable for local delivery on the nasal mucosa. These include peptides and proteins with anti-inflammatory, antimicrobial, or antigenic properties. For instance, Ooi et al. (8) have recently reviewed the role of the sinonasal epithelium in chronic rhinosinusitis, focusing on toll-like receptors, antimicrobial peptides (cathelicidins and defensins), and surfactant proteins, and the potential use of such innate immune peptides as treatments against fungi, biofilms, and superantigens. As another example, IN IL-12 promoted bacterial clearance and extended time to death against pulmonary challenge with *Francisella tularensis novicida*, with this benefit being further increased when IN gentamicin was coadministered (9). Further discussion of the important area of IN vaccine delivery is beyond the scope of the current chapter, and for more information, the reader is referenced to recent reviews (10–12).

SYSTEMIC DELIVERY

IN administration provides an alternative route for the systemic delivery of drugs that are more conventionally delivered orally or (for poorly orally absorbed compounds such as peptides and proteins) by injection. Attributes of IN systemic delivery include no first-pass metabolism, a relatively large surface area for drug absorption, noninvasiveness to maximize patient comfort and compliance, rapid drug onset, and potential for alternate PK and PD effects. Typically, IN delivery provides for rapid onset of drug levels in the blood, with a relatively short time for drug absorption due to mucocilliary clearance from the nasal cavity. Further, detailed discussion of systemic drug delivery via the IN route is provided in the case examples below, and additional information can be found in various literature reviews (13–19).

ACUTE VS. CHRONIC ADMINISTRATION

When determining which delivery route is optimal for a product, it is essential to consider the drug's possible dosing regimen, including whether the intended use is chronic or acute. For an acute indication, the benefit of patient comfort and compliance provided by IN dosing (as compared with injections) may not be a key factor. Thus, it is important to note that IN administration offers more than just comfort and compliance. One example is the case of an emergency room setting, in which avoiding the potential of accidental needle sticks is desired

(20). An example of a peptide drug that has been explored for IN delivery as an alternative to injections is glucagon (21–24). IN glucagon provides for a rapid drug onset, critical for use in treating insulin-induced hypoglycemia. Furthermore, IN dosing provides for facile administered by a family member and is safer and faster than oral glucose in an unconscious subject. Additional examples in this context include IN epinephrine (25), cardiovascular agents such as nitroglycerin (26), and the peptide desmopressin (27).

The dosing frequency of currently marketed IN products ranges from a relatively infrequent dosing, (e.g., weekly dosing for Nascobal® spray to treat vitamin B12 deficiencies), to a dosing regimen of multiple times daily, (e.g., two sprays per nostril two to three times daily for Atrovent® nasal spray, which is indicated for the symptomatic relief of rhinorrhea associated with allergic and nonallergic perennial rhinitis). IN dosing may be especially suited for chronic applications of a drug that is not orally bioavailable and would be given to a needle-naive patient population. An example of such a drug is insulin, which will be discussed later.

CENTRAL NERVOUS SYSTEM DELIVERY

Whether IN drugs are capable of targeting the central nervous system (CNS) is currently an area of significant interest, as reviewed elsewhere (28,29). Improved delivery to the brain via IN administration has been reported for therapeutic peptides and proteins (30–35), as well as some low-molecular-weight drugs (18,36–40). However, it should be noted that there are also cases for which no evidence was found for preferential delivery to the brain via IN dosing (41–44). Consequently, the capability of preferential brain delivery for IN dosing may be drug specific or may depend on the study methods employed. Indeed, van den Berg and Merkus (44) concluded from a comprehensive assessment of available literature in this area (~100 papers) that the majority of studies had not employed sound experimental design and of those that met the authors' criteria, only two studies in rats were able to provide results that can be seen as an indication for direct transport from the nose to the CNS.

As a related example, PK data for IN galantamine (45) demonstrate that the ratio of CSF to plasma drug levels in dogs reaches the same value after one hour irrespective of the dosing route (IN vs. oral). It was noted that there was a more rapid onset of drug levels for IN delivery both in the plasma and in the CSF for IN administration compared with oral dosing (Fig. 1).

In addition to the potential for "nose-to-brain" delivery, drugs can proceed from "nose to systemic circulation to brain." This pathway necessitates that the drug readily permeates the blood-brain barrier (BBB). For this to be accomplished, the drug (or prodrug) must demonstrate acceptable passive or active transport across the TJ barriers of the BBB. For instance, a transporter for insulin across the BBB has been described by Banks (46). Others have reported on delivering insulin to the brain via the IN route (47–50), however, the data are not definitive to support any preferential brain transport via the IN route.

NASAL PHYSIOLOGY

Various aspects of nasal physiology and mechanisms, such as nasal anatomy, resistance, airflow, and the nasal cycle may have a potential impact on IN delivery. Relevant reviews can be found elsewhere (51,52). Briefly, the nasal cavity is separated by the nasal septum (comprised of cartilage and bone), with each half opening at the face (via the nostrils). In addition, the nasopharynx provides a connection to the oral cavity. The olfactory region, the respiratory region, and the anterior and posterior vestibules are the three major areas of the nasal cavity. The lateral walls contain a folded structure, referred to as the conchae. This structure is further divided into the superior, median, and inferior turbinates, giving a total surface area, in humans, of approximately 150 cm^2.

Epithelial tissue in the nasal cavity is highly vascularized, providing a promising conduit for drug delivery. The cellular composition of the nasal

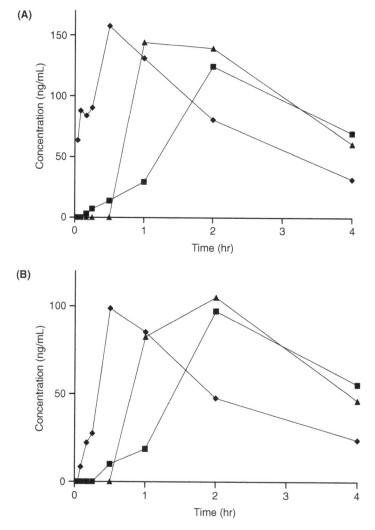

FIGURE 1 Galantamine pharmacokinetics in dogs given intranasal formulation (*filled diamonds*), oral solution (*filled triangles*), and oral tablet (*filled squares*). The data are shown for galantamine measured in (**A**) plasma; (**B**) the CSF; and (**C**) as the ratio of drug measured in CSF to plasma.

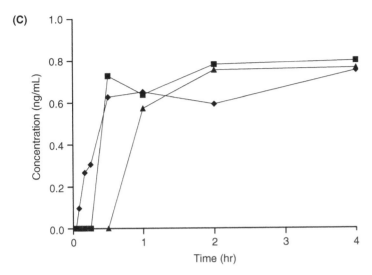

FIGURE 1 (*Continued*)

epithelial tissue consists primarily of basal cells, goblet cells, ciliated columnar cells, and nonciliated columnar cells, with varying proportions throughout different regions of the nasal cavity. Basal cells are poorly differentiated and operate as stem cells to replace other epithelial cells. Goblet cells contain an abundance of secretory granules filled with mucin and produce secretions that form the mucus layer. Ciliated cells aid in the transport of mucus to the nasopharynx.

EFFECT OF NASAL INFLAMMATION
It is commonly asked whether inflammation of the nasal mucosa (e.g., patients with rhinitis) can affect drug bioavailability. Several studies suggest that IN drug PK and/or PD are not affected by the presence of rhinitis. These studies include the examination of IN formulations of both peptide drugs [e.g., desmopressin (53) and buserelin (54)] and low-molecular-weight compounds [e.g., zolmitriptan (55), butorphanol (56), and dihydroergotamine (57)].

EFFECT OF DOSING TECHNIQUE, DOSAGE FORM, AND FORMULATION
There are a variety of dosage forms and dosing techniques that can be explored for IN delivery. In the context of animal studies, dosing by direct instillation into the nares (e.g., by micropipette) is most common, although application of a nasal spray is possible in larger animal species. A number of studies in animal models and in humans have compared different dosing techniques and dosing sites for both small-molecule and peptide drugs.

It has been reported by Liversidge et al. (58) that the dosage form (drop vs. spray) and administration site (dorsal or ventral nostril surface) profoundly affected distribution and clearance of a γ-emitting technitium 99m (^{99m}Tc)-labeled diethylenetriamine pentaacetic acid solution in dogs. The slowest nasal clearance was observed for dorsally administered drops, whereas administration of drops to the ventral surface or sprays to either dorsal or ventral surface results in rapid clearance and low deposition in the turbinates. In another investigation, distribution and nasal clearance of ^{99m}Tc-labeled albumin was determined in humans randomized in a crossover clinical study design to receive nasal spray versus drops (59). It was found that the 50% mean clearance time from the nasopharynx for nasal spray was about 50 minutes, with no drug detected in the lungs. The clearance of drops was found to be highly variable, making direct comparison with the nasal spray data difficult.

Some studies in humans report higher bioavailability achieved for nasal spray versus drops for both low-molecular-weight drugs, for example, fluticasone (60), and peptides, for example, insulin formulated in the presence of sodium glycocholate (61). However, this may not always be the case. For instance, in a study of IN nicotine (62), it was found that there was no significant difference in rate or extent of absorption following administration of drops to the nasal conchae or nasal septum compared with delivering via nasal spray.

There are numerous studies comparing the effectiveness of delivering various types of nasal formulations. Conclusions of such studies vary, likely an indication of the important role of not only dosage form but also formulation composition. For example, liquid and powders provided similar bioavailability of IN dihydroergotamine in rabbits (63). In contrast, IN elcatonin in the powder form provided significantly greater absorption compared with the liquid dosage form (64), whereas a human study of budesonide demonstrated higher bioavailability by spray pump compared with nasal powder (65). Quay et al. (66) have reported on the comparison of a gel versus liquid formulation of the low-molecular-weight drug cyanocobalamin and found bioequivalence between the two different dosage forms despite the markedly different formulations and areas within the nasal cavity where each is applied.

As can be expected, the formulation composition has a profound impact on the PK and PD of IN drugs. A full discussion of this area is beyond the scope of the present chapter. Besides the important topic of permeation enhancers (discussed below), mucoadhesives represent an excipient class worthy of mention [reviews can be found elsewhere (67,68)]. It has been reported that these agents can improve PK and PD performance of IN drugs, for instance, apomorphine in rabbits (69). As an illustration related to peptide drugs, studies in preclinical models have investigated the effect on bioavailability and/or extension of PK profile in the presence of mucoadhesive liposomes (70), particulate systems (71), inserts (72), and gels (73). The latter approach has also been explored in a human study (74). Further discussion of the important area of modulating the IN PK profile is given later in this chapter.

DRUG PHYSICOCHEMICAL PROPERTIES

The physicochemical properties of the drug can have a profound impact on the performance of the drug delivered by the IN route. For a comprehensive review in this area, the reader is referred to elsewhere (19). For the purposes of the current chapter, a paramount consideration for peptides and proteins is their relatively large molecular weight, which limits their transepithelial permeation. Even so, IN delivery of peptides and small proteins can be dramatically improved by employing permeation enhancers that are capable of opening TJs and promoting paracellular transport. It is critical that the permeation enhancers employed exhibit low nasal tissue cytotoxicity, maintain good cell viability, and provide for reversibility of TJs, as described elsewhere (19,75,76).

ROLE OF PERMEATION ENHANCERS

Generally, drug absorption decreases with increasing molecular weight and increases in the presence of permeation enhancers. For example, Schipper et al. (77) reported that for the small peptide adrenocorticotropic hormone 4 to 9 (MW = 906 Da), the IN bioavailability increased from $\sim 10\%$ to $\sim 17\%$ in rabbits and from $\sim 15\%$ to $\sim 65\%$ in rats in the presence of a permeation enhancer. Salmon calcitonin (sCT), a larger peptide (MW = 3432 Da), exhibits lower bioavailability when dosed via the IN route, for example, $\sim 8\%$ in rats (78) and $\sim 3\%$ in humans (79). In rats, the IN bioavailability of sCT achieved in the presence of a permeation enhancer was increased to 27% (78). As another peptide example, the bioavailability of human parathyroid hormone 1 to 34 (PTH_{1-34}) was about 2% in humans [on the basis of PD data (80)], whereas values in the range of 5% to 8% and 12% to 15% were observed in the presence of TJ modulating excipients by Brandt et al. (81). As a representative small protein, the bioavailability of human interferon-α B/D hybrid (MW $\sim$ 19,000) was reported to be $\sim 2.9\%$ without permeation enhancers in rabbits (82). This was compared with a bioavailability for interferon-β (MW $\sim$ 22,500 Da) of $\sim 19\%$ to 21% with permeation enhancers [note that this bioavailability was relative to IM injection and was based on PD, rather than PK data (83)].

There is currently keen interest in development of a variety of different classes of enhancers that promote increased drug permeation by the transcellular and/or paracellular route [see recent reviews (75,76,84–86)]. For instance, Zhang et al. (87) employed electron spin resonance and confocal laser scanning microscopy to study the permeation of recombinant hirudin-2 across rat nasal tissue. It was found that chitosan appeared to promote the paracellular pathway, ammonium glycyrrhizinate invoked transcellular transport, and hydroxyl-propyl-β-cyclodextrin was associated with both mechanisms. As another example, Arnold et al. (88) investigated the alkylglycoside tetradecylmaltoside and described the in vitro internalization of coadministered therapeutic proteins including insulin, leptin, and somatotropin, consistent with a transcellular transport mechanism; with this excipient at high excipient concentration, cellular morphological changes were observed.

SCREENING TOOLS FOR INTRANASAL FORMULATION DEVELOPMENT

The successful development of IN products necessitates robust screening tools predictive of PK performance in humans. Early in the development process, the capability for high-throughput screening (HTS) is important, particularly for peptide and protein formulation development requiring optimization among many product attributes such as permeation enhancement, drug stability, and other properties. The discussion below provides an overview of the in vitro and in vivo models available to guide IN product development.

IN VITRO TISSUE MODELS

While the performance of a new drug or delivery formulation is most critically evaluated using in vivo models, initial screening and development are more efficiently accomplished using a well-defined and controlled in vitro cell culture model. In vitro screening for transport of a drug across an epithelial barrier requires growth of a confluent cell monolayer on a suitable porous support membrane and differentiation into polarized epithelium with robust intercellular TJs. Ideally, these cultures mimic nasal epithelium in their composition of ciliated and nonciliated columnar cells, mucus-producing goblet cells, and basal cells. However, HTS efforts for formulation optimization may be associated with concerns of cost and reproducibility of the cell model.

Human epithelial cell lines are popular choices for oral, nasal, and respiratory drug delivery studies. Models such as Caco-2 are widely used as an oral (i.e., intestinal) delivery model in assays for drug metabolism and transport (89). Most nasal delivery studies utilize Calu-3 (90) or 16HBE14o- (91) cell lines, which readily differentiate into polarized epithelial cells on porous support membranes but are derived from respiratory epithelial cells and may not mimic nasal epithelium [for more in-depth review, see Merkle et al. (92)]. Recently, culture methods for the nasal epithelial cell line RPMI 2650 have been optimized for TJ formation in monolayers (93). When grown at an air-liquid interface, RPMI 2650 differentiates to form confluent monolayers with significantly higher transepithelial resistance (TEER) than cells cultured at a liquid interface. TEER is an important characteristic since its reduction can be used as a formulation-screening tool, that is, an indication of TJ opening. Formation of robust TJs was further indicated by the ability to resist permeation of paracellular flux markers such as mannitol. Additionally, RPMI 2650 cell line has been shown to be useful in drug metabolism studies (92).

Primary respiratory epithelial cell cultures can be differentiated to form pseudostratified, mucociliary cultures. While cost, tissue availability, and reproducibility are major barriers for HTS in primary cell cultures, there are well-established methods for the culture of primary human bronchial/tracheal epithelial cells, as well as companies that provide reproducible, ready-to-use, differentiated cultures in formats suitable for HTS (MatTek Corporation, Ashland, Massachusetts, U.S.A.; Epithelix Sàrl, Plan-Les-Ouates, Genève, Suisse). Further, the recent development of hTERT-immortalized primary nasal cells (94), which allows for extended lifespan while preserving karyotype, holds promise for a differentiated nasal cell model, which preserves the key biological characteristics of the nasal mucosa while allowing for HTS.

IN VIVO ANIMAL MODELS

While in vitro tissue models are highly valuable for HTS, it is also important to employ in vivo models of IN PK and PD. To this end, various animal species and dosing methodologies can be utilized, with varying complexity, cost, and relevance to human dosing (95). Table 1 describes various animal models explored and their nasal surface areas and nasal dose volumes compared with dosing humans.

It is important to note that various factors can affect IN absorption in animals compared with humans, such as nasal cavity anatomical differences, experimental design including whether to perform a surgical preparation, dosing

TABLE 1 Nasal Surface Area and Nasal Dose Volume Comparison Between Species

Species	Weight (kg)	Unilateral surface area (cm^2)	Volume to be administered (μL)	Dose[a] (μL/kg)
Human	70	80	100	
Monkey	7	31	40	6
Mouse	0.03	1.5	3	100
Rat	0.25	7	12.5	50
Rabbit	3	30	45	15
Dog	10	110	150	15
Sheep	60	164	200	3.5

[a]Approximate dose given.

technique, and use of anesthesia. There are nasal cavity interspecies differences; for example, the conchae complexity in the nasal cavity is different between human and other animal species, which may affect nasal absorption (96). The use of anesthesia is also a factor. Anesthesia is known to give rise to a decrease in mucociliary clearance rate and has been shown to enhance the IN drug absorption (95,97). Therefore, it is preferable to conduct such in vivo studies without use of anesthesia to avoid overestimation of IN drug absorption. Likewise, any surgical interventions, such as esophageal suturing or sealing the nasopalatine duct with glue (98), may make in vivo results difficult to generalize to humans.

Among the various animal models, the nasal anatomy of the monkey is the closest to humans, with a single-scroll conchae. Therefore, it is possible to dose the monkey employing a commercially available nasal spray pump [but at about half the human dose volume based on unilateral nasal surface area differences (Table 1)]. However, some species such as cynomolgus have small nares, which can make dosing with a human nasal spray pump difficult. Crossover study design is possible, and serial blood samples can be drawn. This may be the optimal model to evaluate the nasal absorption properties, however, high cost, animal availability, and ethical issues limit their use.

At the other extreme are the rodent models; mouse and rat exhibit nasal double-scroll conchae. Such a difference could potentially affect the comparison of nasal absorption data with that of humans. The nostrils are small and difficult to dose, particularly in the absence of anesthesia. Dose has to be accomplished with a micropipette at very small dose volumes, which can add to experimental variability. It is very difficult to get the amount of blood volume appropriate for assay per time point in one animal, so most studies pool the blood time points among different animals, which may decrease variability. Caution should be exercised when employing rodents as a model for assessing the potential for IN drugs to reach the brain, especially given the disparate relative size of the olfactory region. For instance, in humans the olfactory region covers only about 10 cm^2, corresponding to about 6% of the total nasal surface area (99), as compared with about 40% to 50% for rats and mice (100).

Even given such drawbacks, rodent IN models have utility due to factors such as low cost and availability. For example, we have reported using a rodent PK model for screening IN formulations of the low-molecular-weight drug galantamine (101). Subsequent preclinical studies focused on the ferret model due to its relevance for predicting GI-related side effects in humans, and the data clearly demonstrated that IN dosing dramatically reduced emetic side effects compared with oral dosing (102). As an example, relevant to protein formulation, the rat model has been employed for screening enhancers for IN growth hormone (103).

In between the primate and rodent IN animal models are various intermediate mammalian species such as the rabbit and dog. Both of these species have branching conchae and are easy to dose by the nasal route using a micropipette without anesthesia. Serial blood draws can be taken from an individual animal of these species. PD properties can also readily be measured, for example, glucose levels after administration of IN insulin to rabbits to provide robust formulation screening for IN insulin (104). We have recently published a number of reports describing the use of rabbits as an IN animal model for peptide drugs such as insulin (104), peptide YY_{3-36} (PYY_{3-36}) (105), and carbetocin (106). As an example of interspecies comparison, the bioavailability of PYY_{3-36} in the presence of TJ modulating excipients was found to be about 8% to 16% in rabbits (105,107) and about 19% in humans (108). The case of rabbit to human PK predictability for IN insulin will be discussed in further detail below. The rabbit has also been employed as a PK model for screening for novel TJ modulators (109).

The sheep is yet another mammalian species reported to have utility for IN formulation screening for a variety of low-molecular-weight drugs as well as peptides and proteins (110–113). It should be noted that sheep have a double-scroll conchae, similar to rodents. The nostrils are relatively large; therefore, larger dose volumes can be given compared with dosing other animal species. Crossover design is possible, and serial blood draws can be taken.

PHARMACOKINETIC AND PHARMACODYNAMIC CASE EXAMPLE: IN INSULIN

IN administration of insulin provides an attractive option for a noninvasive diabetes treatment. Currently, marketed prandial (meal time) insulin products are dosed by injection immediately before each meal, posing a barrier among patients and physicians to initiate therapy. The rapid PK profile of IN insulin also offers the potential to reduce postmeal hypoglycemia caused by longer half-life insulin preparations.

To identify promising IN formulations, an in vitro model system was employed to monitor reduction in transepithelial electrical resistance (TER), cellular toxicity, and permeation. From this HTS process, formulations that provided up to 10^5-fold increase in in vitro permeation relative to a simple formulation devoid of enhancers were identified. We have previously reported (104) the assessment of PK parameters following instillation (15 µL/kg) of IN insulin formulations in New Zealand white rabbits. IN insulin provided 9% to 19% bioavailability relative to SC NovoLog® injection (12 IU/kg) on the basis of data out to four hours, with a T_{max} between 13 and 59 minutes and 21% to 67% intersubject coefficient of variance (CV), compared with a T_{max} of 18 minutes for SC injection with a 74% CV. All insulin-containing IN formulations decreased plasma glucose levels.

Human PK and PD effects of IN insulin formulations were assessed in phase 1 clinical trials in healthy volunteers (114). Absorption of three doses of IN insulin was compared with SC insulin aspart (NovoLog, 0.12 IU/kg) or inhaled

Exubera® inhalation powder (3 mg) in 12 healthy subjects 18 to 45 years old with normal BMI (20–26). IN insulin was dosed as a single spray in one nostril. Plasma insulin and glucose were measured at 12 time points for up to six hours after dosing, and insulin C-peptide was measured in each sample to correct for the subject's endogenous insulin secretion. The PK parameters T_{max}, C_{max}, and area under the curve (AUC) were determined. In this study, all doses of IN insulin were well tolerated and there were no clinically significant changes in vital signs (systolic or diastolic blood pressure and heart rate), ECG, or physical examination. This data demonstrated that IN insulin provided the fastest rise in plasma concentrations, followed by the fastest return to baseline when compared with SC NovoLog (Fig. 2A). Human PK and PD parameters following IN administration of four distinct formulations were assessed. For the various IN insulin formulations, the bioavailability (relative to NovoLog) based on AUC_{last}

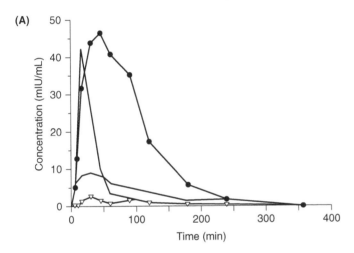

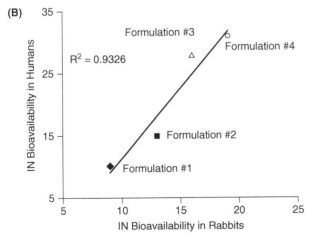

FIGURE 2 (**A**) Insulin pharmacokinetics in humans for intranasal formulation (*solid line*) and NovoLog (*filled circles*). (**B**) Comparison of bioavailability of various IN insulin formulations dosed in rabbits and humans. *Source*: From Ref. 114.

(0–4 hours) of IN insulin was in the range of 10% to 30%, and observed T_{max} was in the range of 30 minutes or lower, compared with about 50 minutes for NovoLog. IN insulin achieved a maximum glucose fall faster ($\sim$56 minutes) than SC Novolog ($\sim$88 minutes) or Exubera ($\sim$100 minutes). The bioavailability of various IN insulin formulations dosed in humans was compared with the data generated from rabbit studies (Fig. 2B), and there was a good correlation between the two species.

Recently, we have reported positive phase 2 data for IN insulin (115), comparing the PK and PD effect versus insulin aspart administered by subcutaneous (SC) injection (NovoLog, Novo Nordisk Inc., Princeton, New Jersey, U.S.A.) on postprandial glycemic control and evaluating of the incidence of hypoglycemia within four hours post dosing. Each patient ate a standard breakfast and then underwent a series of drug challenges including placebo, and optimized insulin aspart dose and an optimized IN insulin dose. Both IN insulin and insulin aspart provided statistically significant postmeal glucose reduction at 60 and 90 minutes compared with usual oral antidiabetic therapy (Fig. 3). IN insulin was noninferior to insulin aspart for postmeal change from baseline glucose levels at 60 and 90 minutes. The mean change from baseline at the 60-minute time point was 56.9 mg/dL and 62.9 mg/dL for IN insulin and insulin aspart, respectively. The percent CVs (intersubject variability) for PK (blood insulin levels) and PD (postmeal glucose) were similar for IN insulin and insulin aspart.

The T_{max} for IN insulin was significantly faster ($p < 0.001$) than that for insulin aspart, with blood levels for IN insulin seen as fast as five minutes post dose. The median time to peak concentrations for IN insulin was 30 minutes, compared with that of 90 minutes for insulin aspart. IN insulin was well

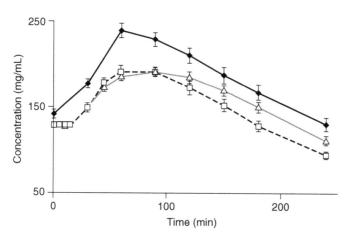

FIGURE 3 Insulin pharmacodynamics in humans for IN formulation (*black line*), insulin aspart (*gray line*), and usual therapy (*dash line*). With respect to postmeal glucose at the 60- and 90-minute postmeal time points, the study demonstrated that both insulin aspart and IN insulin were significantly better than placebo at lowering postmeal glucose and that IN insulin was non-inferior to insulin aspart. IN insulin results in statistically significant postmeal glucose reduction compared with usual therapy at 60 and 90 minutes. The glucose reduction following IN insulin is noninferior to that following insulin aspart at 60 minutes and 90 minutes. *Abbreviation*: IN, intranasal. *Source*: From Ref. 115.

tolerated with only one adverse event reported (one patient sneezed 19 minutes after dosing) out of 146 nasal administrations.

There was a statistically significant decrease in the incidence of patients experiencing hypoglycemia (blood glucose <70 mg/dL) for IN insulin (1 of 29) compared with that for insulin aspart (6 of 29) ($p < 0.05$). These data indicate that, when compared with insulin aspart, IN insulin provides equivalent glycemic control with lower risk of hypoglycemia.

SUMMARY AND FUTURE DIRECTIONS

Some of the earliest permeation enhancers, including surfactants such as sodium dodecyl sulfate and bile salts, have shown limited success at enhancing delivery of peptides and proteins, and it appears that there is a direct correlation between potency and local toxicity [for review, see Johnson and Quay (75)]. Efforts to develop the next generation of IN permeation enhancers have focused on maximizing permeation and controlling PK of macromolecular drugs such as peptides and proteins while minimizing local toxicity. Recently, a systematic screen of surfactants identified concentrations that were both safe and efficacious (116,117), suggesting that a better understanding of the structure-function relationships for this category of enhancers may yield useful permeation enhancers. Further, by optimizing combinations and concentrations, a new repertoire of synergistic enhancer formulations were identified that improved performance without increasing toxicity in in vitro Caco-2 cell model (118).

Recent studies have revealed that TJs are not static but undergo continual remodeling with ZO-1, occludin and claudin entering and leaving the junction with distinct dynamic behavior (119). These findings open up the possibility of rapid TJ remodeling by permeation enhancers that target the functional domains of the junctional proteins or by affecting signaling through pathways that regulate TJ assembly and disassembly. For instance, cholesterol-solubilizing agents have been shown to increase paracellular permeability, possibly through release of lipid-derived second messengers, which could affect the phosphorylation state of TJ proteins (120). More recently, Chen-Quay et al. (121) developed a screen to identify lipids that modulated TJ properties. Lipids from three structural groups were identified (glycosylated sphingosines, oxidized lipids, and ether lipids) that rapidly and reversibly reduced TER by up to 95% and enhanced permeation of 3-kDa dextran with minimal toxicity. The identified lipids are present in lipid raft domains in cell membranes, which have been shown to be sites of TJ assembly. Therefore, it is possible that in addition to more direct effects on signaling pathways, perturbation of local membrane domains could promote TJ disassembly or endocytosis.

Peptide-based permeation enhancers are attractive candidates as they share physiochemical, diffusion, and release properties at the epithelial surface with peptide and protein drugs. Most peptide enhancers, such as zonula occluden toxin (Zot), have been described for oral drug delivery [for review see Maher et al. (122)], with few studies existing for nasal delivery. Attempts to design peptides that competitively bind to extracellular domains of occludin or claudins and thus disrupt TJ assembly demonstrate some promise (123–125). For instance, Mrsny et al. (126) have identified a critical claudin-1 extracellular loop motif that increased paracellular gastric permeability when administered orally to rats (126). As another example, Chen et al. (109) have described an 18–amino acid amphipathic peptide, PN159, capable of reducing TER and increasing permeation of peptide drugs with low toxicity in primary bronchial/tracheal cell cultures. Further, in vivo studies of intranasally dosed rabbits demonstrated PN159 increased permeation of PYY_{3-36} by 50 to 70 fold. PN159 was shown to be chemically stable under storage conditions relevant to IN formulations. These results demonstrate that such peptide enhancers have utility for enhancing nasal delivery of peptide drugs.

REFERENCES

1. Johnson, PH, Cui K, Costantino HR, et al. Exploiting tight junctions for delivery of drugs. Genet Eng News 2004; 24(1):7–10.
2. Bloebaum RM. The role of pharmacotherapy; sound treatment can improve your patient's quality of life. J Respir Dis 2002; 23:370–376.
3. Milford CA, Mugliston TA., Lund VJ, et al. Long-term safety and efficacy study of intranasal ipratropium bromide. J Laryngol Otol 1990; 104:123–125.
4. van Drunen C, Meltzer EO, Bachert C, et al. Nasal allergies and beyond: a clinical review of the pharmacology, efficacy, and safety of mometasone furoate. Allergy 2005; 60(suppl 80):5–19.
5. Stanaland BE. Once-daily budesonide aqueous nasal spray for allergic rhinitis: a review. Clin Ther 2004; 26:473–492.
6. Lumry W, Hampel F, LaForce C, et al. A comparison of once-daily triamcinolone acetonide aqueous and twice-daily beclomethasone dipropionate aqueous nasal sprays in the treatment of seasonal allergic rhinitis. Allergy Asthma Proc 2003; 24:203–210.
7. Salib RJ, Howarth PH. Safety and tolerability profiles of intranasal antihistamines and intranasal corticosteroids in the treatment of allergic rhinitis. Drug Saf 2003; 26:863–893.
8. Ooi EH, Wormald PJ, Tan LW. Innate immunity in the paranasal sinuses: a review of nasal host defenses. Am J Rhinol 2008; 22:13–19.
9. Pammit MA, Budhavarapu VN, Raulie EK, et al. Intranasal interleukin-12 treatment promotes antimicrobial clearance and survival in pulmonary *Francisella tularensis subsp. novicida* infection. Antimicrob Agents Chemother 2004; 48:4513–4519.
10. Sullivan VJ, Mikszta JA, Laurent P, et al. Noninvasive delivery technologies: respiratory delivery of vaccines. Expert Opin Drug Deliv 2006; 3:87–95.
11. Giudice EL, Campbell JD. Needle-free vaccine delivery. Adv Drug Deliv Rev 2006; 58:68–89.
12. Slütter B, Hagenaars N, Jiskoot W. Rational design of nasal vaccines. J Drug Target 2008; 16:1–17.
13. Hussain AA. Intranasal drug delivery. Adv Drug Deliv Rev 1998; 29:39–49.
14. Behl CR, Pimplaskar HK, Sileno AP, et al. Effects of physicochemical properties and other factors on systemic nasal drug delivery. Adv Drug Deliv Rev 1998; 29:89–116.
15. Behl CR, Pimplaskar HK, Sileno AP, et al. Optimization of systemic nasal drug delivery with pharmaceutical excipients. Adv Drug Deliv Rev 1998; 29:117–133.
16. Illum L. Nasal drug delivery—possibilities, problems and solutions. J Control Release 2003; 87:187–198.
17. Song Y, Wang Y, Thakur R, et al. Mucosal drug delivery: membranes, methodologies, and applications. Crit Rev Ther Drug Carrier Syst 2004; 21:195–256.
18. Costantino HR, Sileno AP, Johnson PH. Pharmacokinetic attributes of intranasal delivery: case studies and new opportunities. Drug Deliv 2005; 3:8–11.
19. Costantino HR, Illum L, Brandt G, et al. Intranasal delivery: physicochemical and therapeutic aspects. Int J Pharm 2007; 337:1–24.
20. Wolfe TR, Bernstone T. Intranasal drug delivery: an alternative to intravenous administration in selected emergency cases. J Emerg Nurs 2004; 30:141–147.
21. Freychet L, Rizkalla SW, Desplanque N, et al. Effect of intranasal glucagon on blood glucose levels in healthy subjects and hypoglycaemic patients with insulin-dependent diabetes. Lancet 1988; 18:1364–1366.

22. Pontiroli AE, Calderara A, Perfetti MG, et al. Pharmacokinetics of intranasal, intramuscular and intravenous glucagon in healthy subjects and diabetic patients. Eur J Clin Pharmacol 1993; 45:555–558.
23. Stenninger E, Aman J. Intranasal glucagon treatment relieves hypoglycaemia in children with type 1 (insulin-dependent) diabetes mellitus. Diabetologia 1993; 36:931–935.
24. Hvidberg A, Djurup R, Hilsted J. Glucose recovery after intranasal glucagon during hypoglycaemia in man. Eur J Clin Pharmacol 1994; 46:15–17.
25. Bleske BE, Rice TL, Warren EW, et al. Effect of dose on the nasal absorption of epinephrine during cardiopulmonary resuscitation. Am J Emerg Med 1996; 14:133–138.
26. Landau AJ, Eberhardt RT, Frishman WH. Intranasal delivery of cardiovascular agents: an innovative approach to cardiovascular pharmacotherapy. Am Heart J 1994; 127:1594–1599.
27. Pontiroli AE. Peptide hormones: review of current and emerging uses by nasal delivery. Adv Drug Deliv Rev 1998; 29:81–87.
28. Illum L. Is nose-to-brain transport of drugs in man a reality? J Pharm Pharmacol 2004; 56:3–17.
29. Vyas TK, Shahiwala A, Marathe S, et al. Intranasal drug delivery for brain targeting. Curr Drug Deliv 2005; 2:165–175.
30. Frey WH, Liu J, Chen X, et al. Delivery of 125I-NGF to the brain via the olfactory route. Drug Deliv 1997; 4:87–92.
31. Dufes C, Olivier JC, Gaillard F, et al. Brain delivery of vasoactive intestinal peptide (VIP) following nasal administration in rats. Int J Pharm 2003; 255:87–97.
32. Lerner EN, van Zanten EH, Stewart GR. Enhanced delivery of octreotide to the brain via transnasal iontophoretic administration. J Drug Target 2004; 12:273–280.
33. Thorne RG, Pronk GJ, Padmanabhan V, et al. Delivery of insulin-like growth factor-I to the rat brain and spinal cord along olfactory and trigeminal pathways following intranasal administration. Neuroscience 2004; 127:481–496.
34. Banks WA, During MJ, Niehoff ML. Brain uptake of the glucagon-like peptide-1 antagonist exendin (9-39) after intranasal administration. J Pharmacol Exp Ther 2004; 309:469–475.
35. Ross TM, Martinez PM, Renner JC, et al. Intranasal administration of interferon beta bypasses the blood-brain barrier to target the central nervous system and cervical lymph nodes: a non-invasive treatment strategy for multiple sclerosis. J Neuroimmunol 2004; 151:66–77.
36. Sakane T, Akizuki M, Yamashita S, et al. The transport of drug to the cerebrospinal fluid directly from the nasal cavity: the relation to the lipophilicity of the drug. Chem Pharm Bull 1991; 39:2456–2458.
37. Kao HD, Traboulsi A, Itoh S, et al. Enhancement of the systemic and CNS specific delivery of L-Dopa by the nasal administration of its water soluble prodrugs. Pharm Res 2000; 17:978–984.
38. Chow HH, Anavy N, Villalobos A. Direct nose-brain transport of benzoylecgonine following intranasal administration in rats. J Pharm Sci 2001; 90:1729–1735.
39. Al-Ghananeem AM, Traboulsi AA, Dittert LW, et al. Targeted brain delivery of 17 beta-estradiol via nasally administered water soluble prodrugs. AAPS PharmSciTech 2002; 3:E5.
40. Barakat NS, Omar SA, Ahmed AA. Carbamazepine uptake into rat brain following intra-olfactory transport. J Pharm Pharmacol 2006; 58:63–72.
41. van den Berg MP, Merkus P, Romeijn SG, et al. Uptake of melatonin into the cerebrospinal fluid after nasal and intravenous delivery: studies in rats and comparison with a human study. Pharm Res 2004; 21:799–802.
42. van den Berg MP, Verhoef JC, Romeijn SG, et al. Uptake of estradiol or progesterone into the CSF following intranasal and intravenous delivery in rats. Eur J Pharm Biopharm 2004; 58:131–135.
43. Yang Z, Huang Y, Gan G, et al. Microdialysis evaluation of the brain distribution of stavudine following intranasal and intravenous administration to rats. J Pharm Sci 2005; 94:1577–1588.

44. Merkus FW, van den Berg MP. Can nasal drug delivery bypass the blood-brain barrier?: questioning the direct transport theory. Drugs R D 2007; 8:133–144.
45. Costantino HR, Leonard AK, Brandt G, et al. Intranasal administration of acetylcholinesterase inhibitors. BMC Neurosci 2008; 9(suppl 3):S6.
46. Banks WA. The source of cerebral insulin. Eur J Pharmacol 2004; 490:5–12.
47. Benedict C, Hallschmid M, Schmitz K, et al. Intranasal insulin improves memory in humans: superiority of insulin aspart. Neuropsychopharmacology 2007; 32:239–243.
48. Hallschmid M, Benedict C, Schultes B, et al. Towards the therapeutic use of intranasal neuropeptide administration in metabolic and cognitive disorders. Regul Pept 2008; 149(1–3):79–83.
49. Benedict C, Kern W, Schultes B, et al. Differential sensitivity of men and women to anorexigenic and memory-improving effects of intranasal insulin. J Clin Endocrinol Metab 2008; 93:1339–1344.
50. Reger MA, Watson GS, Green PS, et al. Intranasal insulin administration dose dependently modulates verbal memory and plasma amyloid-beta in memory impaired older adults. J Alzheimers Dis 2008; 13:323–331.
51. Mygind N, Dahl R. Anatomy, physiology and function of the nasal cavities in health and disease. Adv Drug Deliv Rev 1998; 29:3–12.
52. Jones N. The nose and paranasal sinuses physiology and anatomy. Adv Drug Deliv Rev 2001; 51:5–19.
53. Greiff L, Andersson M, Svensson J, et al. Absorption across the nasal airway mucosa in house dust mite perennial allergic rhinitis. Clin Physiol Funct Imaging 2002; 22:55–57.
54. Larsen C, Niebuhr Jorgensen M, Tommerup B, et al. Influence of experimental rhinitis on the gonadotropin response to intranasal administration of buserelin. Eur J Clin Pharmacol 1987; 33:155–159.
55. Dowson AJ, Charlesworth BR, Green J, et al. Zolmitriptan nasal spray exhibits good long-term safety and tolerability in migraine: results of the INDEX trial. Headache 2005; 45:17–24.
56. Shyu WC, Pittman KA, Robinson DS, et al. The absolute bioavailability of transnasal butorphanol in patients experiencing rhinitis. Eur J Clin Pharmacol 1993; 45:559–562.
57. Humbert H, Cabiac MD, Dubray C, et al. Human pharmacokinetics of dihydroergotamine administered by nasal spray. Clin Pharmacol Ther 1996; 60:265–275.
58. Liversidge GG, Wilson CG, Sternson WL, et al. Nasal delivery of a vasopressin antagnoist in dogs. J Appl Physiol 1988; 64:377–383.
59. Hardy JG, Lee SW, Wilson CG. Intranasal drug delivery by spray and drops. J Pharm Pharmacol 1985; 37:294–297.
60. Daley-Yates PT, Baker RC. Systemic bioavailability of fluticasone propionate administered as nasal drops and aqueous nasal spray formulations. Br J Clin Pharmacol 2001; 51:103–105.
61. Pontiroli AE, Alberetto M, Pajetta E, et al. Human insulin plus sodium glycocholate in a nasal spray formulation: improved bioavailability and effectiveness in normal subjects. Diabetes Metab 1987; 13:441–443.
62. Johansson CJ, Olsson P, Bende M, et al. Absolute bioavailability of nicotine applied to different nasal regions. Eur J Clin Pharmacol 1991; 41:585–588.
63. Marttin E, Romeijn SG, Verhoef JC, et al. Nasal absorption of dihydroergotamine from liquid and powder formulations in rabbits. J Pharm Sci 1997; 86:802–807.
64. Ishikawa F, Katsura M, Tamai I, et al. Improved nasal bioavailability of elcatonin by insoluble powder formulation. Int J Pharm 2001; 224:105–114.
65. Thorsson L, Borgâ O, Edsbäcker S. Systemic availability of budesonide after nasal administration of three different formulations: pressurized aerosol, aqueous pump spray, and powder. Br J Clin Pharmacol 1999; 47:619–624.
66. Quay SC, Aprile PC, Go ZO, et al. Cuanocobalamin low viscosity aqueous formulations for intranasal delivery. United States Patent 7,229,636, issued June 12, 2007.
67. Ugwoke MI, Verbeke N, Kinget R. The biopharmaceutical aspects of nasal mucoadhesive drug delivery. J Pharm Pharmacol 2001; 53:3–21.
68. Edsman K, Hägerström H. Pharmaceutical applications of mucoadhesion for the non-oral routes. J Pharm Pharmacol 2005; 57:3–22.

69. Ugwoke MI, Exaud S, Van Den Mooter G, et al. Bioavailability of apomorphine following intranasal administration of mucoadhesive drug delivery systems in rabbits. Eur J Pharm Sci 1999; 9:213–219.

70. Jain AK, Chalasani KB, Khar RK. Muco-adhesive multivesicular liposomes as an effective carrier for transmucosal insulin delivery. J Drug Target 2007; 15:417–427.

71. Vila A, Sánchez A, Tobío M, et al. Design of biodegradable particles for protein delivery. J Control Release 2002; 78:15–24.

72. McInnes FJ, O'Mahony B, Lindsay B, et al. Nasal residence of insulin containing lyophilised nasal insert formulations, using gamma scintigraphy. Eur J Pharm Sci 2007; 31:25–31.

73. Varshosaz J, Sadrai H, Heidari A. Nasal delivery of insulin using bioadhesive chitosan gels. Drug Deliv 2006; 13:31–38.

74. D'Souza R, Mutalik S, Venkatesh M, et al. Insulin gel as an alternate to parenteral insulin: formulation, preclinical, and clinical studies. AAPS PharmSciTech 2005; 6: E184–E189.

75. Johnson PH, Quay SC. Advances in nasal drug delivery through tight junction technology. Expert Opin Drug Deliv 2005; 2:281–298.

76. Johnson PH, Frank D, Costantino HR. Discovery of tight junction modulators: significance for drug development and delivery. Drug Discov Today 2008; 13:261–267.

77. Schipper NG, Verhoef JC, De Lannoy LM, et al. Nasal administration of an ACTH(4-9) peptide analogue with dimethyl-beta-cyclodextrin as an absorption enhancer: pharmacokinetics and dynamics. Br J Pharmacol 1993; 110:1335–1340.

78. Matsuyama T, Morita T, Horikiri Y, et al. Enhancement of nasal absorption of large molecular weight compounds by combination of mucolytic agent and nonionic surfactant. J Control Release 2006; 110:347–352.

79. Novartis, .Miacalcin®. Available at: http://www.pharma.us.novartis.com/product/ pi/pdf/miacalcin_nasal.pdf.

80. Matsumoto T, Shiraki M, Hagino H, et al. Daily nasal spray of hPTH(1-34) for 3 months increases bone mass in osteoporotic subjects: a pilot study. Osteoporos Int 2006; 17:1532–1538.

81. Brandt G, Spann BM, Sileno AP, et al. Teriparatide Nasal Spray. Pharmacokinetics and safety vs. subcutaneous teriparatide in healthy volunteers. American Association of Clinical Endocrinologists 15th Annual Meeting & Clinical Congress, April 28, 2006, Chicago, IL.

82. Bayley D, Temple C, Clay V, et al. The transmucosal absorption of recombinant human interferon-alpha B/D hybrid in the rat and rabbit. Pharm Pharmacol 1995; 47:721–724.

83. Vitkun SA, Cimino L, Sileno A, et al. A comparative study of a nasal formulation of interferon beta-1a versus Avonex®. Presented at: the 56th American Academy of Neurology Annual Meeting, San Francisco, CA, April 29, 2004.

84. Sharma S, Kulkarni J, Pawar AP. Permeation enhancers in the transmucosal delivery of macromolecules. Pharmazie 2006; 61:495–504.

85. Di Colo G, Zambito Y, Zaino C. Polymeric enhancers of mucosal epithelia permeability: synthesis, transepithelial penetration-enhancing properties, mechanism of action, safety issues. J Pharm Sci 2007; 97:1652–1680.

86. Maggio ET. Intravail: highly effective intranasal delivery of peptide and protein drugs. Expert Opin Drug Deliv 2006; 3:529–539.

87. Zhang YJ, Zhang Q, Yang J, et al. Promoting mechanism of enhancers and transport pathway of large hydrophilic molecular across nasal epithelium studied by ESR and CLSM technologies. Yao Xue Xue Bao 2007; 42:1195–1200.

88. Arnold JJ, Ahsan F, Meezan E, et al. Correlation of tetradecylmaltoside induced increases in nasal peptide drug delivery with morphological changes in nasal epithelial cells. J Pharm Sci 2004; 93:2205–2213.

89. Sambuy Y, DeAngelis I, Ranaldi G, et al. The Caco-2 cell line as a model of the intestinal barrier: influence of cell and culture-related factors on Caco-2 cell functional characteristics. Cell Biol Toxicol 2005; 21:1–26.

90. Grainger CI, Greenwell LL, Lockley DJ, et al. Culture of Calu-3 cells at the air interface provides a representative model of the airway epithelial barrier. Pharm Res 2006; 23:1482–1490.

91. Ehrhardt C, Kneuer C, Fiegel J, et al. Influence of apical fluid volume on the development of functional intercellular junctions in the human epithelial cell line 16HBE14o-: implications for the use of this cell line as an in vitro model for bronchial drug absorption studies. Cell Tissue Res 2002; 308:391–400.

92. Merkle HP, Ditzinger G, Lang SR, et al. In vitro cell models to study nasal mucosal permeability and metabolism. Adv Drug Deliv Rev 1998; 29:51–79.

93. Bai S, Yang T, Abbruscato TJ, et al. Evaluation of human nasal RPMI 2650 cells grown at an air-liquid interface as a model for nasal drug transport studies. J Pharm Sci 2008; 97:1165–1178.

94. Kurose M, Kojima T, Koizumi JI, et al. Induction of claudins in passaged hTERT-transfected human nasal epithelial cells with an extended life span. Cell Tissue Res 2007; 330:63–74.

95. Illum L. Animal models for nasal delivery. J Drug Target 1996; 3:717–724.

96. Gizurarson S. The relevance of nasal physiology to the design of drug absorption studies. Adv Drug Deliv Rev 1993; 11:329–347.

97. Mayor SH, Illum L. An investigation of the effect of anaesthetics on the nasal absorption of insulin in rats. Int J Pharm 1997; 149:123–129.

98. Hirai S, Yashiki T, Matsuzawa T, et al. Absorption of drugs from nasal mucosa of rat. Int J Pharm 1981; 7:317–325.

99. Bear MF, Connors BW, Pradiso MA. Neuroscience: Exploring the Brain. 2nd ed. Baltimore: Lippincott Williams and Wilkins, 2001:269.

100. Gross EA, Swenberg JA, Fields S, et al. Comparative morphometry of the nasal cavity in rats and mice. J Anat 1982; 135:83–88.

101. Kays Leonard A, Sileno AP, MacEvilly C, et al. Development of a novel high-concentration galantamine formulation suitable for intranasal delivery. J Pharm Sci 2005; 94:1736–1746.

102. Kays Leonard A, Sileno AP, Brandt GC, et al. In vitro formulation optimization of intranasal galantamine leading to enhanced bioavailability and reduced emetic response in vivo. Int J Pharm 2007; 335:138–146.

103. O'Hagan DT, Chirchley H, Farraj NF, et al. Nasal absorption enhancers for biosynthetic human growth hormone in rats. Pharm Res 1990; 7:772–776.

104. Cohen AS, Sileno AP, Peddakota LR, et al. *In vitro* and *in vivo* screening of intranasal insulin formulations. Presented at: the 2006 American Association of Pharmaceutical Scientists Annual Meeting and Exposition, San Antonio, TX.

105. Kleppe M, Deshpande A, Go Z, et al. Development of an intranasal formulation of the Y2R agonist peptide YY 3-36. Presented at: the 2003 NAASO Annual Meeting, Ft. Lauderdale, FL.

106. Peddakota L, Leonard A, Sileno A, et al. *In vitro* and *in vivo* screening of intranasal carbetocin formulations. Presented at: the 2008 AAPS Annual Biotechnology Conference, Toronto.

107. Foerder C, MacEvilly C, Haugaard D, et al. Quantitative determination of peptide YY 3-36 in plasma by radioimmunoassay. Presented at: the 2004 AAPS National Biotechnology Conference, Boston, MA.

108. Brandt G, Park A, Wynne K, et al. Nasal peptide YY 3-36. Phase 1 dose ranging and safety studies in healthy human subjects. Presented at: the 2004 ENDO Conference, New Orleans, LA.

109. Chen SC, Eiting K, Cui K, et al. Therapeutic utility of a novel tight junction modulating peptide for enhancing intranasal drug delivery. J Pharm Sci 2006; 95:1364–1371.

110. Critchley H, Davis SS, Farraj NF, et al. Nasal absorption of desmopressin in rats and sheep. Effect of a bioadhesive microsphere delivery system. J Pharm Pharmacol 1994; 46:651–656.

111. Illum L, Davis SS, Pawula FM, et al. Nasal administration of morphine-6-glucuronide in sheep – a pharmacokinetic study. Biopharm Drug Dispos 1996; 17:717–724.

112. Gill IJ, Fisher AN, Farraj N, et al. Intranasal absorption of granulocyte-colony stimulating factor (G-CSF) from powder formulations in sheep. Eur J Pharm Sci 1998; 6:1–10.
113. Dyer AM, Hunchcliffe M, Watts P, et al. Nasal delivery of insulin using novel chitosan based formulation. A comparative study in two animal models between simple chitosan formulations and chitosan nanoparticles. Pharm Res 2002; 19:998–1008.
114. Cohen AS, Forseth KT, Sileno AP, et al. Assessing pharmacokinetics and pharmacodynamics of an intranasal formulation of human insulin. Presented at: the 2007 American Association of Pharmaceutical Scientists Annual Meeting and Exposition, San Diego, CA.
115. Brandt G, Sileno A, Cohen AS, et al. Intranasal insulin. Phase 2 glucose tolerance study in type 2 diabetics. Presented at the 2008 American Diabetes Association 68th Scientific Sessions, San Francisco, CA.
116. Whitehead K, Karr N, Mitragotri S. Safe and effective permeation enhancers for oral drug delivery. Pharm Res 2008; 25(8):1782–1788.
117. Whitehead K, Karr N, Mitragotri S. Mechanistic analysis of chemical permeation enhancers for oral drug delivery. Pharm Res 2008; 25:1412–1419.
118. Whitehead K, Karr N, Mitragotri S. Discovery of synergistic permeation enhancers for oral drug delivery. J Control Release 2008; 128:128–133.
119. Shen L, Weber CR, Turner JR. The tight junction protein complex undergoes rapid and continuous molecular remodeling at steady state. J Cell Biol 2008; 181:683–695.
120. Francis SA, Kelly JM, McCormack J, et al. Rapid reduction of MDCK cell cholesterol by methyl-beta-cyclodextrin alters Eur J Cell Biol 1999; 78:473–484.
121. Chen-Quay S-C, Eiting KT, Li Aw, et al. Identification of tight junction modulating lipids. J Pharm Sci 2008; 98:606–619.
122. Maher S, Brayden DJ, Feighery L, et al. Cracking the junction: update on the progress of gastrointestinal absorption enhancement in the delivery of poorly absorbed drugs. Crit Rev Ther Drug Carrier Syst 2008; 25:117–168.
123. Kondoh M, Masuyama A, Takahashi A, et al. A novel strategy for the enhancement of drug absorption using a claudin modulator. Mol Pharmacol 2005; 67:749–756.
124. Everett RS, Vanhook MK, Barozzi N, et al. Specific modulation of airway epithelial tight junctions by apical application of an occludin peptide. Mol Pharmacol 2006; 69:492–500.
125. Herman RE, Makienko EG, Prieve MG, et al. Phage display screening of epithelial cell monolayers treated with EGTA. Identification of peptide FDFWITP that modulates tight junction activity. J Biomol Screen 2007; 12:1092–1101.
126. Mrsny RJ, Brown GT, Gerner-Smidt K, et al. A key claudin extracellular loop domain is critical for epithelial barrier integrity. Am J Pathol 2008; 172:905–915.

9 Oral Delivery of Protein and Peptide Therapeutics: PK/PD and Metabolism Considerations

Puchun Liu and Steven Dinh

Noven Pharmaceuticals, Inc., Miami, Florida, U.S.A.

INTRODUCTION

The success of protein and peptide therapeutics is revolutionizing the biotech and pharmaceutical market, spurring the creation of next-generation products with reduced immunogenicity, improved safety, and greater effectiveness. Endogenous proteins and peptides play an important role in the regulation and integration of life processes and act with high specificity and potency (1). Because of the rapid progress in biotechnology, as well as in gene technology, the industry is capable of producing a large number of potential therapeutic peptides and proteins in commercial quantities. Despite this remarkable success, protein and peptide drugs continue to suffer from drawbacks, particularly in regard to their mode of delivery, which is primarily by injection. Improvements in protein and peptide drug delivery using less invasive or noninvasive approaches would increase patient compliance and expand many drug market opportunities. In addition, some new formulations may be patentable and can be used to extend the life cycle management of a drug. For these reasons, pharmaceutical and biotechnology companies are researching and developing new delivery methods for protein and peptide drugs (2).

Oral drug delivery is the most common and convenient form of drug administration that can provide patient compliance and acceptance. In addition to the advantages of patient compliance, the cost associated with an oral dosage form is usually more competitive than that associated with other dosage forms, from the lower cost of manufacturing to the benefit of no direct involvement of the health care provider. Perhaps more importantly, a growing body of data suggests that for certain polypeptides such as insulin, the oral delivery route may be more physiological. The availability of safe and effective oral protein and peptide formulations would provide a major advancement in treating a variety of diseases that currently require chronic parenteral administrations. The success achieved by Sandoz with oral cyclosporine formulations remains one clear example of what can be achieved. Some of the oral protein and peptide therapeutics under current clinical development are reported in Table 1. This partial list of molecules includes salmon calcitonin (sCT), human insulin, human parathyroid hormone 1–34 (PTH1–34), human growth hormone (hGH), glucagon-like peptide 1 (GLP-1), and peptide YY_{3-36} (PYY). Many other protein and peptide drug candidates, some of which are already available as parenterals, have undergone significant oral formulation efforts in the laboratory and testing in animal models.

Oral delivery of therapeutic protein and peptides drugs is a challenge primarily due to their poor bioavailability that originates from the enzymatic degradation of these molecules in the gastrointestinal (GI) tract and the difficulty of transporting a relatively large molecule across the intestinal barrier.

192

TABLE 1 Oral Protein and Peptide Therapeutics in Clinical Development

Drug	Product description	Therapeutic indication	Developer	Originator	Clinic status	Update date
Calcitonin	Salmon calcitonin (Eligen)	Osteoporosis	Novartis/Emisphere	Emisphere	Phase III	Feb 13, 2009
	Salmon calcitonin (Eligen)	Osteoarthritis	Novartis/Emisphere	Emisphere	Phase III	Feb 13, 2009
	Salmon calcitonin	Osteoporosis	Unigene	Unigene	Phase II	Feb 03, 2009
	Salmon calcitonin (Axcess)	Osteoporosis	Bone Medical	Bone Medical	Phase II	Nov 06, 2008
Insulin	Insulin (Eligen)	Diabetes	Emisphere	Emisphere	Phase II	June 24, 2008
	IN-105 (conjugated insulin)	Diabetes	Nobex	Nobex	Phase II	Sep 22, 2008
PTH	PTH1–34	Osteoporosis	Bone Medical	Bone Medical	Phase II	Nov 06, 2008
	PTH1–34 (Eligen)	Osteoporosis	Novartis/Emisphere	Emisphere	Phase I	Nov 25, 2008
hGH	hGH (Eligen)	Growth hormone deficiency	Novartis/Emisphere	Emisphere	Phase II	Dec 09, 2008
GLP-1	GLP-1 (Eligen)	Diabetes	Emisphere	Emisphere	Phase I	Nov 07, 2008
PYY	Peptide YY$_{3-36}$ (Eligen)	Obesity	Emisphere	Emisphere	Phase I	Nov 07, 2008

Abbreviations: PTH, parathyroid hormone; GLP-1, glucagon-like peptide-1; hGH, human growth hormone.

Researchers are approaching this problem from several different angles, such as by modifying the composition of the drug, by developing novel drug delivery technologies, and by combining these two approaches. The potential advantages of breakthrough technologies in this area justify continued efforts to identify and optimize both pharmaceutical and biological approaches for improving and maximizing the oral bioavailability of protein and peptide drugs. The opportunity for success remains, but it is more important to focus the effort by selecting an appropriate protein and peptide candidate and combining it with well-designed formulations and dosage forms. By continuously stretching the limit that may have prevented oral protein and peptide delivery from achieving its full potential, the merits of each application have to be examined individually in terms of therapeutic rationale, market potential, and technical feasibility. The positive and negative effects need to be weighed carefully before large expenditures for developmental work are committed. Another important issue related to drug stability (degradation, aggregation, and conformation changes, etc.) in the manufacturing process and during the storage of the drug product has remained largely unattended. The formulation development, scale-up, and production problems may vary for different proteins and peptides and drug delivery systems and for each combination of a drug with a particular delivery approach.

This chapter summarizes the progress in the oral delivery of protein and peptide therapeutics. In addition to pharmaceutical approaches to improve bioavailability, the section "Product Strategies with Special PD/PK Considerations" is presented. This section highlights the determination of the therapeutic indication with known drug action target, and selection of the meaningful PK profile and efficacy-relevant PK parameter(s) to evaluate bioavailability. An approach to analyze the cause(s) of the bioavailability variations is also included using the development of oral insulin as a case study.

PRODUCT STRATEGIES WITH SPECIAL PD/PK CONSIDERATIONS
Therapeutic Indication: Acute Vs. Chronic Conditions (Single Vs. Repeat Dose)

Protein and peptide drugs may be used for both acute and chronic conditions, such as in the use of glucagons for rescuing hypoglycemic episodes versus application of glucagons for maintaining a basal level. The criticality of variability on PK performance is different for different indications. In an acute situation, a single-dose drug exposure is critical, while repeat dosing has a cumulative effect that triggers a PD response in chronic therapy, as exemplified by the cumulative effect of hGH to affect a change in insulin-like growth factor I (IGF-1). Orally administered protein and peptide drugs for chronic conditions [e.g., sCT for osteoporosis (OP) and osteoarthritis (OA), insulin and GLP-1 for type 2 diabetes, PTH1-34 for OP, hGH for growth hormone (GH) deficiency, and PYY for obesity] may have some flexibility on the criticality of variability on PK performance.

Insulin is a hormone that is used to treat patients with type 1, type 2, and gestational diabetes. The variability in onset and duration of action following insulin absorption is the biggest confounder of efforts to mimic physiological insulin secretion, and hypoglycemia is the most common adverse effect of insulin therapy (3). Patients with type 1 diabetes require insulin therapy because the β cells of their pancreas are no longer manufacturing sufficient amounts of insulin to control their blood sugar levels. The lack of insulin can affect the body in two ways: first, high blood sugar can cause damage to the eyes, heart, and other organs; and second, poor protein synthesis can lead to a general weakening of the body. In this case, "physiological insulin replacement" therapy is a must to mimic normal insulin secretion for both basal and "bolus" (prandial) conditions. In type 2 diabetes, the body produces insulin, but the cells do not respond appropriately. Type 2 diabetes can disproportionately affect the sedentary, obese, and elderly. In this case, "nonphysiological insulin replacement" therapy that does not necessarily need to mimic normal β-cell secretion is adequate.

Action Target: Systemic Vs. Portal PD/PK Profiles

In pharmacology textbooks, a typical sequence of the drug-body interactions is described by PK first, followed by PD, where the PK profile is the drug concentration in the systemic circulation. This sequence is reversed, however, if the targeted organ where the pharmacological action occurs is before the systemic circulation (e.g., GI or liver) after oral administration. For example, in the case of insulin where the target organ is the liver, the meaningful PK profile after oral delivery is in the portal vein rather than in the systemic circulation. Physiological insulin that is secreted by the pancreas enters the portal circulation and inhibits hepatic production. As a result, it undergoes significant metabolism in the liver (about 50%). The ratio of plasma insulin in the portal circulation to the peripheral circulation is approximately 2. In experiments using rodents, approximately two-thirds of insulin absorbed in the portal vein was used by the liver to reduce hepatic glucose production. The measured mean insulin portal to systemic concentration ratio is about 3, while the glucose level remained approximately the same in both portal and systemic circulation (4).

Oral administration of insulin mimics the physiological pathway, where, after absorption from the GI tract, insulin enters into the portal vein and targets the liver. Hence, oral insulin not only replaces the needle but can also be used as a targeted drug delivery system. Potential clinical advantages of the first-pass hepatic insulin extraction include: (*i*) less incidence of hypoglycemia, (*ii*) reduction in long-term complications related to hyperinsulinemia, (*iii*) improvement in hepatic glucose control especially during nighttime, and (*iv*) enhanced patient compliance.

PK Parameter: Area Under the Curve Vs. Peak Concentration

Many endogenous hormones exhibit physiological pulsatile secretion. Similar to neurological signaling, this natural circadian rhythm is closely related to its biological mode of action. For those proteins and peptides, the peak concentration (C_{max}) is the efficacy-relevant PK parameter. Insulin is secreted in a high-frequency pulsatile manner in humans. These pulses are delivered directly into the portal vein, the hormone undergoing extraction and dilution before delivery into the systemic circulation.

Pulsatile insulin secretion has an interval (periodicity) of approximately five minutes, with the amplitude being affected by the systemic glucose load. Oscillations of the insulin concentration observed in the portal vein are typically five-fold greater than that observed in the arterialized vein. Enhanced insulin release in response to hyperglycemia is achieved by amplification of these high-frequency pulses (5). This pattern of insulin secretion has been shown to be abnormal in subjects with type 2 diabetes and in patients at risk of developing type 1 diabetes. As the magnitude of insulin pulses is decreased in patients with type 2 diabetes (6), it is likely that the liver of these patients is exposed to even more attenuated oscillations in insulin pulse amplitude, which may contribute to the hepatic insulin resistance characteristic of this disease.

hGH is synthesized and secreted from the anterior pituitary gland in a pulsatile manner throughout the day, with surges occurring at three- to five-hour intervals. The largest and most predictable of these hGH peaks occurs about an hour after onset of sleep (7). Analysis of the pulsatile profile of hGH indicates that its concentration is less than 1 ng/mL for basal levels, while the peaks are approximately 10 to 20 ng/mL (8). This therapeutic hGH level is intended to raise the level of IGF-1, which mediates many of the systemic actions of hGH. A study in monkeys showed that hGH C_{max} rather than area under the curve (AUC) correlated with IGF-1 increase for both SC and oral dosing (9).

For sCT, an extensive amount of research has been conducted, in which effects on food intake and circadian variation are well-established parameters (10,11). Pre-dinner dosing was more effective in comparison with morning or bedtime dosing. These investigations and data are instrumental in understanding and utilizing natural circadian rhythms for optimizing drug efficacy.

PHARMACEUTICAL APPROACHES TO IMPROVING ORAL BIOAVAILABILITY
Overview

Figure 1 represents a typical drug concentration–time PK profile after oral administration. The C_{max} and the time to achieve the peak concentration (T_{max}) can be estimated from this profile. For the purpose of simplicity, a one-compartment model leads to the following expressions for T_{max} and C_{max} (the same concept applies to other compartment or noncompartment models):

$$T_{max} = \frac{\ln(k_a/k_e)}{k_a - k_e} \qquad (1)$$

$$C_{max} = \frac{\text{Dose} \times F}{V_d} e^{-k_e T_{max}} \qquad (2)$$

where k_e and k_a are the first-order elimination and absorption constants, respectively; V_d is the volume distribution; Dose is the oral dose; and F is the absolute bioavailability (expressed in fraction).

In turn, the bioavailability F is given by

$$F = f_{abs}(1 - f_{GI})(1 - f_H) \qquad (3)$$

where f_{abs} is the fraction of drug absorbed into the GI wall, f_{GI} is the fraction metabolized in the GI (preabsorption in lumen and intracellular), and f_H is the fraction extracted during the first-pass hepatic metabolism.

Despite the high level of activity in peptide-based drug research, imparting good bioavailability while maintaining pharmacological efficacy is one of the key challenges that have hindered the development of peptides into useful

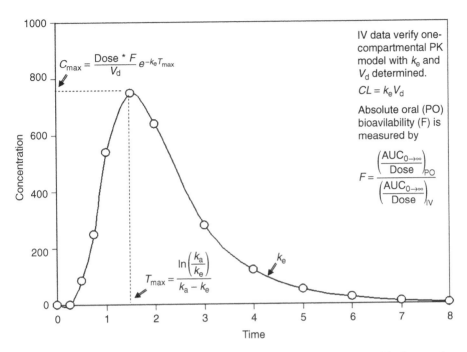

FIGURE 1 A representative pharmacokinetic profile of oral protein and peptide therapeutics, showing one-compartment model.

therapeutic products. The oral administration of peptide and protein drugs faces two formidable problems. The first is the protection against the metabolic barriers of the GI tract. The entire GI tract and the liver are designed to metabolize and break down proteins and peptides into smaller fragments of 2 to 10 amino acids using a variety of proteolytic enzyme (proteases). There are four major types of enzymes of concern: aspartic proteases (pepsin, rennin), cystine proteases (papain, endopeptidase), metallo proteases (carboxypeptidase-A, ACE), and serine proteases (thrombin, trypsin). The second problem is the absence of a carrier system for absorption of peptides with more than three amino acids. Pharmaceutical approaches to address drug permeation across these barriers that have been successful with traditional smaller organic drug molecules cannot be readily developed into effective protein and peptide formulations. Designing and formulating a protein and peptide drug delivery for the GI tract has been a formidable challenge because of these unfavorable physicochemical properties of enzymatic degradation and poor membrane permeability.

The development of oral drug delivery technologies to improve the absorption of peptide and protein drugs is one of the greatest challenges in drug development. The next section reviews the following approaches to deliver therapeutic protein and peptide drugs: (*i*) chemical modifications of protein and peptide drugs, (*ii*) formulation additives to enhance the GI absorption of protein and peptide drugs, and (*iii*) dosage form design that protects protein and peptide drugs from GI metabolism. In addition, combination of these approaches can further maximize the overall drug bioavailability. Those approaches have been described in a number of excellent review articles that readers are referred to for additional details (1,12–27).

Chemical Modifications of Protein/Peptide Drugs

Chemical modification of protein and peptide drugs can improve their enzymatic stability and/or GI membrane permeability. This approach can also be used to reduce immunogenicity. Lipidization, which is the covalent conjugation of a hydrophobic moiety, can increase the lipophilicity of protein and peptide molecules, whereas conjugation with polyethylene glycol (PEG) can improve solubility and offer protection from enzymatic degradation. As an example, hexyl-insulin-monoconjugate 2 (HIM2) is a conjugated insulin developed by Nobex. A single amphiphilic oligomer is covalently linked to the free amino group on the Lys-β 29 residue of human insulin via an amide bond (28). As demonstrated in phase II studies using the conjugated insulin, the bioavailability was increased from an increase in GI absorption, a lowering in enzyme degradation, and a reduction in protein self-association.

Recently, there are some interesting approaches to make a new generation of therapeutic proteins (new biological entities). The first one is Nautilus' approach to develop a naked protein with minimal structural change (a single amino acid variation) (29). Using its proprietary semirational technology, Nautilus has designed specific point mutations aimed at protecting the therapeutic protein molecule from degradation by a wide range of proteases. In most cases, a single point mutation is sufficient to confer high resistance to proteolysis. These changes are designed so that they do not alter the specific activity of the protein (i.e., their potency) with the respective native ones. This has been demonstrated to make these molecules highly protected in vivo in skin, blood, and other biological fluids. As a consequence, these molecules are not degraded rapidly in the GI tract and can be readily absorbed into the bloodstream. In addition, their half-life and in vivo activity profile were significantly increased.

A second interesting development is an unusual class of proteins called β-peptides (30). Most protein drugs in use and in development, such as antibodies that target cancer cells, are made from α-amino acids, which are the building blocks of naturally occurring proteins. These new β-peptides are metabolized much more slowly than α-proteins in the body. Advances have been made that enable the control of the β-peptides' three-dimensional structure and activity similar to that of α-proteins made by the body.

Formulation Additives Enhancing Protein/Peptide Drugs' GI Absorption

Designing and formulating protein and peptide drugs for delivery across the GI tract require a multitude of strategies, such as enhancing transcellular permeation. Numerous classes of compounds with diverse chemical properties have been reported to enhance the intestinal absorption of protein and peptide drugs (14,31). The key criterion is to ensure the safety and effectiveness of drug delivery, and an understanding of the mechanism of absorption enhancement. Emisphere has developed a series of "carriers," broadly referred to as the Eligen technology, that enable the oral delivery of protein and peptide drugs. These carriers are low-molecular-weight organic compounds that interact weakly and noncovalently with the drug to increase the ability of the drug to cross the GI epithelium. The application of these carriers to achieve increased oral absorption of important protein and peptide therapeutics has been demonstrated in clinical studies that are at various stages of development. Interest in this approach has also prompted other companies to search for GRAS (generally regarded as safe) materials to enhance the oral delivery of protein and peptide drugs.

While the absorption of insulin dosed orally in a rodent model was demonstrated to take place throughout the entire GI tract (32), the optimal site of absorption was further evaluated and shown to be formulation dependent in a dog model (33). The effects of dosing site and formulation parameters (including a carrier) on the GI absorption of insulin were evaluated in a conscious, jejunum-cannulated dog model. After correcting for inter-animal variations, oral uptake of insulin was shown to be more favorable in the jejunum using solution formulations containing three different insulin/carrier dose ratios.

The transport enhancement across the GI mucus layer is reported to be important for some hydrophobic proteins and peptides. The stomach antrum has the thickest "firmly adherent" layer with the largest pH gradient, while the jejunum has the thinnest firmly adherent layer with a small pH gradient. Mucoadhesive polymeric systems provide an intimate contact to the mucin layer of the mucosal epithelium and can thus result in an increase in oral bioavailability. Additionally, the residence time of the delivery system at the site of drug absorption is increased. These polymers decrease the drug clearance rate from the absorption site and thereby increase the time available for absorption (34).

The strategy that involves modulating the permeability of the tight junction is not without safety concerns. Once tight junctions have been opened, transport is enhanced not only for drugs but also for potentially toxic or unwanted molecules present in the GI tract. Because many proteins and peptides are used for the treatment of chronic conditions, the safety of this approach merits further investigation.

Dosage Form Design Protecting Protein/Peptide Drugs from GI Metabolism

Cyclosporine is reported to have a relatively high oral bioavailability (>30%) and provides an example where the drug is relatively stable from metabolic degradation in the GI tract. In contrast, drugs that are susceptible to being degraded in the GI tract would typically result in low bioavailability, with a high degree of variability. As such, strategies to protect the protein and peptide drugs from degradation in the GI tract can further enhance oral bioavailability. Various strategies have been employed to address this issue. Instead of "inhibiting GI enzymes," a more pragmatic approach is to protect the drugs from GI metabolism without altering the natural flora of enzymes in the GI tract. Side effects such as systemic intoxication and disturbed digestion of food proteins are often associated with the use of enzyme inhibitors. Here, the use dosage forms developed for small organic drug molecules may be applicable. These include liposomes, microspheres, nanoparticles, emulsions, hydrogels, lipid vesicles, particulates, and capsule mucoadhesives. The challenge is to incorporate the protein and peptide drugs into these vehicles.

While small intestinal cells secrete a variety of enzymes, including aminopeptidases, this part of the GI tract does not significantly contribute to the digestive process compared with events in the stomach or those resulting from pancreatic enzymes. Enteric-coated dosage forms are especially useful to transit through the stomach. It has been found that insulin delivery to the mid-jejunum protects insulin from gastric and pancreatic enzymes, and release from the dosage form is enhanced by intestinal microflora (35). Dosage forms such as nanoparticulates and microparticulates, mucoadhesives, or microspheres that are targeted to a part of the gut where proteolytic activities are relatively low can protect the protein and peptide drug from luminal proteolytic degradation and release the drug at the most favorable site for absorption. The protein and peptide drugs encapsulated in the nanoparticles are less sensitive to enzyme degradation through their association with polymers and have been demonstrated to have better absorption through GI tract than their native counterparts. Micro- and nanoencapsulation techniques, such as the w/o/w multiple emulsion technologies, have evolved to allow for the incorporation of sensitive proteins to resist the harsh environment of the mucosa. In addition, biodegradable polymers have been used with well-known degradation properties. For instance, a dry emulsion formulation was enteric coated using a pH-responsive polymer hydroxypropyl methylcellulose phthalate (HPMCP) (36). The release behavior of encapsulated insulin was found to be responsive to external pH and the presence of lipase under simulated GI conditions. These microspheres restrict the release of proteins to a more favorable region of the GI tract. Another approach to inhibit enzyme activity is to alter the pH to inactivate the local digestive enzymes. Pepsin in stomach is active at low pH, but it can be rapidly inactivated above pH 5 (37). A sufficient amount of a pH-lowering buffer that lowers the local intestinal pH to values below 4.5 can deactivate trypsin, chymotrypsin, and elastase.

CASE STUDIES
Oral Salmon Calcitonin: The Lead in Product Development
sCT is approved for the treatment of OP and other diseases that are involved in the acceleration of bone turnover. Preliminary evidence suggested that sCT might provide therapeutic benefits that could translate into chondroprotective effects in progressive degenerative joint diseases, such as OA and rheumatoid arthritis (38). Novartis is applying the Eligen technology to develop an oral dosage form of sCT for OP and OA indications. Clinical developments of these products are currently in phase III.

In the first randomized clinical study, the results showed the antiresorptive safety and efficacy of oral Eligen formulation in 277 postmenopausal women who participated in a three-month study (39). The results showed effective, well-tolerated, oral absorption, marked inhibition of bone resorption with minimal alteration of formation, and the reproducibility of responses over a three-month period. In addition, the level of serum concentrations achieved could evoke pronounced biological responses in bone tissue, as indicated by marked decreases in biomarkers of bone resorption. Interestingly, with this oral formulation, an inhibition in bone resorption without the secondary effect on bone formation was also reported. If dosed optimally, this approach might alter bone turnover favorably through the primary inhibition of bone resorption. Bagger et al. assessed the efficacy of a three-month study that investigated the effect of oral sCT on cartilage degradation on the basis of the changes in the urinary excretion of C-terminal telopeptide of collagen type II (CTX-II) (40). In addition, they investigated whether the response of oral sCT to urinary CTX-II depended on the baseline level of cartilage turnover. The clinical study was a randomized, double-blind, and placebo-controlled study that included 152 Danish postmenopausal women with ages from 55 to 85 years. The results demonstrated that oral sCT induced a significant dose-dependent decrease in 24-hour urinary CTX-II excretion. Similar dose-dependent responses were found in 24-hour urinary CTX-I and CTX-II corrected for creatinine excretion at month 3. The study demonstrated that orally delivered sCT, in addition to its pro-nounced effect on bone resorption, also reduced cartilage degradation and can provide additional therapeutic benefits of chondroprotection. Women with high cartilage turnover can significantly benefit from oral sCT treatment.

Very recently, a report has been published on the investigation of diurnal variation in bone resorption and the effect of dosing regimen on the oral sCT bioavailability and PK profile in healthy postmenopausal women (41). The study concluded that pre-dinner dosing under fasting conditions was more effective in achieving absorption and in the suppression of serum CTX com-pared with morning or bedtime dosing. The results of this investigation are instrumental in understanding and utilizing natural circadian rhythms for optimizing drug efficacy.

Oral Human Insulin: Expectations Vs. Challenges

Among protein and peptide drugs, the oral delivery of insulin has received the widest attention. The development of oral insulin has long been regarded as the Holy Grail of drug delivery, as developers search for a replacement to daily injections, particularly for young diabetic children who often have problems with injections. An oral insulin product would have tremendous benefits by decreasing the number of injections for diabetic patients and reducing the incidence of side effects (42,43).

The development of oral insulin has been at different stages for different companies and covered a broad spectrum, from preclinical testing to phase II clinical trials (44). A notable advancement is the completion of phase II trials in two oral insulin products: Nobex's oral HIM2 (45) and Emisphere's oral Eligen insulin pill (46). The carrier-enhanced insulin permeability across the GI mem-brane barrier makes the drug absorption possible, while the drug available for transport is site and time dependent primarily because of the enzyme degra-dation. The protective effect of the carrier on the drug to prevent degradation, either by the drug interacting with itself or with the enzyme, is relatively weak and variable because of the local dynamic environment at the site of absorption. Addition of excipients (preferably GRAS materials) may work with the carrier synergistically to reduce variability and further enhance drug absorption. Although the carrier can deliver the drug from any region of GI tract, insulin is quickly metabolized in the stomach after oral dosing and only a small fraction of the drug is available for further absorption in the small intestine (47).

One of the notable observations from the delivery of insulin using Eligen in both monkey and human studies was the relatively small variability in T_{max} and the large variability in C_{max} from the insulin PK profiles. On the basis of equations (1–3), an analysis was conducted, and the results are summarized in Table 2. First, the small variation in T_{max} suggests that variability associated with drug absorption (k_a) is small, and the contribution to T_{max} from drug elimination (k_e) is independent of absorption. Hence, the large variability in C_{max}

TABLE 2 Analysis of Oral Eligen Insulin's PK Profiles: Small Variability in T_{max} and Large Variability in C_{max}

PK parameters, bioavailability	Contributing factors	Drug dose dependent[a]	Carrier dose dependent[a]	Remarks
$t_{1/2}$	k_e	No	No	
T_{max}	k_e	No	No	
	k_a	No	Yes	Small variability in T_{max} suggests small variability in k_a at the fixed carrier dose.
C_{max} or $AUC_{0 \to \infty}$	k_e	No	No	
	k_a	No	Yes	(see above)
	V_d	No	No	
	Dose	Yes	No	
	F	No	Yes	Large variability in C_{max} is due to large variability in F at the fixed doses of both the drug and the carrier.
F	f_H	No	No	
	f_{abs}	No	Yes	With small variability in k_a at the fixed carrier dose (see T_{max}), a large variability in drug concentration at the absorption site can cause large variability in f_{abs}, and then F.
	f_{GI}	No	Yes	The carrier's weak and time/position-dependent enzyme inhibition at the absorption site may provide large variability on the drug concentration for GI transport.

[a]Assumed no saturation kinetics with relatively low concentrations of the drug and the carrier at absorption sites.
Abbreviation: GI, gastrointestinal.

is then due to a large variability in F, since other parameters (V_d, Dose) are not related. Finally, it can be concluded that a large variability in drug concentration at the absorption site can lead to the large variation in f_{abs} and F. In addition, the weak, time- and position-dependent enzyme inhibition by the carrier at the site of absorption suggests that the value of F is small and more sensitive to variability. Because of the enzymatic degradation, only a small fraction of drug (with large inter- and intra-variability) is available for absorption even if permeability is enhanced by the carrier. Overall, insulin is subjected to acid-catalyzed degradation in the stomach, luminal degradation by pepsin, pancreatic trypsin and/or α-chymotrypsin, and intracellular degradation (48). The cytosolic enzyme that degrades insulin is the insulin-degrading enzyme. Insulin can be available for absorption only if the activity of the enzyme attack is significantly reduced or eliminated (49).

Oral Human Growth Hormone: Feasibility Assessment from Preclinical to Clinical

hGH is indicated primarily as a replacement therapy that can benefit both children and adults with GH deficiency. Currently, all approved GH products require subcutaneous (SC) or intramuscular (IM) injections that are administered daily or thrice weekly. The therapeutic GH blood level of 10 to 20 ng/mL is intended to raise the level of IGF-1 that in turn mediates many of the actions of GH to the normal range. The feasibility of oral hGH was assessed in preclinical and clinical studies using Eligen technology.

Initial preclinical studies conducted in rats and in primates showed that hGH blood concentrations rose quickly after oral dosing and was rapidly eliminated (50). In rats, the PK profiles from all doses showed that T_{max} was less than 15 minutes. The bioavailability of oral hGH relative to SC dosing in rats from these simple solution formulations was 5% on the basis of C_{max}. The bioavailability (relative to SC) was 3% in primates on the basis of the C_{max}. T_{max} was 30 to 45 minutes in studies conducted in monkeys using capsule formulations. The bioactivity of GH was further investigated in hypophysectomized rats, with their pituitary gland surgically removed. In a 10-day growth study, a significant increase in weight gain of hypophysectomized rats after repeated oral dosing of hGH combined with a carrier (compared with the carrier alone) indicated that hGH was biologically active (51).

The type of dosage form can affect the efficiency of absorption. Aqueous solutions and powder-in-capsule formulations of hGH in combination with a carrier were compared in a primate study (52). The results showed that capsule formulations were more efficient than solutions in delivering hGH. However, the T_{max} from the capsules was approximately 15 minutes later than that from solutions. In addition, the delivery of hGH from solutions at pH 12 was shown to be indifferent than that from capsules. However, solutions at lower pH (10 and 8.4) delivered significantly lower amounts of hGH than the capsules.

In a four-week repeat dosing, oral hGH in combination with a carrier was well tolerated, and both bioavailability and bioactivity were demonstrated in cynomolgus monkeys. The toxicokinetics (TK) of the carrier, the pharmacokinetics of hGH, and the pharmacodynamics of IGF-1 were determined (9). The PK profile of oral hGH using Eligen is similar to physiologically pulsatile secretion of hGH, with a rapid rise in hGH concentration ($T_{max} \leq 30$ minutes), which is then followed by a relatively quick elimination phase ($t_{1/2} \approx 30$ minutes). No accumulation of either the carrier or hGH was observed after the repeat four-week dosing study. The rise in IGF-1 was 80% by SC administration and 25% by oral delivery.

The oral delivery of hGH was also evaluated in a single-dose, double-blind, double-dummy, randomized, dose-escalating crossover study with nine healthy subjects. The clinical results showed that administration of the solid dosage form was well tolerated, and physiologically relevant hGH concentrations were achieved (53). The T_{max} was shorter and less variable from the oral delivery of hGH ($T_{max} = 0.6 \pm 0.1$ hour) than that from SC delivery ($T_{max} = 5.4 \pm 2.9$ hours). The half-life from oral delivery of hGH ($t_{1/2} = 0.4 \pm 0.1$ hour) was similar to IV and differed from SC delivery ($t_{1/2} = 2.8 \pm 0.5$ hours). The bioavailability of hGH relative to SC was 2%, with a C_{max} of 12 ng/mL.

A phase I single-arm study was carried out in GH-deficient adult patients who were on hGH treatment (54). All patients received four 100 mg of hGH tablets per day during seven days (one in the morning, one in the evening, and two at bedtime). After oral dosing, hGH was rapidly absorbed and excreted, normally within two hours. There was evidence of greater hGH absorption on day 1 compared with day 7. The absorption of hGH was generally lower in the evening dose compared with the morning and bedtime doses. Following treatment with hGH/carrier tablets, there was a statistically significant increase in IGF-1 levels on treatment day 7 when compared with values at the end of the seven-day washout period. Patients who showed IGF-1 response were those who had no endogenous GH secretion and/or had the highest absorption of hGH.

CONCLUDING REMARKS

The primary reasons for the low oral bioavailability of protein and peptide drugs are presystemic enzymatic degradation and poor drug permeation across the intestinal mucosa. Attempts to improve the oral bioavailability of protein and peptide drugs have ranged from chemical modification, formulations with absorption enhancers, and special dosage forms that protect the drug from enzymatic degradation. To date, success to reduce drug degradation has been limited. The development of technologies to overcome this limitation would provide a significant breakthrough for product development of this class of important drugs.

Although the challenges are significant to develop oral protein and peptide products that meet regulatory requirements, intra- and inter-reproducibility, and manufacturing cost, the potential of this emerging field is promising. Oral sCT products for both OP and OA are in phase III clinical development. In addition, the development of oral insulin would provide significant benefits to diabetes care.

At the current pace of research coupled with data from recent developments, the idea of oral protein and peptide drug delivery is more convincing than ever before. Better scientific understanding of breakthrough technologies to improve the product developability (improved bioavailability, product stability, and tolerability, etc.) justifies continued efforts to identify and optimize opportunities for innovation. There is no doubt that delivery of protein and peptide drugs will continue to be a vibrant area of research. Successful oral drug delivery technologies will bring more protein and peptide therapeutics to fruition and make them available to serve the needs of the patients.

REFERENCES

1. Rick S. Oral protein and peptide drug delivery. In: Wang B, Siahaan T, Soltero R, eds. Drug Delivery: Principles and Applications. Hoboken, NJ: Wiley Interscience, 2005:189–215.
2. Dubin CH. Proteins & peptides: dependent on advances in drug delivery? Drug Deliv Technol 2009; 8(3):36–41.
3. DeWitt DE. Outpatient insulin therapy in type 1 and type 2 diabetes mellitus. JAMA 2003; 289(17):2254–2264.
4. Liu P, Zhang H, Dinh S. Oral delivery of insulin: in situ portal perfusion in a rat model. DTS Annual Meeting, San Francisco, CA, November 2003. Diabetes Technology Society.
5. Song SH, McIntyre SS, Shah H, et al. Direct measurement of pulsatile insulin secretion from the portal vein in human subjects. J Clin Endocrinol Metab 2000; 85:4491–4499.
6. Laedtke T, Kjems L, Porksen N, et al. Overnight inhibition of insulin secretion restores pulsatility and the proinsulin/insulin ratio in type 2 diabetes. Am J Physiol 2000; 279:E520–E528.
7. Takahashi Y, Kipnis DM, Daughaday WH. Growth hormone secretion during sleep. J Clin Invest 1968; 47:2079–2090.

8. Nindl BC, Hymeret WC, Deaver DR, et al. Growth hormone pulsatility profile characteristics following acute heavy resistance exercise. J Appl Physiol 2001; 91:163–172.
9. Liu P, Shepard T, Dinh S. Oral delivery of human growth hormone in monkeys using a carrier: 4-week repeat dose toxicokinetics/pharmacokinetics/pharmacodynamics. CRS International Symposium on Controlled Release of Bioactive Materials, Honolulu, July 2004. Controlled Release Society.
10. Bjarnaso NH, Henriksen EE, Alexandersen P, et al. Mechanism of circadian variation in bone resorption. Bone 2002; 30:307–313.
11. Christgau S. Circadian variation in serum crosslaps concentration is reduced in fasting individuals. Clin Chem 2000; 46:431.
12. Fix JA. Oral controlled release technology for peptides: status and future prospects. Pharm Res 1996; 13(12):1760–1764.
13. Jain NK. Oral protein drug delivery. In: Jain NK, ed. Advances in Controlled and Novel Drug Delivery. New Delhi: CBS Publishers, 2001:232–254.
14. Sood A, Panchagnula R. Peroral Route: an opportunity for protein and peptide drug delivery. Chem Rev 2001; 101(11):3275–3304.
15. Adessi C, Sotto C. Converting a peptide into a drug: strategies to improve stability and bioavailability. Curr Med Chem 2002; 9(9):963–978.
16. Shah RB, Ahsan F, Khan MA. Oral delivery of proteins: progress and prognostication. Crit Rev Ther Drug Carrier Syst 2002; 19(2):20–114.
17. Lambkin I, Pinilla C. Targeting approaches to oral drug delivery. Expert Opin Biol Ther 2002; 2:67–73.
18. Lee HJ. Protein drug oral delivery: the recent progress. Arch Pharm Res 2002; 25(5):572–584.
19. Mahato RI, Narang AS, Thoma L, et al. Emerging trends in oral delivery of peptide and protein drugs. Crit Rev Ther Drug Carrier Syst 2003; 20(2–3):153–214.
20. Hamman JH, Enslin GM, Kotze AF. Oral delivery of peptide drugs: barriers and developments. BioDrugs 2005; 19(3):165–177.
21. Arhewoh IM, Ahonkhai EI, Okhamafe AO. Optimising oral systems for the delivery of therapeutic proteins and peptides. Afr J Biotechnol 2005; 4(13):1591–1597.
22. Mustata G, Dinh S. Approaches to oral drug delivery for challenging molecules. Crit Rev Ther Drug Carrier Syst 2006; 23(2):111–135.
23. Kumar TR, Soppimath K, Nachaegari SK. Novel delivery technologies for protein and peptide therapeutics. Curr Pharm Biotechnol 2006; 7:40–76.
24. Morishita M, Peppas NA. Is the oral route possible for peptide and protein drug delivery. Drug Discov Today 2006; 11(19–20):905–910.
25. Semalty A, Semalty M, Singh R, et al. Properties and formulation of oral drug delivery systems of protein and peptides. India J Pharm Sci 2007; 69(6):741–747.
26. Shaji J, Patole AV. Protein and peptide drug delivery: oral approaches. India J Pharm Sci 2008; 70(3):269–277.
27. Shingh R, Singh S, Lillard JW. Past, present, and future technologies for oral delivery of therapeutic proteins. J Pharm Sci 2008; 97(7):2497–2523.
28. Kipnes M, Dandona P, Tripathy D, et al. Control of postprandial plasma glucose by an oral insulin product (HIM2) in patients with type 2 diabetes. Diabetes Care 2003; 26:421–426.
29. Martin P. Beyond the next generation of therapeutic proteins, BTi, October 2006. Available at: http://www.biotech-online.com/uploads/tx_ttproducts/datasheet/beyond-the-next-generation-of-therapeutic-proteins.pdf.
30. Bourzac K. Protein drugs with more power. Technology Review published by MIT, February 23, 2007.
31. Aungst BJ. Intestinal permeation enhancers. J Pharm Sci 2000; 89:429–442.
32. Sarubbi D, Maher J, Liu P, et al. Gastric absorption of oral insulin in rats. AAPS Annual Meeting, Toronto, October 2002. American Association of Pharmaceutical Scientists.
33. Liu P, Liao J, Kidron M, et al. Effects of dosing site, dosage form, and drug/carrier dose ratio on oral insulin delivery in a conscious dog model. CRS International Symposium on Controlled Release of Bioactive Materials, Vienna, July 2006. Controlled Release Society.

34. Bernkop-Schnürch A, Krauland AH, Leitner VM, et al. Thiomers: potential excipients for non-invasive peptide delivery systems. Eur J Pharm Biopharm 2004; 58:253–263.
35. Toorisaka E, Hashida M, Kamiya N, et al. An enteric-coated dry emulsion formulation for oral insulin delivery. J Control Release 2005; 107:91–96.
36. Sakuma S, Hayashi M, Akashi M, et al. Design of nanoparticles composed of graft copolymers for oral peptide delivery. Adv Drug Deliv Rev 2001; 47:21–37.
37. Lin Y, Fusek M, Lin X, et al. pH dependence of kinetic parameters of pepsin, rhizopuspepsin, and their active-site hydrogen bond mutants. J Biol Chem 1992; 267(26): 18413–18418.
38. Karsdal MA, Sondergaard BC, Arnold M, et al. Calcitonin affects both bone and cartilage: a dual action treatment for osteoarthritis? Ann N Y Acad Sci 2007; 1117:181–195.
39. Tankó LB, Bagger YZ, Alexandersen P, et al. Safety and efficacy of a novel salmon calcitonin (sCT) technology-based oral formulation in healthy postmenopausal women: acute and 3-month effects on biomarkers of bone turnover. J Bone Miner Res 2004; 19:1531–1538.
40. Bagger YZ, Tankó LB, Alexandersen P, et al. Oral salmon calcitonin induced suppression of urinary collagen type II degradation in postmenopausal women: a new potential treatment of osteoarthritis. Bone 2005; 37:425–430.
41. Karsdal MA, Byrjalsen I, Riis BJ, et al. Investigation of the dilurnal variation on bone resorption for optimal drug delivery and efficacy in osteoporosis with oral calcitonin. BMC Clin Pharmacol 2008; 8:12.
42. Gowthamarajan K, Kulkarni GT. Oral insulin—fact or fiction? Possibilities of achieving oral delivery for insulin. Resonance 2003; 8(5):38–46.
43. Korytkowski M. When oral agents fail: practical barriers to starting insulin. Int J Obesity 2006; 26(suppl 3):S18–S24.
44. Werle M. Innovations in oral peptide delivery—a report. Future Drug Delivery June 2006, Touch Briefings, London.
45. Still JG. Development of oral insulin: progress and current status. Diabetes Metab Res Rev 2002; 18(suppl 1):S29–S37.
46. Arbit E, Majuru S, Gomez-Orellana I. Oral delivery of biopharmaceuticals using the Eligen technology. In: McNally EJ, Hastedt JE, eds. Protein Formulation and Delivery. New York: Informa Healthcare USA Inc., 2008:285–303.
47. Chen W, Wang H, Fotso J, et al. A novel method of applying an ELISA for determination of insulin in rat gastro-intestinal (GI) fluids. AAPS Annual Meeting, San Antonio, October 2006. American Association of Pharmaceutical Scientists.
48. Ghilzai NMK. Oral insulin delivery strategies using absorption promoters, absorption enhancers, and protease inhibitors. Pharm Technol 2006; 30:88–98.
49. Agarwal V, Khan MA. Current status of the oral delivery of insulin. Pharm Technol 2001; 25:76–90.
50. Liu P, Kalbag S, Maher J, et al. Oral delivery of recombinant human growth hormone. CRS International Symposium on Controlled Release of Bioactive Materials, Seoul, July 2002. Controlled Release Society.
51. Maher J, Havel H, Sarubbi D, et al. Evaluation of recombinant human growth hormone (hGH) bioactivity with a growth assay after oral dosing of hGH in hypophysectomized rats. AAPS Annual Meeting, Toronto, October 2002. American Association of Pharmaceutical Scientists.
52. Liu P, Khan A, Dinh S. Oral delivery of recombinant human growth hormone in primates: aqueous solution and powder-in-capsule formulations. CRS International Symposium on Controlled Release of Bioactive Materials, Glasgow, July 2003. Controlled Release Society.
53. Dinh S, Liu P, Arbit E, et al. A Phase I, double-blind, randomized, dose escalating crossover study of an oral formulation of human growth hormone in healthy male volunteers. CRS International Symposium on Controlled Release of Bioactive Materials, Honolulu, July 2004. Controlled Release Society.
54. Dinh S, Liu P. Human growth hormone formulations. United States Patent Application 2008/0095837, April 24, 2008.

10

Drug Interaction Studies of Therapeutic Biologics: Protein and Peptide Actions and Interactions That Alter Their Pharmacokinetics, Pharmacodynamics, and Metabolism

Iftekhar Mahmood
Office of Blood Review and Research, Center for Biological Evaluation and Research, Food and Drug Administration, Bethesda, Maryland, U.S.A.

INTRODUCTION

Drug interaction can alter the pharmacokinetics and/or pharmacodynamics of a drug. In pharmacokinetic drug interactions, the concentrations of one or more drugs are altered by another drug. This change in concentration in a given drug may be due to changes in absorption, distribution, metabolism, or elimination (1). The pharmacodynamic drug interactions can be either negative (toxic effects) or positive (therapeutic benefit) and may be due to various mechanisms including receptor interaction and changes in effector mechanisms (1). For example, the combination therapy of ribavirin and interferon (IFN)-α2b in patients with chronic hepatitis C provided improved therapeutic benefit than either treatment alone, and the safety profiles of combination therapy were similar to the monotherapy treatment (2).

Biological products such as therapeutic proteins and monoclonal antibodies are becoming widely popular for the management and cure of many diseases. These products are termed as macromolecules because of their size (>1000 Da). Therapeutic proteins are naturally occurring substances in the body, each having a unique amino acid sequence and thus distinct physicochemical properties. These differences in properties affect the folding, formulation, and stability of each protein. Indeed, long-term stability issues associated with many protein therapeutics can pose unique challenges for pharmacokinetic studies. Administration of exogenous proteins can influence the stimulation or feedback mechanism of endogenous proteins, thus estimation of pharmacokinetic parameters may become difficult (3).

To properly understand the mechanism of interaction of therapeutic proteins or monoclonal antibodies with other drugs (small as well as macromolecules), it is important to know the pharmacokinetic characteristics (especially metabolism and elimination) of these macromolecules.

Absorption

Because of gastrointestinal enzymatic degradation of the protein molecules, the two most frequently used routes of administration are intravenous (IV) and subcutaneous (SC). However, for high-molecular-weight proteins, intramuscular or SC routes of administration may not be even possible. For example, tissue plasminogen activators (molecular weight of 65,000) cannot readily cross the endothelial cell membrane; therefore, these have to be administered by IV bolus or infusion (3).

Distribution

Once a drug reaches the systemic circulation, the process of distribution and elimination begins. After reaching the bloodstream, a protein molecule is distributed to cellular target sites and the interstitial space via the vascular space across the microvascular wall and cell membranes (4). This biodistribution of the macromolecules depends on the physicochemical properties of the molecule as well as the physicochemical properties and structure of the capillaries responsible for the passage of the molecule from the systemic circulation to the interstitial fluid (4). Many macromolecules are distributed into the lymphatic system following SC administration. With increasing molecular weight of proteins, the lymphatic system becomes the predominant pathway for absorption and distribution of macromolecules (4).

Although plasma protein binding of macromolecules is generally not considered to be an issue that is sufficiently important to be automatically evaluated, there are studies that do indicate that there are binding proteins for macromolecules (5). For example, six specific binding proteins were identified for insulin-like growth factor 1 (IGF-1), with IGFBP-3 (insulin-like growth factor binding protein-3) being the most important binding protein. The elimination half-life of bound IGF-1 was comparatively longer (three to four hours) than that of unbound IGF-1 (10 minutes). For growth hormone (GH), at least two binding proteins have been identified: one with high binding affinity and the other with low affinity. The clearance of unbound GH is 10 times faster than that of bound GH (5). There are many other drugs such as IFN, interleukins (ILs), and tumor necrosis factor (TNF) for which specific binding proteins have been identified (5).

Metabolism

Interaction between two drugs that affect the metabolism of one or both can be clinically important. Drugs can be inducers or inhibitors of metabolizing enzymes. Pharmacokinetic interactions for small molecules can be biotransformation and/or transporter based.

Biotransformation-based drug-drug interactions may occur because of the presence of cytochrome P (CYP)450 system (1). High concentrations of these enzymes are located in the liver and small intestine. The concentrations of CYP450 system can be altered by inhibition and induction and can vary from person to person. Many drugs may increase or decrease the activity of various CYP isozymes, which may result in adverse drug interactions, since changes in CYP enzyme activity may affect the metabolism and clearance of various drugs.

Transporter-based drug-drug interactions play an important role in the processes of drug absorption, distribution, and excretion (6). Because of these characteristics, transporters are involved in clinically significant transporter-mediated drug interactions. The ultimate result of such interactions is alteration in the efficacy or safety of a given drug (the substrate). Transporter-mediated drug interaction with macromolecules has not been established.

It is widely believed that therapeutic proteins are metabolized by the same catabolic pathways as endogenous proteins and are broken down into amino acid fragments (7). Generally, the metabolic products of proteins are not considered as a safety risk. Compared with conventional small-molecule drugs, characterizing the metabolites of therapeutic proteins is a much more difficult task. These difficulties arise because of the lack of suitable analytical method(s) and abundance of potential sites of metabolism (because of the complex structure of therapeutic proteins). Most proteins are catabolized by proteolysis via enzymes distributed throughout the body. Proteolytic activity at a site of injection can lead to protein degradation after SC administration (7). Furthermore, the rate and extent of production of the metabolites will depend on the route of administration. Although therapeutic proteins and monoclonal antibodies may not be metabolized by the CYP450 system, they can inhibit or stimulate the CYP450 system in the body (8).

Elimination

Clearance of protein drugs from the systemic circulation begins with passage across the capillary endothelium cells (9). This endothelial passage depends on the size, shape, and charge of the protein molecule and the structural and physicochemical properties of the capillaries (8). Chemical and enzymatic processes can lead to the degradation of proteins and peptides. The enzyme proteases are responsible for proteolysis, for accelerating hydrolysis and other chemical degradation processes (10).

Proteins and macromolecules are mainly cleared by renal filtration and non-p450 liver metabolism. Receptor-mediated clearance is another important mechanism of elimination for therapeutic proteins (5).

Renal Excretion

The kidneys play an important role in the clearance of proteins and amino acids. Many proteins with molecular weight less than 30 kDa are filtered by the glomerulus and excreted (11). Peptides and proteins less than 5 kDa are filtered efficiently, and their glomerular rate equals the glomerular filtration rate observed in humans (120 mL/min). As the molecular weight of proteins increases (>30 kDa), the capacity for glomerular filtration decreases. Besides molecular weight, charge and size of proteins are also important for glomerular filtration (12). After glomerular filtration, peptides can be excreted unchanged in the urine or degraded to the products that are excreted in the urine (9). Polypeptides and proteins can also be actively reabsorbed by the proximal tubules through a process known as luminal endocytosis and then hydrolyzed by the digestive enzymes in the lysosomes to peptide fragments and amino acids (8). The amino acids are then reabsorbed by a carrier-mediated, energy-dependent transport mechanism. The net result is that only a small fraction of intact protein is eliminated unchanged in the urine (9).

Hepatic Elimination

The liver plays an important role in the removal of proteins from the systemic circulation. Uptake of peptides and proteins from the blood into hepatocytes occurs by two mechanisms (9): receptor-mediated endocytosis (RME) and nonselective pinocytosis

RME can lead to saturable clearance or nonlinear pharmacokinetics for peptide and protein drugs (3). In RME, circulating proteins are taken up by specific hepatic receptors (9). Insulin and epidermal growth factor are examples of RME. Receptor-mediated uptake for some proteins is so extensive that the clearance of such macromolecules can equal the liver blood flow (4).

Pinocytosis, on the other hand, is a nonspecific, nonsaturable, noncarrier-mediated form of membrane transport; this type of transport mechanism involves vesicular uptake of bulk fluid into cells from the surrounding medium (13). Proteins taken up by this mechanism are internalized according to their concentration within plasma (9). Polymer conjugates, some antigen-antibody complexes, some glycoproteins, and pancreatic proteins are examples of proteins cleared from plasma by pinocytosis (9).

Some proteins can also be cleared from the systemic circulation by biliary excretion. Insulin and epidermal growth factor are examples of therapeutic proteins, which are excreted in the bile (14). These proteins may undergo partial proteolysis prior to biliary excretion (14).

PHARMACOKINETIC DRUG INTERACTION STUDIES
Unlike small molecules, there are only few reported drug-drug interaction studies for protein drugs and monoclonal antibodies. Some examples of drug interaction studies for therapeutic proteins are presented below.

EFFECT OF TYPE I INTERFERONS ON THE CYP450 SYSTEM
IFN-α and IFN-β are considered type I IFNs and belong to the larger helical cytokine superfamily that includes GHs, ILs, several colony-stimulating factors, and several other regulatory molecules. Type I IFNs help the immune response by inhibiting viral replication within cells of the body; these proteins are mainly metabolized in the liver and kidneys. Since type I IFNs can affect the CYP450 system, a wide variety of interaction studies with proteins of this class have been conducted. The following are some examples of type I IFN's impact on the CYP450 system and on drug interactions.

Interferon-α
Okuno et al. (15) studied the effect of IFN-α on drug-metabolizing activity in the human liver. Twelve patients with chronic hepatitis B received 6×10^6 IU of IFN-α intramuscularly for four weeks. IFN-α treatment reduced the O-demethylase and O-deethylase activities in the liver by 31% and 33%, respectively. The extent of reduction in the enzymatic activities widely varied. For example, in 2 patients, no inhibition of either enzyme was observed, whereas O-demethylase and O-deethylase activities in the remaining 10 patients varied from 9% to 61% and 16% to 62%, respectively.

In another study, Pageaux et al. (16) investigated the effect of therapeutic doses of IFN-α on CYP450 1A2 and 3A in patients with chronic hepatitis C. Eighteen patients with chronic hepatitis C received 3 million units (MU) of IFN-α three times a week. The results of the study indicated that IFN-α had no effect on the enzymatic activities of CYP1A2 and CYP3A.

Interaction with P-gp Substrates
Reguiga et al. (17) studied the in vivo effect of IFN-α on P-glycoprotein (P-gp) activity in rats. Digoxin, a P-gp substrate, was chosen to evaluate the effect of

IFN-α on the bioavailability of digoxin. Human recombinant IFN-α was given to rats (five to seven rats per group) daily for eight days at different doses (Intron A, 1×10^6, 2×10^6, or 4×10^6 IU/kg, SC), whereas pegylated IFN (peg-IFN)-α was given at 29 μg/kg SC three times a week. Rats were then given digoxin (32 μg/kg) IV or orally. The pharmacokinetics of IV administered digoxin was not modified by IFN-α, but a dose-dependent increase in the bioavailability of orally administered digoxin was noted (Table 1). Peg-IFN-α did not modify digoxin bioavailability. The authors concluded that IFN-α induced a significant dose-dependent inhibitory effect on intestinal P-gp activity that resulted in increased bioavailability of digoxin.

In an another study, Reguiga et al. (18) investigated the effect of recombinant human interferon-α (rhIFN-α) and peg-IFN-α on the pharmacokinetics of docetaxel (Taxotere), a P-gp substrate, in the rat. Sprague Dawley rats were SC pretreated with either rhIFN-α for eight days (INTRON A, 4 MIU/kg once daily) or with peg-IFN-α (ViraferonPeg, 60 mg/kg on days 1, 4, and 7). Rats received a single dose of ^{14}C docetaxel (20 mg/kg) either orally or as an IV bolus injection. Nonpeg- and peg-IFN-α increased docetaxel bioavailability (Table 2). The absorption of docetaxel was also delayed by these IFNs. IV pharmacokinetics of docetaxel was not altered by treatment with these IFNs. The results of the aforementioned two studies could have important clinical relevance since IFN-α is widely used in cancer and antiviral therapy and could affect P-gp substrates such as anticancer drugs (vincristine and doxorubicin) or antiviral drugs (indinavir and efavirenz); increased bioavailability of P-gp substrates given with IFN-α may lead to an increase in the efficacy or toxicity.

TABLE 1 Pharmacokinetic Parameters Following IV and Oral Administration of Digoxin to Rats

Parameters	Control	1 MIU/kg	2 MIU/kg	4 MIU/kg	Peg-IFN-α
Intravenous					
C_0 (μg/L)	14.1 ± 3.2	12.9 ± 1.7	14.4 ± 3.0	16.3 ± 4.4	18.0 ± 5.9
AUC (min·μg/L)	2016 ± 338	2068 ± 254	2087 ± 422	2390 ± 369	2169 ± 369
Oral					
C_{max} (μg/L)	1.08 ± 0.30	0.93 ± 0.12	1.06 ± 0.08	1.22 ± 0.48	1.21 ± 0.11
AUC (min·μg/L)	286 ± 111	286 ± 50	392 ± 83	550 ± 97	354 ± 101
% F	14.2	13.8	18.8	23	16.3

Source: From Ref. 17.

TABLE 2 Pharmacokinetic Parameters Following IV and Oral Administration of Docetaxel to Rats

Parameters	Control	rhIFN-α	Pegylated rhIFN-α
Intravenous			
C_0 (μg/L)	139.3 ± 27.8	153.1 ± 22.5	159.4 ± 34.1
AUC (min·μg/L)	0.101 ± 0.005	0.097 ± 0.005	0.108 ± 0.011
Half-life (min)	207.4 ± 46.8	160.7 ± 63.2	152.9 ± 98.6
Oral			
C_{max} (μg/L)	7.4 ± 2.5	17.0 ± 4.0	18.6 ± 5.5
AUC (min·μg/L)	0.012 ± 0.004	0.036 ± 0.01	0.033 ± 0.009
Half-life (min)	14.48 ± 4.01	11.84 ± 1.92	13.37 ± 3.84
% F	10.4	37.2	30.2

Source: From Ref. 18.

Recombinant Human Interferon-αA

Williams et al. (19) studied the effect of recombinant human IFN-αA on the pharmacokinetics of antipyrine. Following a single intramuscular dose of IFN-αA in nine patients, antipyrine clearance was found to be decreased. The decrease was variable, ranging from 5% to 47%.

In another study, the same authors (20) studied the effect of IFN-αA (single intramuscular dose) on theophylline clearance in five patients with stable chronic active hepatitis B and four healthy subjects. Like antipyrine, theophylline clearance was reduced and varied from 31% to 81% in eight of nine patients. In one patient, no change in theophylline clearance was observed.

Effect of IFN-α-Ribavirin Biotherapy on Cyrochrome P450

Becquemont et al. (21) studied the effect of IFN-α and ribavirin combination therapy on the activities of CYP1A2, CYP2D6, CYP3A4, and N-acetyltransferase-2 (NAT2) after one month of treatment. There were 14 patients with chronic hepatitis C in the study. The patients received three MU of IFN-α three times a week and 600 mg ribavirin twice a day in five patients and 500 mg ribavirin twice a day in nine patients (who were less than 75 kg of body weight). Before the initiation of the therapy, the patients also received 80 mg dextromethorphan and 140 mg caffeine. The results of this study are summarized below:

- CYP3A4 activity increased almost by threefold just before treatment to one month of IFN-α and ribavirin therapy. However, the increase in CYP3A4 activity was not unidirectional. There were seven patients in whom CYP3A4 activity increased from 112% to 1677%, whereas in six patients, CYP3A4 activity decreased by 47% to 67% from pretreatment values.
- CYP2D6 activity also increased almost by threefold, but, like CYP3A4 activity, the increase was not unidirectional. CYP2D6 activity increased from 120% to 322% in nine patients and decreased from 42% to 93% in five patients.
- CYP1A2 and NAT2 activities remained unchanged from the pretreatment period to one month after treatment. Smoking status did not alter CYP1A2 activity before and after treatment.

Overall, the study indicates that IFN-α and ribavirin combination therapy can decrease or increase the activities of CYP3A4 and CYP2D6 in patients with chronic hepatitis C. Therefore, individual therapeutic drug monitoring may become necessary in this patient population if they are taking any drug that is a substrate for either CYP3A4 or CYP2D6.

Pegylated Interferon-α2a

Peg-IFN-α2a is a covalent conjugate of recombinant IFN-α2a. Treatment with peg-IFN-α2a once weekly for four weeks in healthy subjects resulted in inhibition of P450 1A2 and a 25% increase in theophylline area under the curve (AUC). On the basis of the study, it was suggested that theophylline serum levels should be monitored and appropriate dose adjustments considered for patients given both theophylline and peg-IFN-α2b (package insert). There was, however, no effect of peg-IFN-α2b on the pharmacokinetics of drugs that are metabolized by CYP2C9, CYP2C19, CYP2D6, or CYP3A4.

TABLE 3 Probe Drugs in Pittsburg Mixture, the Respective Doses, and Enzymes

Probe Drug	Dose (mg)	Enzymes
Caffeine	100	CYP1A2
Mephenytoin	100	CYP2C19
Debrisoquine	10	CYP2D6
Chlorzoxazone	250	CYP2E1
Dapsone	100	CYP2C8/9

Source: From Ref. 23.

Sulkowski et al. (22) evaluated the effect of peg-IFN-α2a on the pharmacokinetics of methadone in patients with chronic hepatitis C undergoing methadone maintenance therapy for at least three months. The subjects ($n = 24$) received 180 µg SC peg-IFN-α2b once weekly for four weeks and continued their methadone regimen. The results of the study indicated that treatment with peg-IFN-α2b once weekly for four weeks was associated with methadone levels that were 10% to 15% higher than that at baseline. The clinical significance of this finding is not known.

Interferon-α2b

Islam et al. (23) studied the effect of high-dose IFN-α2b (INTRON A) on the activities of CYP enzymes. Seventeen patients with high-risk melanoma received IFN-α2b IV at a dose of 20 MU/m^2/day for 5 days/week for 4 weeks (induction phase) followed by SC administration of IFN-α2b at a dose of 10 MU/m^2/day for 3 days/week for 48 weeks (maintenance phase). In vivo CYP enzyme activities were measured by administrating the "Pittsburgh mixture" (Table 3). The "five-drug mixture" was given orally simultaneously. The results of the study indicated that high-dose IFN-α2b had different degrees of effect on CYP enzymes. There was a 60% inhibition in CYP1A2 activity, whereas no effect on CYP2E1 activity was noted. Significant inhibition of 1A2 and 2D6 were noted immediately after the first IFN-α2b dose, whereas significant inhibition of CYP2C19 was found on day 26. CYP inhibition led to the side effects (fever and neurological toxicity) of IFN-α2b-treated patients.

Effect of Interferon-α2b on Methadone Pharmacokinetics

Methadone, a racemic mixture, is primarily metabolized by CYP3A4, secondarily by CYP2D6, and to a lesser extent by CYP1A2 and CYP2B6 (18). Gupta et al. (24) evaluated the effects of multiple doses of peg-IFN-α2b (PEGASYS) on the steady-state pharmacokinetics of methadone in patients with hepatitis C. Twenty adults with hepatitis C virus infection received peg-IFN-α2b (1.5 µg/kg/wk) SC for four weeks and maintained their normal methadone regimen (approximately 40 mg/day). There was approximately 15% increase in the exposure of individual isomers as well as total methadone after four weekly doses of peg-IFN-α2b, as compared with the exposure observed before peg-IFN-α2b administration. According to the authors, this increase in methadone exposure may not be of any clinical significance, although this study further supports the concept that peg-IFN-α2b can modulate metabolic enzyme functions associated with drug clearance.

Berk et al. (25) studied the effect of peg-IFN-α2b (1.5 mg/kg given SC, 2 doses given one week apart) on steady-state pharmacokinetics of methadone (40–200 mg/day given orally) in nine patients with hepatitis C virus and HIV. Overall, a 24% and 17% increase in mean methadone C_{max} and $AUC_{(0-24)}$ was observed after two weeks of peg-IFN-α2b administration. No dosage adjustment of methadone was suggested by the authors when it was given with peg-IFN-α2b in patients with hepatitis C virus and HIV.

Interferon-β

Okuno et al. (26) studied the effect of IFN-β on the drug-metabolizing activity in the human liver. Seven patients with chronic hepatitis C were given IFN-β at doses of 3×10^6 to 9×10^6 IU/day for eight weeks. IFN-β treatment reduced the O-demethylase and O-deethylase activities in the liver by 53% and 58%, respectively. The extent of reduction in enzymatic activities widely varied, and the magnitude of enzymatic inhibition was not dose dependent. In the same study, Okuno et al. also evaluated the theophylline pharmacokinetics as an indicator of mixed-function oxidase activity. The extent of reduction in the O-demethylase and O-deethylase activities correlated well with the decreased clearance and increased half-life of theophylline.

Hellman et al. (27) studied the effects of IFN-β on the CYP2C19 and CYP2D6 activities in patients with multiple sclerosis. There were 10 Caucasian patients with multiple sclerosis who received IFN-β1a (AVONEX) 30 mg weekly intramuscularly, Rebif 22 or 44 mg thrice weekly SC, or IFN-β1b (Betaferon) 250 mg every other day SC. The urinary S/R mephenytoin ratio (for determination of CYP2C19 activity) and debrisoquine metabolic ratios (for determination of CYP2D6 activity) were used to characterize the activity of the two CYPs. The results of the study indicated that one-month IFN-β treatment did not alter CYP2C19 or CYP2D6 activities.

Mechanism of Type I Interferon Inhibition of CYP450 System

The mechanism by which type I IFNs inhibit the CYP450 system is not well understood. It has been postulated that increased degradation of CYP450 apo-protein may be the reason for depressed enzymatic activity (28). It is also possible that type I IFNs increased the synthesis of xanthine oxidase, which in turn produces superoxide that destroys CYP450 (29). Some investigators (30–32) have suggested that high antiviral or antitumor activity of these IFNs through the biochemical pathways that mediate these effects may depress the CYP450 system.

Clinical Consequences of the Effects of Interferons on the CYP450 System

The aforementioned studies provide some evidence that type I IFNs may inhibit or induce hepatic drug metabolism in humans and there may be some potential of toxic drug-drug interactions. However, the studies also indicate that inhibition or induction may not be of any clinical significance mainly because of the high variability and bidirectionality (inhibition in some subjects and induction in some). Because of limited number of interaction studies and since these studies were only short-term studies, it is very difficult to make an outright conclusion about the clinical consequences of these studies.

EFFECT OF CYTOKINES ON CYP450 SYSTEM

Razzak et al. (33) examined the role of five cytokines (IL-1β, IL-4, IL-6, TNF-α, and IFN-γ) on the expression of CYP1A2, CYP2C, CYP2E1, CYP3A, and epoxide hydrolase in primary human hepatocyte cultures. The results of the study indicated that among these cytokines, IL-1β, IL-6, and TNF-α were the most potent inhibitors of P450 enzymes. After three days of treatment, both mRNA levels and enzyme activities were inhibited by at least 40%. IFN-γ suppressed CYP1A2 and CYP2E1 mRNA levels but had no effect on CYP3A and epoxide hydrolase mRNAs. On the other hand, IL-4 increased the CYP2E1 activity up to fivefold, suggesting that different regulatory mechanisms may be involved. This change in drug-metabolizing capacity may be due to a downregulation of P450 gene transcription and RNA modulation.

Interleukin-10 and Prednisone

Combination therapy of IL-10 and the corticosteroid prednisone may be beneficial, and the combined use of these two drugs may result in mutual dose reduction. Chakraborty et al. (34) studied pharmacokinetic interactions of a single oral dose of prednisone and SC recombinant human IL-10 in 12 healthy male volunteers. Single doses of IL-10 (8 μg/kg), IL-10 with prednisone (15 mg orally), placebo with prednisone, or placebo were administered on four separate occasions with at least three-week washout periods. The results of the study indicated that prednisone had no effect on the pharmacokinetics of IL-10 or vice versa (Tables 4 and 5). Both prednisone and prednisone/IL-10 treatments caused marked suppression of endogenous cortisol concentrations; however, IL-10 alone was sufficient to significantly increase the 24-hour AUC of endogenous cortisol by 20%.

TABLE 4 Pharmacokinetic Parameters of Prednisolone with and Without IL-10

Parameters	Placebo + prednisolone	IL-10 + prednisolone
$t_{1/2}$ (hr)	2.42 ± 0.38	2.73 ± 0.44
CL/F (L/hr)	15.2 ± 2.2	14.5 ± 3.1
V_c/F (L)	46.4 ± 11.8	37.6 ± 19.7

Abbreviation: IL, interleukin.
Source: From Ref. 34.

TABLE 5 Pharmacokinetic Parameters of IL-10 with and Without Prednisolone

Parameters	IL-10	IL-10 + prednisolone
C_{max} (ng/mL)	3.20 ± 2.06	2.85 ± 1.77
$t_{1/2}$ (hr)	5.63 ± 2.64	5.68 ± 3.10
CL/F (mL/hr/kg)	231.6 ± 87.5	242.3 ± 83.1
V_c/F (mL/kg)	139.8 ± 107.4	154.2 ± 112.7

Abbreviation: IL, interleukin.
Source: From Ref. 34.

EFFECT OF RECOMBINANT TUMOR NECROSIS FACTOR ON CYP450 SYSTEM

Ghezzi et al. (35) studied the effect of recombinant TNF on liver CYP450 and related drug metabolism enzymes in male CD1 mice. Treatment of mice with rTNF caused a marked depression of activities of CYP450, ethoxycoumarin deethylase, and arylhydrocarbon hydroxylase in the liver and other organs such as intestine, lung, and adrenal. This effect was maximal 24 to 48 hours after treatment and was dose dependent. Depression of liver drug-metabolizing enzymes was also observed in the endotoxin-resistant C3H/HeJ strain of mice, thus ruling out that this effect may be due to minor endotoxin contamination of rTNF. These data indicate that depression of liver drug metabolism might be an important side effect of rTNF when given with drugs that are metabolized by CYP450 system.

INTERACTION WITH MONOCLONAL ANTIBODIES

Antibodies are glycoproteins produced by an individual in response to the invasion of a xenobiotic or a foreign molecule in the body. These foreign molecules known as antigens are generally produced by or are a component of an invading organism such as a bacterium or virus. Pathogens are not the only reason against which antibodies are produced; the external administration of naturally occurring endogenous chemical substances such as proteins, carbohydrates, and organic compounds can also initiate the production of antibodies (36).

The pioneering work of Kohler and Milstein in 1975 led to modern-day interest in the monoclonal antibodies. These antibodies have become important research tools in the diagnosis and therapy of human disease. In recent years, by manipulating antibody genes, the redesigned antibody molecules are becoming not only very important for the therapy of human diseases but are also playing an important role in diagnostics and research (37).

Antibodies are also known as immunoglobulins. Higher animals have five classes of immunoglobulins, known as IgG, IgM, IgA, IgE, and IgD (36). The IgG class is the most abundant of these immunoglobulins. Each IgG molecule consists of four polypeptide chains, two heavy chains of approximately 50 kDa, and two light chains of 25 kDa. Both these chains are held together by four disulfide bridges. The amino acid sequences are located in both these chains. In addition to amino acid sequences, each heavy chain of IgG molecule contains N-linked carbohydrate moiety that can be critical to the antibody's effecter function. Each light chain comprises a variable domain and a constant domain, whereas each heavy chain consists of one variable domain and three constant domains. The heavy and light chain variable domains form the antigen-binding site. In humans, IgD, IgE, and IgG are present as monomeric structures with a molecular weight of 150 to 200 kDa (36).

Antibody clearance mechanisms are complex (37). Antibodies may be eliminated via excretion or catabolism. Since the size of immunoglobulins is large, minimal amounts of intact immunoglobulins are filtered by the kidneys. Low-molecular-weight antibody fragments are, however, filtered in the kidney but are not excreted, being reabsorbed and metabolized by proximal tubular cells. Some immunoglobulins are also excreted in the bile (approximately 3%), but the vast majority of immunoglobulins are eliminated by catabolism (37). Receptor-mediated drug disposition studies have demonstrated nonlinearity in the pharmacokinetics of several monoclonal antibodies such as omalizumab and herceptin (37).

Adalimumab-Methotrexate

Adalimumab is a recombinant human IgG1 monoclonal antibody specific for human TNF that is indicated for treatment of rheumatoid arthritis (RA). Repeated administration of adalimumab had no effect on the pharmacokinetics of methotrexate (MTX). On the other hand, MTX reduced adalimumab clearance after single and multiple dosing by 29% and 44%, respectively (package insert). The reduced clearance of adalimumab given with MTX is due to reduced formation of antibodies rather than a direct interaction between adalimumab and MTX.

Etanercept

Etanercept is a dimeric fusion protein that is indicated for the treatment of RA, juvenile RA, ankylosing spondylitis, psoriatic arthritis, and psoriasis. Drug interaction studies of etanercept have been conducted with several drugs. The summary of these drug-drug interaction studies is as follows.

Etanercept-Paclitaxel

Administration of paclitaxel in combination with etanercept resulted in a twofold decrease in etanercept clearance in a nonhuman primate study and in a 1.5-fold increase in etanercept serum levels in clinical studies (package insert).

Etanercept-Digoxin

Etanercept did not alter the pharmacokinetics of digoxin. On the other hand, digoxin reduced the C_{max} and AUC of etanercept by 4.2% and 12.5%, respectively. The clinical significance of this change is most likely irrelevant. There were no clinically relevant changes in the electrocardiograms of patients receiving the combination of etanercept and digoxin. The combined administration of etanercept and digoxin also did not increase the adverse events as compared with monotherapy for each drug (38).

Etanercept-Warfarin

Etanercept did not affect the PK and PD of warfarin. Warfarin also did not alter the pharmacokinetics of etanercept (39).

Etanercept-Methotrexate

The concurrent administration of MTX in patients with RA did not alter the pharmacokinetics of etanercept (40).

Herceptin (Trastuzumab)-Paclitexal

Herceptin (Trastuzumab) is a recombinant DNA-derived humanized monoclonal antibody for the treatment of cancer. Furtlehner et al. (41) studied the effect of herceptin on the pharmacokinetics of paclitexal in 10 patients. The results of the study indicated that the C_{max} and AUC values of paclitexal were about 25% and 9% lower, respectively, when given with herceptin. Herceptin also caused a lower C_{max} (22%) of the main metabolite 6-OH paclitexal, but the AUC of 6-OH paclitexal was not different with and without herceptin. The C_{max} and AUC of minor metabolites of paclitexal were not different in both dosing regimens. Overall, herceptin did not alter the PK profile of paclitexal.

Cetuximab-Irinotecan

Cetuximab (ERBITUX) is a recombinant chimeric monoclonal antibody against the epidermal growth factor receptor with an approximate molecular weight of 152 kDa. Ettlinger et al. (42) investigated the impact of cetuximab on the pharmacokinetics of irinotecan. Irinotecan has been approved as a potent drug against metastatic colorectal cancer and is used in combination with 5-fluorouracil and leucovorin. Patients with advanced colorectal cancer received irinotecan (350 mg/m^2) every third week and cetuximab as a loading dose (400 mg/m^2) on day 2, followed by a weekly maintenance dose (250 mg/m^2). The results of the study indicated that cetuximab has no clinically relevant effect on the pharmacokinetics of irinotecan or its metabolites. The effect of irinotecan on the pharmacokinetics of cetuximab was not studied.

Daclizumab and Mycophenolate Mofetil

Daclizumab is a humanized monoclonal antibody with specificity for the α chain of the IL-2 receptor (CD25). Daclizumab has been shown to decrease the incidence of acute rejection in renal transplantation (43). In a study (43), 75 renal transplant recipients were randomized 2:1 to receive either daclizumab 1 mg/kg or placebo pretransplantation every other week, for a total of five doses. All patients also received cyclosporine, steroids, and mycophenolate mofetil (MMF). MMF is a potent immunosuppressive agent that facilitates hematopoietic engraftment (44). Mycophenolic acid (MPA), the active metabolite of MMF, interferes with cell proliferation by inhibiting inosine monophosphate dehydrogenase type II, which subsequently blocks de novo purine synthesis in T and B lymphocytes (45). Levels of MPA, its glucuronide metabolite, and daclizumab were measured after dosing on days 28 and 56. The results of the study indicated that concomitant administration of daclizumab and MMF had no effect on the pharmacokinetics of MPA. The $AUC_{(0-8)}$ of MPA on days 28 and 56 was similar in patients with and without daclizumab. However, $AUC_{(0-8)}$ of MPA glucuronide on both days 28 and 56 was about 18% lower in daclizumab-treated patients than that in placebo group.

Basiliximab and Cyclosporine

Basiliximab is a chimeric mouse-human monoclonal antibody to the IL-2Rα receptor of T cells. It is used to prevent rejection in organ transplantation, especially in kidney transplants. Basiliximab reduces the incidence and severity of acute rejection in kidney transplantation without increasing the incidence of opportunistic infections (46). In a retrospective study (47), 39 pediatric renal transplant patients received an initial oral daily dose of 500 mg/m^2 cyclosporine. Thereafter, the dose was adjusted daily to achieve cyclosporine trough levels of 200 to 300 μg/L. In 24 patients, 20 mg basiliximab was added to induction treatment at days 0 and 4 after transplantation (10 mg if <35 kg body weight). The remaining 15 patients served as controls. Cyclosporine trough concentrations, dose requirements, and side effects were measured daily for the first 30 days and twice weekly up to 90 days after transplantation.

Within the first 10 days, substantially less cyclosporine was required in the basiliximab group than that in controls (mean cyclosporine dose 412 mg/m^2 vs. 443 mg/m^2), but this dosing strategy resulted in higher trough concentrations (mean 258 μg/L vs. 228 μg/L) despite daily dose reduction. Twenty-two of

24 children who received basiliximab reached cyclosporine trough concentrations above 300 µg/L, compared with 9 of 15 controls. At days 28 to 50, patients who received basiliximab showed a substantial decline in whole blood concentrations of cyclosporine (21 of 24 patients had <170 µg/L vs. 3 of 15 controls) despite unaltered dose. All seven acute rejections in the basiliximab group occurred during this time, and these patients showed an especially high-dose requirement to achieve the target concentration of cyclosporine. The authors suggested that as CD25 saturation fades at days 28 to 50, cyclosporine concentrations decline and 20% higher doses are required to maintain adequate trough concentrations. Furthermore, an IL-2 receptor-mediated alteration of the CYP450 system caused the systemic drug interaction. The authors proposed that with the use of basiliximab, the initial cyclosporine dose in children should be limited to 400 mg/m^2 or less to avoid unnecessary toxic side effects.

PHARMACODYNAMIC DRUG INTERACTION STUDIES

Pharmacodynamic drug interactions can lead to additive, synergistic, or antagonistic effects for a particular drug (1). Pharmacodynamic interaction studies may also provide essential information regarding dosage adjustment for drug(s) given concomitantly. Some examples of pharmacodynamic drug interaction studies involving macromolecules are presented below.

Pegylated Interferon-α2b and Methadone

To determine the effect of peg-IFN-α2b on opiate withdrawal symptoms, Berk et al. (25) evaluated their patients with HIV and chronic Hepatitis C virus infection using subjective opiate withdrawal scale (SOWS) and objective opiate withdrawal scale (OOWS) assessment methods at baseline and 7, 14, and 21 days after the administration of the first dose. Weekly clinical evaluation for signs and symptoms of methadone withdrawal of peg-IFN-α2b was conducted. The results of this study indicated that SOWS and OOWS scores of methadone were not statistically different with and without peg-IFN-α2b. Therefore, the authors recommended that methadone dosage adjustment, when given with peg-IFN-α2b, is not necessary in patients with HIV and chronic Hepatitis C virus infection.

Prednisolone and Interleukin-10

A pharmacodynamic interaction study between prednisolone and IL-10 was conducted by Chakraborty and Jusko (48). To determine the nature and intensity of interaction between these two compounds, the authors used isobolograms along with parametric competitive and noncompetitive interaction models. The isobolographic model indicated that the effect of prednisolone and IL-10 is additive in suppressing lymphocyte proliferation. The competitive interaction model revealed the interaction to be slightly synergistic, whereas the noncompetitive interaction model indicated a small degree of antagonism. This study highlights the importance of different modeling methods and the outcome of these methods, which may be model dependent. Therefore, one should be cautious in the interpretation of data when using divergent models.

Xemilofiban and Abciximab

Platelet glycoprotein (GP) IIb/IIIa receptor antagonists prevent thrombotic occlusions in patients with acute coronary syndromes and reduce ischemic complications of coronary angioplasty. In a clinical trial, Kereiakes et al. (49) evaluated the pharmacodynamic effects of xemilofiban, an oral nonpeptide GP IIb/IIIa antagonist, with or without concomitant abciximab therapy. Xemilofiban is a human-mouse chimeric monoclonal antibody that functions as a GP IIb/IIIa antagonist currently used in percutaneous coronary interventions to avoid platelet activation, thrombosis, and inflammation. Of the 74 patients enrolled in a placebo-controlled, dose-ranging study of xemilofiban, 17 patients received IV abciximab as a bolus (0.25 mg/kg) and 12-hour (0.6 mg/hr) IV infusion. Both the magnitude and the duration of response to xemilofiban were enhanced by prior abciximab treatment, but this effect was no longer evident after one week. Although the sample size was small and the patients were not randomly assigned to receive abciximab, it appears that the combination therapy of xemilofiban and abciximab may provide inhibition of platelet aggregation for an additional period of time and may be beneficial in deriving doses of orally administered compounds.

Heparin and Abciximab/Reviparin and Abciximab

In a randomized and placebo-controlled parallel group design, Klinkhardt et al. (50) studied the pharmacodynamic effect of unfractionated heparin or a low-molecular-weight heparin, reviparin, on efficacy outcomes for abciximab or tirofiban. Abciximab is a platelet aggregation inhibitor mainly used during and after coronary artery procedures like angioplasty to prevent platelets from sticking together and causing thrombus (blood clot) formation within the coronary artery. Its mechanism of action is inhibition of glycoprotein IIb/IIIa. The pharmacodynamic effects measured were bleeding time, fibrinogen binding at the GP IIb/IIIa receptor, expression of the platelet secretion marker CD62, and ADP (20 μM) and collagen (5 μg/mL) induced platelet aggregation. The results of the study showed that unfractionated heparin attenuated platelet aggregation and fibrinogen binding induced by abciximab or tirofiban, but reviparin did not exert any effect on their pharmacodynamic characteristics. According to the authors, the study suggests an advantage of reviparin over unfractionated heparin when given with abciximab or tirofiban.

Heparin + Aspirin and YM337

Graff et al. (51) investigated the pharmacodynamic effect of unfractionated heparin and aspirin on YM337, a Fab fragment humanized monoclonal antibody of the platelet GP IIb/IIIa receptor. Eighteen healthy male volunteers between 18 and 40 years of age were enrolled in the study. Each volunteer also received 325 mg aspirin or matching placebo for three days. The pharmacodynamic effects measured were bleeding time, expression of the platelet secretion marker CD62, and ADP (20 μM) as well as collagen (5 μg/mL) induced platelet aggregation. The results of this study indicated that unfractionated heparin and YM337 have strong synergistic effects on bleeding time, whereas coadministration of aspirin has strong inhibitory effects of YM337 on collagen-induced platelet aggregation.

PERSPECTIVES AND FUTURE DIRECTIONS

Conducting pharmacokinetic drug-drug interaction studies of therapeutic proteins and peptides are complicated. Because of long half-lives of some of these molecules, such as monoclonal antibodies, conventional crossover studies with adequate washout periods can be difficult to design and conduct. With the advent of therapeutic biological products, the use of these products in clinical conditions such as leukemia, lymphoma, solid tumor cancers, organ transplants, and autoimmune diseases is rapidly growing. The patients with aforementioned disease states may also require concurrent use of other medications. As a result, there is always a potential for drug interactions; therefore, a more routine investigation of pharmacokinetic drug-drug interaction studies for new therapeutic biological products is necessary.

In a recent paper, Seitz and Zhou (52) have emphasized the importance of drug-drug interaction studies for therapeutic monoclonal antibodies and state the following:

> To successfully keep pace with each technological and therapeutic advancement, industry sponsors should routinely conduct thorough evaluations of the drug-drug interaction potential of investigational monoclonal antibodies. The knowledge gained will expand our overall understanding of identified metabolic pathways and pharmacokinetic mechanisms and simultaneously ensure the safety of all patients who may require concurrent pharmacotherapy involving a therapeutic monoclonal antibody.

Although drug interaction studies of therapeutic proteins and peptides are not as common as those of conventional drugs, this trend is changing. It is becoming important to evaluate the impact of therapeutic proteins and peptides on the pharmacokinetics and pharmacodynamics of conventional drugs because more and more conventional drugs are given with therapeutic proteins and peptides to improve therapies. The current trend is to evaluate the effect of therapeutic proteins and peptides on the PK/PD of conventional drugs, but the reverse rarely occurs. It is important that drug interaction studies be conducted in both directions. Considering that IFNs, as examples of a class of therapeutic proteins, do inhibit or induce drug-metabolizing enzymes (although long-term clinical effect of such inhibition or induction is not known), it is important to evaluate the effect of other classes of therapeutic proteins (cytokines, ILs, monoclonal antibodies) on drug-metabolizing enzymes.

REFERENCES

1. Gibaldi M. Pharmacokinetic variability–drug interactions. In: Biopharmaceutics and Clinical Pharmacokinetics, 3rd ed. Philadelphia: Lea and Febiger, 1984:257–285.
2. Khakoo S, Glue P, Grellier L, et al. Ribavirin and interferon alpha-2b in chronic hepatitis C: assessment of possible pharmacokinetic and pharmacodynamic interactions. Br J Clin Pharmacol 1998, 46(6), 563–570.
3. Mahmood I. Pharmacokinetics and pharmacodynamic considerations in the development of therapeutic proteins. In: Mahmood I, ed. Clinical Pharmacology of Therapeutic Proteins. Rockville: Pine House Publishers, 2006:123–174.
4. Braeckman R. Pharmacokinetics and pharmacodynamics of protein therapeutics. In: Reid ER, ed. Peptides and Protein Drug Analysis. New York: Marcel Dekker, 2000:633–669.
5. Mohler MA, Cook JE, Baumann G. Binding proteins of protein therapeutics. In: Ferraiolo BL, Mohler MA, Gloff CA, eds. Protein Pharmacokinetics and Metabolism. New York: Plenum Press, 1992:35–71.
6. Lin JH. Transporter-mediated drug interactions: clinical implications and *in vitro* assessment. Expert Opin Drug Metab Toxicol 2007; 3(1):81–92.

7. Kozlowski A, Charles SA, Harris JM. Development of pegylated interferons for the treatment of chronic hepatitis C. BioDrugs 2001; 15(7):419.8–429.8.
8. Ferraiolo BL, Mohler MA. Goals and analytical methodologies for protein disposition studies. In: Ferraiolo BL, Mohler MA, Gloff CA, eds. Protein Pharmacokinetics and Metabolism. New York: Plenum Press, 1992:1–33.
9. Kompella A, Lee VHL. Pharmacokinetics of peptide and protein drugs. In: Lee VHL, ed. Peptide and Protein Drug Delivery. New York: Marcel Dekker, 1991:391–484.
10. Ho RJ, Gibaldi M. Pharmacology, toxicology, therapeutic dosage formulations, and clinical response. In: Ho RJ, Gibaldi M, eds. Biotechnology and Biopharmaceuticals: Transforming Proteins and Genes into Drugs. Hoboken: John Wiley & Sons, 2003:97–123.
11. Takakura Y, Fujita T, Hashida M, et al. Disposition characteristics of macromolecules in tumor-bearing mice. Pharm Res 1990; 7(4):339–346.
12. Bocci V. Catabolism of therapeutic proteins and peptides with implications for drug delivery. Adv Drug Deliv Rev 1990; 4(2):149–169.
13. Preusch PC. Equilibrative and concentrative drug transport mechanisms. In: Atkinson AJ, Daniels CE, Dedrick RL, et al., eds. Principles of Clinical Pharmacology. New York: Academic Press, 2001:201–222.
14. LaRusso NF. Proteins in bile: how they get there and what they do. Am J Physiol 1984; 247(3 pt 1):G199–G205.
15. Okuno H, Kitao Y, Takasu M, et al. Depression of drug-metabolizing activity in the human liver by interferon-a. Eur J Clin Pharmacol 1990; 39(4):365–367.
16. Pageaux GP, le Bricquir Y, Berthou F, et al. Effects of interferon-alpha on cytochrome P-450 isoforms 1A2 and 3A activities in patients with chronic hepatitis C. Eur J Gastroenterol Hepatol 1998; 10(6):491–495.
17. Ben Reguiga M, Bonhomme-Faivre L, Orbach-Arbouys S, et al. Modification of the p-glycoprotein dependent pharmacokinetics of digoxin in rats by human recombinant interferon-alpha. Pharm Res 2005; 22(11):1829–1836.
18. Ben Reguiga M, Bonhomme-Faivre L, Farinotti R. Bioavailability and tissue distribution of docetaxel, a P-glycoprotein substrate, are modified by interferon-α in rats. J Pharm Pharmacol 2007; 59(3):401–408.
19. Williams SJ, Farrell GC. Inhibition of antipyrine metabolism by interferon. Br J Clin Pharmacol 1986; 22(5):610–612.
20. Williams SJ, Baird-Lambert JA, Farrell GC. Inhibition of theophylline metabolism by interferon. Lancet 1987; 2(8565):939–941.
21. Becquemont L, Chazouilleres O, Serfaty L, et al. Effect of interferon alpha-ribavirin bitherapy on cytochrome P450 1A2 and 2D6 and N-acetyltransferase-2 activities in patients with chronic active hepatitis C. Clin Pharmacol Ther 2002; 71(6):488–495.
22. Sulkowski M, Wright T, Rossi S, et al. Peginterferon alpha-2a does not alter the pharmacokinetics of methadone in patients with chronic hepatitis C undergoing methadone maintenance therapy. Clin Pharmacol Ther 2005; 77(3):214–224.
23. Islam M, Frye RF, Richards TJ, et al. Differential effect of IFNa-2b on the cytochrome P450 enzyme system: a potential basis of IFN toxicity and its modulation by other drugs. Clin Cancer Res 2002; 8(8):2480–2487.
24. Gupta SK, Sellers E, Somoza E, et al. The effect of multiple doses of peginterferon alpha-2b on the steady-state pharmacokinetics of methadone in patients with chronic hepatitis C undergoing methadone maintenance therapy. J Clin Pharmacol 2007; 47(5):604–612.
25. Berk SI, Litwin AH, Arnsten JH, et al. Effects of pegylated interferon alpha-2b on the pharmacokinetic and pharmacodynamic properties of methadone: a prospective, nonrandomized, crossover study in patients coinfected with hepatitis C and HIV receiving methadone maintenance treatment. Clin Ther 2007; 29(1):131–138.
26. Okuno H, Takasu M, Kano H, et al. Depression of drug-metabolizing activity in the human liver by interferon-beta. Hepatology 1993; 17(1):65–69.
27. Hellman K, Roos E, Österlund A, et al. Interferon-b treatment in patients with multiple sclerosis does not alter CYP2C19 or CYP2D6 activity. Br J Clin Pharmacol 2003; 56(3):337–340.

28. Gooderham NJ, Mannering GJ. Depression of cytochrome P-450 and alterations of protein metabolism in mice treated with the interferon inducer polyriboinosinic acid. polyribocytidylic acid. Arch Biochem Biophys 1986; 250(2):418–425.
29. Ghezzi P, Saccardo B, Bianchi M. Induction of xanthine oxidase and heme oxygenase and depression of liver drug metabolism by interferon: a study with different recombinant interferons. J Interferon Res 1986; 6(3):251–256.
30. Singh G, Renton KW, Stebbing N. Homogenous interferon from E. coli depresses hepatic cytochrome P-450 and drug biotransformation. Biochem Biophys Res Commun 1982; 106(4):1256–1261.
31. Parkinson A, Lasker J, Kramer MJ, et al. Effects of three recombinant human leukocyte interferons on drug metabolism in mice. Drug Metab Dispos 1982; 10(6):579–585.
32. Renton KW, Singh G, Stebbing N. Relationship between the antiviral effects of interferons and their abilities to depress cytochrome P-450. Biochem Phannacol 1984; 33(23):3899–3902.
33. Abdel-Razzak Z, Loyer P, Fautrel A, et al. Cytokines down-regulate expression of major cytochrome P-450 enzymes in adult human hepatocytes in primary culture. Mol Pharmacol 1993; 44(4):707–715.
34. Chakraborty A, Blum RA, Suzette M, et al. Pharmacokinetic and adrenal interactions of IL-10 and prednisone in healthy volunteers. J Clin Pharmacol 1999; 39(6):624–635.
35. Ghezzi P, Saccardo B, Bianchi M. Recombinant tumor necrosis factor depresses cytochrome P450-dependent microsomal drug metabolism in mice. Biochem Biophys Res Commun 1986; 136(1):316–321.
36. Liu XY, Pop LM, Vitetta ES. Engineering therapeutic monoclonal antibodies. Immunol Rev 2008; 222(4):9–27.
37. Mahmood I, Worobec A. Therapeutic monoclonal antibodies. In: Mahmood I, ed. Clinical Pharmacology of Therapeutic Proteins. Rockville: Pine House Publishers, 2006:357–411.
38. Zhou H, Parks V, Patat A, et al. Absence of a clinically relevant interaction between etanercept and digoxin. J Clin Pharmacol 2004; 44(11):1244–1251.
39. Zhou H, Patat A, Parks V, et al. Absence of a pharmacokinetic interaction between etanercept and warfarin. J Clin Pharmacol 2004; 44(5):543–550.
40. Zhou H, Mayer PR, Wajdula J, et al. Unaltered etanercept pharmacokinetics with concurrent methotrexate in patients with rheumatoid arthritis. J Clin Pharmacol 2004; 44(11):1235–1243.
41. Furtlehner A, Schueller J, Jarisch I, et al. Disposition of paclitaxel (Taxol) and its metabolites in patients with advanced breast cancer (ABC) when combined with trastuzumab (Hercpetin). Eur J Drug Metab Pharmacokinet 2005; 30(3):145–150.
42. Ettlinger DE, Mitrhauser M, Wadsak W, et al. In vivo disposition of irinotecan (CPT-II) and its metabolites in combination with the monoclonal antibody cetuximab. Anticancer Res 2006; 26(2B):1337–1342.
43. Pescovitz MD, Bumgardner G, Gaston RS, et al. Pharmacokinetics of daclizumab and mycophenolate mofetil with cyclosporine and steroids in renal transplantation. Clin Transplant 2003; 17(6):511–517.
44. Bacigalupo A. Management of acute graft-versus-host disease. Br J Haematol 2007; 137:87–98.
45. Bullingham RE, Nicholls AJ, Kamm BR. Clinical pharmacokinetics of mycophenolate mofetil. Clin Pharmacokinet 1998; 34:429–455.
46. Waldman TA. Immunotherapy: past, present and future. Nat Med 2003; 9:269–277.
47. Strehlau J, Pape L, Offner G, et al. Interleukin-2 receptor antibody-induced alterations of ciclosporin dose requirements in paediatric transplant recipients. Lancet 2000; 356(9238):1327–1328.
48. Chakraborty A, Jusko WJ. Pharmacodynamic interaction of recombinant human interleukin-10 and prednisolone using in vitro whole blood lymphocyte proliferation. J Pharm Sci 2002; 91(5):1334–1342.
49. Kereiakes DJ, Runyon JP, Kleiman NS, et al. Differential dose-response to oral xemilofiban after antecedent intravenous abciximab: administration for complex coronary intervention. Circulation 1996; 94(5):906–910.

50. Klinkhardt U, Graff J, Westrup D, et al. Pharmacodynamic characterization of the interaction between abciximab or tirofiban with unfractionated or a low molecular weight heparin in healthy subjects. Br J Clin Pharmacol 2001; 52(3):297–305.
51. Graff J, Klinkhardt U, Westrup D, et al. Pharmacodynamic characterization of the interaction between the glycoprotein IIb/IIIa inhibitor YM337 and unfractionated heparin and aspirin in humans. Br J Clin Pharmacol 2003; 56(3):321–326.
52. Seitz K, Zhou H. Pharmacokinetic drug-drug interaction potentials for therapeutic monoclonal antibodies: reality check. J Clin Pharmacol 2007; 47(9):1104–1118.

The views expressed in this article are those of the author and do not reflect the official policy of the FDA. No official support or endorsement by the FDA is intended or should be inferred.

11

Intersection of Pharmacogenomics with Pharmacokinetics and Pharmacodynamics

Cecile M. Krejsa
Applied Pharmacology LLC, Seattle, Washington, U.S.A.

Emmanuel Monnet
Neurology and Autoimmune and Inflammatory Diseases, Stratified Medicine–Exploratory Medicine, Medical Sciences and Innovation, Merck Serono International, Geneva, Switzerland

David Cregut
Sequence/Structure Bioinformatics, Drug Discovery Informatics, Merck Serono International, Geneva, Switzerland

THE PROMISE OF PHARMACOGENOMICS

Pharmacogenomics (Pgx) may be defined as the application of genomic or genetic biomarkers to predict patient-specific responses to a therapeutic intervention (1). No longer solely concerned with individual differences in drug metabolism and pharmacokinetics (PK), Pgx inquiry now addresses the use of biomarkers associated with variability in drug targets and pharmacodynamic (PD) responses. In addition, protein biomarkers and diagnostics have been developed to identify patient subsets for specialized treatment. The introduction of Pgx-based tools to assist in clinical decision-making has been publicly anticipated by multiple stakeholders: clinicians, patients, insurers, and regulatory agencies. Thus, Pgx strategies are increasingly integrated into drug development paradigms.

Pgx holds great promise for improved selection of drugs and dose adjustment based on patient needs. Regulatory agencies, recognizing an opportunity to identify patients most likely to benefit, have released guidance encouraging the use of Pgx biomarkers in patient diagnostic and treatment regimens (1,2). This has raised expectations for using such tools during the process of pharmaceutical research and development. Understanding patient status with regard to a diagnostic test result would allow indication refinement and more benefit to patients. Because of their inherent specificity, many protein drugs are prescribed on the basis of test results; examples include (historically) replacement therapies and (recently) receptor-targeted antitumor drugs (3).

In the past two decades, increasing numbers of protein therapeutics have entered clinic use and conveyed unprecedented benefits to patients. The commitment of time and expense involved with development usually leads to premium pricing of novel protein drugs. Their safety and efficacy advantages in responsive patients strongly affirm their value. Although protein therapeutics may have superb efficacy in a subset of patients, selection of responders has often been difficult. Protein and peptide therapeutics are intrinsically targeted drugs that specifically alter selected pathways (4). As a class, therapeutic proteins are less subject to off-target activities than conventional drugs, because of their endogenous or highly selective nature. Both the PD and the PK of a protein drug are largely dependent on the nature of the drug target.

Application of Pgx to development of protein therapeutics has lagged behind its adoption in conventional drug development. For small molecules, decades of research on the phenotypes associated with variants in drug transporters and metabolizing enzymes preceded the use of genotyping tools. There was clearly value in generating detailed exposure data and identifying patients who might require dose customization. These goals became easier as validated methods for testing genetic variants in major drug elimination pathways became available (5–7). Drug product labels have also become more informative about available Pgx testing and dose customization protocols to reduce risk to patients.

More recently, pathways associated with PD responses have become a focus of Pgx investigations. It appears that predictive value can be enhanced by considering both exposure (PK) and response (PD) variables, as was demonstrated for the case of coumarin-based anticoagulants (8).

In this chapter, we discuss Pgx with regard to exposure and response variables for protein therapeutics. Our approach is to provide an overview of how Pgx may be used in different phases of pharmaceutical research and discovery, and describe some strategies and statistical considerations for design of Pgx studies. The effect of patient variability on PK of protein drugs is addressed for major protein drug elimination mechanisms. Examples of Pgx effects on PD and clinical responses to treatment are presented. As carriage of specific major histocompatibility complex (MHC) alleles contributes to drug immunogenicity, we discuss the problem of anti-drug antibody responses and methods used to screen candidates for possible MHC interactions. Finally, we predict that future expansion of Pgx research will add value to drug development programs, improve treatment options, and facilitate the customization of patient care.

PHARMACOGENOMICS IN DRUG DEVELOPMENT
Overview of Strategy for Pgx in Biopharma R&D

Application of Pgx in drug development may support three main activities. First, to support identification of new drug targets, biomarker analysis of phenotypically distinct patient groups may be incorporated into the program. Identification of Pgx associations may reveal novel mechanisms of molecular pathology and trigger design of new drug candidates. A second application is to monitor effects of drug candidates at different stages in the program. Early signs of PD effects (before clinical efficacy can be observed) may inform decision-making, elucidate or confirm drug mechanism, and identify target populations for design of later clinical trials. Third, Pgx may support stratification of the patient population to identify responsive patient subsets and to avoid adverse events may be applied in late development. Adoption of a Pgx-based therapeutic strategy may allow a market advantage, patent protection, or a more rapid move to frontline therapy for patients in the responsive subset. This last strategy could also provide a salvage strategy in the case of failed phase II/III trials if a sensitive or responsive patient subset can be identified.

Target Discovery

A major use of high-content technologies is to identify novel targets for therapeutic intervention. Disease association studies using single-nucleotide polymorphism

(SNP) whole-genome arrays can yield numerous potential targets that may be prioritized through pathway interrogation, mechanistic studies, and analysis of clinical specimens for gene or protein expression. In cancer, the development of somatic mutations leads to additional complexity in target discovery as the genomic characteristics of the tumor are known to change over time.

Clinical association data suggesting mechanistic involvement of a target with a certain patient subset should guide the early Pgx strategy. Depending on the strength of association, knowledge of variable target expression might immediately trigger plans to codevelop a diagnostic tool to identify suitable patients for treatment (9). Development of cetuximab for tumors overexpressing epidermal growth factor receptor (EGFR) followed this paradigm. Alternatively, a plan to stratify patients on the basis of a Pgx test result would allow the association with response to being tested prospectively. Either way, a fit-for-purpose diagnostic test for the candidate marker will be needed in early clinical development. In cases where the target is related to a clinical disease end point, qualified assays may already be available.

Patient Selection and Stratification Strategies

Initiating the codevelopment of a Pgx marker with every lead development program may not be feasible. However, it makes sense to consider the potential need for patient selection and stratification early in development so that candidate markers and exploratory assays can be incorporated into the clinical program. Assay development can be a lengthy process, and qualification of tests suitable for clinical decision-making often requires collaboration with an outside vendor. As exploratory assays may be run as research end points, early efforts to establish the value of candidate markers can benefit the program by prioritizing a subset of Pgx markers for further assay development and inclusion in later-phase studies.

The statistical considerations outlined in section "Statistical Considerations" illustrate the dilemma facing Pgx researchers in drug development. Early studies are generally too small to provide adequate power for genome-wide exploratory studies. Later-phase studies may have sufficient sample size, but fewer chances remain to prospectively test the hypotheses generated in late-phase studies. Patient selection or stratification strategies will require a qualified assay, information on the proportion of patients who are "marker positive," and an idea of the expected effect size associated with the marker (10).

A case in point is the recent approval of panitumumab for advanced colorectal cancer (CRC). Retrospective analysis of biopsies from patients included in a pivotal phase III trial showed that patients with mutated forms of the oncogene *KRAS* (43% of the assessable patients) did not respond to panitumumab treatment. As the trial included a crossover design, it is not possible to assess the prognostic value of *KRAS* mutations on overall survival, but the retrospective study showed clearly that panitumumab benefit (tumor response and progression free survival (PFS)) was limited to patients whose tumors had wild-type *KRAS* (11,12). European regulatory authorities have approved a specific indication for CRC patients who test negative for activating *KRAS* mutations. Based on retrospective analyses, the FDA recently made changes to the product labels for cetuximab and panitumumab, recommending against use of these agents in patients with *KRAS* mutations in codon 12 or 13.

Improvements to diagnostic tests may alter the therapeutic paradigm, as illustrated by trastuzumab in treatment of metastatic breast cancer. At the time of registration, the clinical diagnostic test for overexpression of the Her2/neu receptor was based on immunohistochemistry (IHC). Later studies demonstrated that in situ hybridization (ISH), which detects gene amplification, was a more predictive analytical method. A recent algorithm for defining Her2/neu status uses results from either IHC or ISH (13,14). This practical change has allowed treatment of a wider patient subset and resulted in therapeutic benefit to additional patients.

PHARMACOGENOMICS STRATEGIES
Hypothesis-Based Approaches
Certain target- or disease-related genes may be obvious candidates for Pgx investigation. Replacement therapies have clear diagnostic paradigms and are not generally considered for Pgx analysis. Most Pgx studies focus on protein therapies that augment or interfere with existing pathways within a heterogeneous disease setting. For example, oncology agents may be directed against pathways that are upregulated in only a subset of tumors. Likewise, elevation of a specific cytokine may identify a population of autoimmune patients who would benefit from targeted blockade of that factor. Candidates for Pgx analysis might include both the therapeutic target and genes that are mechanistically related to that target.

The drug target must be considered in relation to the heterogeneity of the disease population, as most diseases are polygenic in nature and few therapies are effective in all treated patients. As the size of the indication is usually considered quite early in development, it makes sense to anticipate patient selection strategies and estimate the fraction of the disease population likely to benefit from a targeted approach. For novel pathways, this may require additional epidemiological or clinical association research. We suggest to

- develop a list of candidate markers on the basis of the medical hypothesis for a given target;
- integrate selected markers into early studies, as justified by preclinical data, epidemiology, and the results of exploratory biomarker analyses;
- accrue data over multiple studies to judge the utility of such candidate markers; and
- prioritize confirmed markers for further development, including assay validation.

The value of using a candidate marker approach is that mechanistic hypotheses may be tested. This can simplify statistical analysis and may lend credibility to data generated early in development when patient numbers are especially limited. There is a risk, however, that the candidate marker will be falsely associated with a patient response end point. This can occur for a variety of reasons, including assay failure, population skewing, and an inappropriate threshold for significance. The risk of false association may be amplified by a sense that mechanisms are well understood and that a positive result is "confirmatory." Initial Pgx studies must always be considered exploratory, and their results must be interpreted with caution. Associations must be confirmed using independent population(s) of appropriate size and qualified, fit-for-purpose assays. Statistical approaches to development of Pgx classifiers are discussed in section "Statistical Considerations."

A rich literature is developing on Pgx candidate markers that associate with responses to many different classes of protein drug (3,15). Candidate genes may include upstream factors that poise the patient for differential sensitivity to treatment, the target itself, if variably expressed in the disease, or the presence of polymorphisms in downstream pathways that influence the magnitude of PD responses. Mechanisms involved with protein drug clearance and altered PK have been examined less widely (see sect. "Effect of Patient Variability on Exposure").

Non-Hypothesis-Based Approaches

There is a trend away from hypothesis-based approaches in exploratory Pgx studies. This shift reflects the relative ease in generating and processing large genomics datasets. It may also be attributed to the feeling that candidate gene approaches introduce bias and unnecessarily limit the investigation. Computational methods have been developed to rapidly test associations between response (phenotype) and carriage of SNP in large patient populations. For gene expression profiling and proteomics analysis, data visualizations such as clustering give scientists a means to conceptualize correlations within large multivariate datasets. Other tools for processing large datasets use iterative model-fitting algorithms for factor reduction and the identification of multivariate classifiers.

Technological advances have enabled global SNP and gene expression analysis in patient samples and shifted the focus of Pgx to data-driven discovery. The adoption of chip-based technologies in pharmaceutical development is more or less complete; novel higher-throughput and/or lower cost-per-sample platforms introduced in recent years may spur another quantum leap in the amount of data generated from clinical studies. Many newer platforms can derive high-content data from standard patient samples such as the formalin-fixed paraffin-embedded tissues commonly collected for biopsy.

Existing technologies for RNA expression analysis are quite comprehensive, including about 45,000 gene transcripts, although recent trends are to reduce the number of transcripts on the chips, as many are now considered in silico artifacts (16). Technologies to screen DNA polymorphisms currently measure about one million SNP, covering around 80% to 90% of the genome. Because of genomic complexity, the remaining 10% to 20% will require exponential numbers of markers to correctly cover these regions (17). Copy number variations in DNA have piqued the interest of researchers, as association studies have shown the importance of these variations in disease susceptibility (18–20). Multiple somatic chromosomal aberrations, including altered karyotype, microsatellite instability, gene copy amplification, and translocations can occur in cancer, resulting in functional changes in gene expression (21). Current high-content tools allow measurement of a few hundreds of thousands of replicated genomic regions, while molecular cytogenetics techniques detect structural variations. Epigenetic changes in DNA methylation have recently provided additional characterization of tumors. Existing tools allow screening for a few thousand of these variations in one assay (22).

Non-hypothesis-based approaches are designed to provide comprehensive datasets, which may be analyzed to reveal the underlying structure and identify possible Pgx markers for further study. This is generally possible, and results of such analyses may provide insight to novel mechanisms and/or targets. By way of caution, a recently published simulation using data from breast cancer patients showed that supervised analysis of gene expression profiles from a single phase II study was unlikely to identify Her2/neu expression as a predictor of trastuzumab response because of the low patient numbers and the high number of genes in the analysis (23). As with candidate gene approaches, results of data-driven biomarker discovery approaches must be confirmed through repeated analysis of independent datasets.

Statistical Considerations

Generally speaking, identification of a Pgx marker requires association of that marker with phenotype data from the clinical study. Phenotype data may be binary (positive/negative for the given response) or may have other characteristics. Often continuous phenotype data are converted to binary or ordinal format to enable analyses; however, for some types of data, correlation of markers with the response end points may be a useful exploratory approach. One important consideration is the assay itself; some clinical assays that seem to yield "continuous" data values are validated only for use as binary assays (positive/negative based on a threshold) and may not provide accurate data across a wide concentration range.

Whether a small group of candidate markers is selected or a whole-genome dataset is used, the search for putative classifiers should follow an exploratory/confirmatory paradigm. A marker or multivariate classifier that is identified in one sample set should be tested independently in a different sample set (24). Bootstrap techniques use multiple subsets of a single sample to demonstrate the sensitivity of the derived model to changes in the input data. This type of validation does not constitute an adequate test of the model, since identification of the classifier is based on the composition of initial sample set, and may be affected by sampling bias. Thus, a major hurdle for decision-making based on Pgx association data is the independent replication of results (25). Robust association markers should derive from a sufficient number of examples to provide at least two datasets for analysis, ideally from independently collected samples of the general population. To provide multiple datasets during drug development, Pgx activities must be included early in the program.

In practice, the number of samples required is directly related to the number of variables in the testing algorithm. For candidate Pgx markers, it is possible to generate relevant association data from fairly small numbers of treated patients (i.e., less than 100) whereas whole-genome association algorithms should include much larger sample numbers (200–300 per group). In this case, corrections for multiple testing must be used to reduce the associations found by chance alone.

Methods to adjust for multiple comparisons are usually applied to whole-genome scans for SNP association studies, but there is controversy about using very low significance thresholds because truly associated markers may not be identified. In exploratory studies, a less stringent threshold may lead to the identification of several functionally related markers, such as upstream and downstream markers of a certain pathway. This can lend mechanistic strength to the association, although it is not sufficient to confirm a marker or pathway in the absence of replication in an independent dataset.

If candidate markers have been previously identified (through literature or exploratory studies), a possible tactic is to perform an initial analysis using a predetermined set of candidate markers, with appropriate correction for false discovery rate based on the number of markers to be tested. The whole-genome analysis for non-hypothesis-based associations may follow, with a more stringent significance threshold to limit the false discovery rate.

Gene expression profiles are generally analyzed for hypothesis generation; however, multivariate classifiers such as gene expression signatures, once identified, should be tested with the same stringency as other candidate diagnostics (25). Methods for incorporating composite markers to identify patient subsets on the basis of gene expression have been proposed (26,27). Few diagnostics based on multivariate assays have been developed to guide patient treatment, and none has yet been codeveloped for launch with a protein therapeutic. Nonetheless, the results of gene expression studies may provide insight into the mechanism of action of a novel drug or target and may identify a subset of candidate markers for further analysis.

EFFECT OF PATIENT VARIABILITY ON EXPOSURE

Therapeutic peptides and proteins are eliminated from the body by normal physiological pathways that govern the breakdown of endogenous proteins. Low-molecular-weight proteins are subject to glomerular filtration, which can be reduced by modifications that increase size, such as pegylation or conjugation to larger proteins such as albumin. The PK behavior of larger protein drugs is dependent on target characteristics, specific interactions with receptors, including Fc receptors (FcR), and the actions of proteases within blood and tissue microenvironments. Variations in certain genes involved in these mechanisms may influence the rate of elimination.

Target-Mediated Elimination Pathways

Protein therapeutics that act on cell surface receptors often show target-dependent PK behavior. The fate of the target (endocytosis, degradation, etc.) and the consequences of target engagement (gene induction, modulation, posttranslational modifications) will influence both exposure and response kinetics.

Immunostimulatory cytokines may be subject to receptor-mediated endocytosis, as is the case for interleukin (IL)-2, IL-15, and type 1 interferons (IFNs). Ligand engagement triggers receptor phosphorylation and endocytosis, eliminating the cytokine from the extracellular space. In the case of IL-2, the cytokine is released from the heterotrimeric IL-2 receptor complex during endosomal sorting and is degraded along with the IL-2/15 receptor β subunit (IL-2Rβ) (28). In contrast, IL-15 is subject to trans-endosomal recycling with IL-15Rα, while IL-2Rβ is sorted to lysosomes (29). While patient responses to IL-2 therapy are quite variable, association studies of candidate polymorphisms in the IL-2 receptor pathway have not been published.

The IFN receptors IFNAR1 and IFNAR2 are also subject to ligand-dependent internalization (30). The activation of IFN-mediated signals induces suppressor of cytokine signaling 1 (SOCS-1) and SOCS-3, resulting in the downregulation of IFNAR1 (31,32). The SOCS box domain of SOCS proteins is thought to act by targeting E3-ubiquitin ligases to specific receptor proteins, which initiates their degradation through the proteosome (33). Recent reports also suggest a role for nuclear translocation and transcriptional activation by the internalized IFNAR1 complex following ligand engagement (34). Polymorphisms in *IFNAR1* and *SOCS3*, as well as pretreatment mRNA levels of *SOCS1* and *SOCS3*, have been associated with sustained viral responses following IFN-α therapy for chronic hepatitis C (35–38). Other targets of SOCS family members include receptors for growth hormone (somatotropin), insulin, leptin, and granulocyte colony-stimulating factor (G-CSF) (39–42). The impact of genetic variations in SOCS proteins on the PK and PD of these drugs has not been reported.

Monoclonal antibodies (mAbs) are also subject to intracellular degradation pathways if the target is internalized following mAb binding. Clearance may be altered by the "antigen sink"; in effect, more rapid clearance is observed until doses are high enough to saturate the target (43). Upon binding, efalizumab induces internalization of its target, the T-cell adhesion molecule CD11b (44,45). Efalizumab has nonlinear PK and decreased half-life at low doses, apparently because of target saturation (46).

Trastuzumab and cetuximab target erbB family members and are indicated for tumors that overexpress Her2/neu and EGFR, respectively (47). Vastly different degrees of target expression may be encountered by these drugs on the basis of tumor burden, target expression, and tumor heterogeneity. The effect of this variability on exposure in patents has not been reported. Nonetheless, mechanism-based physiologically based PK/PD models have demonstrated that target number influences receptor-mediated clearance and alters the PK of tumor-targeted mAbs in nonclinical test species (48).

Soluble forms of Her2/neu and EGFR in serum have been monitored in association with patient responses to trastuzumab and chemotherapy in breast cancer and gefitinib in lung cancer (49–52). In particular, a reduction from baseline in serum Her2/neu may be predictive of good response to therapy. Soluble extracellular domains of target receptors may be characteristic of aggressive disease and have been studied as prognostic factors. In paired longitudinal samples from breast cancer patients, serum Her2/neu increased with metastatic disease (53). By acting as a sink for active drug, serum receptor extracellular domains may result in lower drug exposure to the tumor sites. For both trastuzumab and cetuximab, clearance decreased with increasing dose, as expected for saturation of a target-mediated clearance mechanism (54,55). The effect of serum receptor extracellular domains on drug clearance was not studied in ascending dose trials. In the case of trastuzumab, PK analysis of multiple dose studies showed lower trough concentrations of active drug in patients with high levels of circulating extracellular Her2 (55).

The B-cell antigen CD20 is neither shed nor internalized following engagement; mAbs targeting CD20 remain on the cell surface. Through multiple mechanisms, binding of rituximab to CD20 leads to activation of phagocytic cells and destruction of the target cell (56,57). PK studies have shown increased rituximab half-life, attributed to the lower number of CD20 positive cells, following repeated dosing in non-Hodgkin's lymphoma (58). This phenomenon was not observed in case of rheumatoid arthritis patients. The difference has been attributed to a lower body burden of CD20-positive cells in autoimmune patients and hence less dramatic clearance following the initial rituximab dose (59).

These examples illustrate how PK parameters may vary because of variable expression of a target (receptor) as a result of downstream mechanisms triggered by target engagement. Receptor internalization, induction of compensatory mechanisms, and elimination of receptor-bearing cells have been shown to alter drug clearance pathways. Genetic alterations in one or more of these processes could influence the rate of elimination. Whereas Pgx strategies have commonly focused on the targets of protein drugs for patient eligibility and response end points, the link between drug target variability and variable drug exposure has seldom been investigated in Pgx studies.

Fc-Mediated Effects

Molecules that contain an IgG Fc domain have long half-lives because of the MHC-like neonatal Fc receptor (FcRn), a heterodimer composed of FcRn and β_2-microglobulin (60,61). FcRn binds and stabilizes Fc-containing proteins within acidic endosomal vesicles and provides a recycling pathway via transcytosis and exocytosis (62,63). FcRn is widely expressed on vascular endothelial cells, upper airway epithelial cells, myeloid cells, and in kidney, the podocytes and proximal convoluted tubule. In addition to its role in IgG homeostasis, FcRn has been shown to extend the serum half-life of albumin (64). Some Fc engineering efforts have utilized the unique Fc binding functions of FcRn to enhance drug delivery and to extend the half-life of therapeutic mAbs (65–67). The known biological functions and potential therapeutic uses of FcRn have recently been reviewed (68).

Polymorphisms have been reported in the gene encoding FcRn (*FCGRT*) (69,70). Although the coding region appears to harbor only synonymous polymorphisms, the number of alleles sequenced is small, and very few studies have been published on the effect of *FCGRT* polymorphisms. Given the attention to Fc modifications in mAb design, it is surprising that no Pgx investigations have been published for *FGCRT*.

Large variance in IgG levels at birth has been observed, presumably because of variance in FcRn activity. Birth IgG level was not associated with carriage of synonymous SNP in the coding region of *FGCRT*. Furthermore, the rare nonsynonymous coding polymorphisms of *FGCRT* are unlikely to explain this variability (69). One study showed that a variable number of tandem repeats (VNTR) polymorphism in the FcRn promoter region were associated with differential transcriptional activity of *FCGRT*. Monocytes from individuals homozygous for the major allele (VNTR3/3) showed a pH-dependent increase in IgG binding compared with monocytes from VNTR2 carriers (70).

Other investigations focused on the *DMPK* gene in type 1 dystrophia myotonica (DM1), a disease associated with IgG deficiencies. In this case, carriage of a CTG repeat in the 3′ untranslated region of the *DMPK* gene, which might influence expression levels of the nearby *FCTRT*, has been studied. A small study in Japanese DM1 patients showed a correlation between CTG repeat number and serum IgG levels, but a larger study in Swedish patients showed no association between either serum IgG levels or FcRn α chain transcripts (71,72). In cattle, *FCGRT* haplotypes have been associated with the risk of falling into extreme IgG phenotypes at birth. This large, population-based study in cattle suggests that additional investigation of human *FCGRT* polymorphisms may be warranted (73).

Pharmacogenomic studies of immune effector cells FcR are discussed in section "Effects of Upstream and Downstream Processes."

Non-Target-Mediated Elimination Mechanisms

Both intracellular and extracellular proteolytic processes contribute to non-specific elimination of protein and peptide therapeutics. These processes could potentially be affected by genetic polymorphisms or disease-related alterations in gene expression, as has been described for the SOCS-targeted proteolytic degradation of specific receptors (see sect. "Target-Mediated Elimination Pathways"). No Pgx studies have been published on these protein clearance mechanisms.

Antibody-bound protein therapeutics may form immune complexes that are rapidly cleared by the reticuloendothelial systems (phagocytic cells in the skin, liver, and spleen). Genetic variability may influence this elimination pathway, both by favoring formation of anti-drug antibodies (see sect. "Immunogenicity Predictions") and through the FcR-mediated events that initiate clearance of immune complexes (see sect. "Effects of Upstream and Downstream Processes").

Clearance of proteins modified with specific carbohydrate residues may be mediated by the mannose receptor on macrophages or asialoglycoprotein receptor (ASGP-R) on parenchymal hepatocytes. The ASGP-R has been investigated as a nonviral carrier for gene therapy transduction and as a potential method of improving tumor targeting of protein therapeutics and imaging agents (74,75). If a glycosylated protein drug is subject to ASGP-R binding, rapid distribution phase elimination may limit exposures. Recombinant tissue plasminogen activator, and the experimental drugs lenercept and recombinant neutrophil inhibitory factor are subject to this mechanism (76–78). Although the number of glycoproteins in therapeutic development has increased with construction of soluble receptor fusion proteins, Pgx investigations of ASGP-R variants have not yet been published.

Impact of PK-Related Variants on Later Product Development

The identification of Pgx markers predicting drug PK provides an option for dose customization strategies. Addition of subgroups for specialized dose-response analysis within the context of clinical development will add complexity and cost to drug trials. Therefore, the impact of the dose customization should be assessed with regard to clinical outcomes.

In many cases, early clinical studies, which carefully monitor PK, are insufficiently powered to identify potential genetic sources of PK variability. However, the goal to understand exposure-response relationships may lead to more extensive monitoring of PK in later trials, where subject numbers are higher. If variable exposures are associated with adverse events, the goal is to identify the sensitive patient subset by genomic or genetic screening. If variable exposures are clearly associated with responses, the underlying mechanisms could be addressed by patient selection, dose customization, or advanced molecular design to overcome the differential effect (e.g., engineering of Fc constructs to avoid effects of FcR polymorphisms or to improve half-life through the FcRn recycling pathway).

EFFECT OF PATIENT VARIABILITY ON RESPONSE

Most PD markers currently under investigation for predicting treatment response belong to two main categories. The first is related to the disease and its manifestation at the molecular level, and the second, to the therapeutic target and its downstream pathways.

Mechanism- and Disease-Associated PD Markers

PD markers related to the manifestation of disease include those that identify patient subsets prior to therapeutic intervention and those that change in response to treatment and may be useful for defining biological response in patient subsets. Multiple sclerosis (MS) is a disease in which the Pgx (both types) markers have been studied.

Human leukocyte antigen (HLA) variants are known to be risk factors in MS, and the HLA-DRB1*1501 variant has the best association with increased disease risk (79). Carriage of HLA-DRB1*1501 has also been associated with response to treatment with glatiramer acetate but not with IFN-β1b response (80). Other MS-associated Pgx markers are *CTSS* and *PSMB8* genetic variants. The proteins encoded by these genes, cathepsin S and proteasome (prosome, macropain) subunit, β type 8, both play a role in the antigen processing. Genetic variants may trigger disease by processing viral antigens and inducing a specific immune response, leading to attacks of the neuron proteins by the immune system. Notably, *PSMB8* and *CTSS* gene variations have been associated with more favorable response to IFN-β1b therapy (81). The exact relationship between disease markers and response to treatment remains to be explained.

Therapeutic target–related responses may also define patient subsets. In MS, myxovirus (influenza virus) resistance 1 (*MX1*) gene expression provides an example of this type of marker. Patients with low expression of *MX1* following IFN-β1b treatment had a higher risk of new relapses, suggesting inadequate PD response to therapy. Induction of *MX1* is also considered a biomarker for production of neutralizing antibodies, as expression is reduced in antibody-positive patients (82,83).

Target Variability

Genetic variability in drug targets has been a primary hypothesis for Pgx studies of protein therapeutics. In the treatment of growth hormone deficiency, several groups have studied the effect of a genetic variation in the growth hormone receptor gene (*GHR*) on response to growth hormone (somatotropin) treatment (84,85). The *GHR-d3* allele results in production of a truncated GHR protein with deleted exon 3, a common variation that is carried in 50% of patients, with 9% of patients homozygous for *GHR-d3* (86). In children treated with growth hormone for idiopathic short stature, *GHR-d3* was reported to explain up to 40% of the variance in total growth in response to treatment. However, this data has not been consistently replicated (87,88). Different results may be explained by underpowered studies and by different clinical end points used to define actual growth.

Studies of response to treatments that block tumor necrosis factor (TNF) in autoimmune disease have focused on candidate polymorphisms in the *TNF* gene. In the case of rheumatoid arthritis, promoter region variations have been associated with treatment response, and small studies from several authors have repeated this finding (89–91). The most commonly associated variant is a G-to-A variation in position −308; it has been shown that patients with the GG genotype are twice as likely to respond to anti-TNF treatment as those carrying an A in this position. In vitro studies have demonstrated increased production of TNF from peripheral blood mononuclear cells of donors that carried the −308A allele (92).

Whereas many examples demonstrate a direct impact of drug target variability on treatment response, we must underline that it this not always the case. For instance, in the case of MS treatment by IFN-β1b, none of the identified *IFNAR1* variants influences treatment response; this includes coding variants that modify the structure of the receptor (81,93). Similarly, while cetuximab has been indicated for CRC patients whose tumors have EGFR overexpression, recent studies have shown benefit to patients whose tumors were scored as negative for EGFR expression, and there was also lack of benefit for some patients with EGFR-positive tumors (94–97). Variations in the drug target expression may, in some cases, be trumped by other factors, such as downstream pathways (see below).

Effects of Upstream and Downstream Processes

Beyond the interaction of the drug with its target, variations in downstream effector pathways such as signaling proteins and transcription factors may influence the overall action of the treatment. We recently published a comprehensive review of Pgx studies of mechanistically related candidate markers in drug target pathways for protein therapeutics (15). A few examples are provided to illustrate the effects of downstream processes on drug responses.

One example illustrating this effect involves the presence of somatic mutations of *KRAS* in certain tumors. Activating *KRAS* mutations have been associated with lack of response to panitumumab, which targets EGFR upstream of this GTPase (11,12,94). Mutations in *KRAS* may subvert the mechanism of EGFR inhibitors, since downstream signaling is no longer dependent on signals through EGFR. In a similar fashion, loss of function of PTEN has also been associated with poor response to trastuzumab and cetuximab therapy (98–100). The mutation status of downstream signaling molecules in drug target pathways is especially important to consider with oncology agents, since tumor escape mechanisms generally involve coincident or sequential alteration of several growth-regulatory pathways (21).

The suppressors of SOCS system, which modulates certain cell surface receptors by inhibiting downstream signal transduction pathways and targeting receptors for degradation through the proteosome, may have effects upstream or downstream of the therapeutic target (see sect. "Target-Mediated Elimination Pathways"). Specific SOCS family proteins have been shown to regulate signaling by IFN-α, insulin, leptin, G-CSF, and somatotropin (31,32,39–41). Several Pgx studies have found significant associations between *SOCS1* and *SOCS3* genotypes or pretreatment mRNA levels and response to IFN-α treatment in chronic hepatitis C patients (35–38).

By far, the largest literature on Pgx of downstream effectors has described the effects of coding polymorphisms in the FcR genes. While FcR-dependent activities of phagocytic cells and the reticuloendothelial system can indirectly mediate clearance of some therapeutic proteins, most studies have focused on the efficacy side of the equation. Pharmacogenomic studies have repeatedly demonstrated FcR associations with responses to selected therapeutic mAb drugs. mAbs that bind cellular targets may activate FcR on phagocytic cells and NK cells, stimulating the destruction of the targeted cell by antibody-mediated cellular cytotoxicity (ADCC) or by phagocytosis of the opsonized cell.

Several publications describe the association of coding polymorphisms in *FCGR3A* with responses to rituximab in oncology and autoimmunity trials (101–105). Similar findings have been reported for *FCGR2A*, although it may be that these associations result from linkage within the genetic region encoding several FcRs (103). In vitro studies have shown that the *FCGR3A* variant 158V has higher affinity for IgG than the 158F variant, effectively broadening the concentration range in which Fc engagement activates ADCC in NK cells (106,107). Likewise, it has been observed that the anti-TNF mAb drugs may act as cell-targeting agents, since TNF-α is produced in a transmembrane form before release from the cell surface by cleavage (108). Recent studies of *FCGR3A* and infliximab therapy demonstrated that patients homozygous for the high-affinity 158V allele had a greater decrease in C-reactive protein, a PD marker in Crohn's disease, compared with 158F carriers. A follow-up study demonstrated that variations in the *CRP* gene itself did not account for this association. While the association of *FCGR3A* genotype with biological response in serum CRP levels to infliximab was confirmed, no association was shown with clinical response in the two cohorts studied (109–111).

Fewer studies have investigated associations between FcR variants and erbB-targeting mAbs, perhaps because endocytosis of the mAb-receptor complex reduces the window of opportunity for immune effector cell recognition. However, a role for ADCC in trastuzumab and cetuximab responses has been postulated, and recently, several studies have found associations between FcR variants and/or NK cell activity with clinical response to trastuzumab in breast cancer patients (112–115).

Nonfucosylated mAb constructs have been designed to increase cell-mediated killing of tumor cells by binding with higher affinity to FcR and overcoming the variability associated with coding polymorphisms in *FCGR3A* and *FCGR2A* (116). Interestingly, the phagocytic activity of polymorphonuclear cells appears to favor high fucose content of IgG Fc, whereas the ADCC performed by mononuclear effector cells such as NK is enhanced by low fucose content (116,117).

IMMUNOGENICITY PREDICTIONS
Development of Anti-drug Antibody Responses

Protein drugs are known to elicit immune responses in patients. Immunogenicity was first detected a century ago with therapeutic proteins of animal origin, such as bovine and porcine insulin for the treatment of diabetes (118). Although recombinant technologies now produce proteins using fully human sequences, some patients still develop antibodies. Therapeutic proteins are often engineered with regions of "nonnatural" sequence. Novel sequences, or the novel presentation of endogenous sequences, can trigger the development of specific binding or neutralizing anti-drug antibodies. Almost all therapeutic proteins have been shown to be immunogenic; however, the incidence of anti-drug antibody responses varies greatly among products.

The risks associated with drug immunogenicity depend on many factors, including the nature of the drug and the strength and duration of the antibody response. If a drug supplements an autologous protein, anti-drug antibody responses could potentially neutralize the endogenous protein as well, so the risk depends on the redundancy of the pathway targeted. If the drug is a nonautologous protein or peptide, diminished efficacy may result if the antibody response prevents target binding or significantly reduces exposure by accelerating clearance. Strong antibody responses could cause hypersensitivity reactions, which range from mild to life-threatening ones. Several excellent reviews on immunogenicity risk management strategies and bioanalytical approaches to monitoring anti-drug antibody responses have been published recently (119–121).

Development of an antibody response requires the presentation of peptide epitopes to T cells in the context of an activating costimulatory signal. In general, this transaction is mediated by professional antigen-presenting cells (APC). Besides the primary protein sequence, which will govern binding to MHC and the T-cell receptor (TCR), other factors intrinsic to a drug product may affect its immunogenicity: propensity toward self-aggregation into structures that might activate pattern recognition receptors on APC, tendency to transition into non-native forms, and the presence of other factors such as drug product contaminants (122).

Recognition of antigenic peptides by T lymphocytes can be split into two essential steps. First, a peptide bound to MHC is presented by an APC; second, the peptide-MHC complex is recognized by TCR to initiate T-cell activation. This double mechanism for recognition restricts the range of peptide sequences recognized by the system; thus, only a limited fraction of a protein sequence can elicit an immune response.

Peptides derived from therapeutic proteins are presented by MHC class II molecules, which bind peptides resulting from the cleavage of soluble proteins. The MHC class II proteins are composed of two chains encoded by several similar highly polymorphic genes, each of which binds a different range of peptide sequences (123). Degenerate binding sites on MHC proteins allow a relatively small number of different MHC molecules present in a person to recognize many different peptide sequences. This characteristic gives an individual the ability to respond to a wide range of pathogens. The surface of the TCR that interacts with the peptide-MHC class II complex is extremely variable, since the TCR repertoire is produced in part by somatic recombination. Therefore, a particular peptide-MHC complex will be recognized specifically by a particular TCR.

MHC Genetics and Immunogenicity of Therapeutic Proteins

The MHC locus is diverse and redundant in humans, with multiple genes from ancient duplication events. The genes themselves are highly polymorphic. Thus, carriage of several alleles benefits the immune system, as amino acid substitutions in the binding groove of MHC molecules can alter the repertoire of antigens presented to the adaptive immune system.

Because of the highly repetitive nature of the MHC region, high-throughput genotyping methods have not been widely applied to the study of MHC polymorphisms. Specialized SNP chips that target the sequences of MHC alleles with a rich representation are in development. However, at present, genotyping for HLA type uses long-range PCR reactions that can identify different alleles on the basis of primer design and length of the amplicon. These tests are available through clinical laboratories at blood centers and have been used in many disease association studies; however, application of these tests to Pgx investigations has been less widespread.

Association of specific MHC alleles with development of anti-drug antibody responses has been demonstrated for both small-molecule and protein therapeutics. For example, treatment with IFN-β1b for MS leads to generation of binding or neutralizing antibodies in about 10% to 20% of the population. A study of MHC genes showed that MS patients carrying the HLA-DRB1*0701 allele had higher chances (eight out of nine) of generating antibodies against IFN-β1b (124). The HLA-B*5701 allele has been associated with severe drug hypersensitivity reactions to the antiretroviral drug abacavir (125). Further study of anti-drug antibody responses in association with MHC alleles should enhance our ability to predict which patients are prone to developing such responses.

Computational Approaches to Immunogenicity

From the perspective of drug development, avoiding immunogenicity may begin with candidate selection. Accurate prediction of immunogenic peptides early in a project should allow the design of therapeutic proteins with reduced risk of anti-drug antibody responses. Identification of potential T-cell epitopes within therapeutic proteins is therefore an active field of research. Empirical methods to directly measure peptide-MHC binding affinities are now available. Although usually quite powerful, most of these methods have been set up and tested on a limited number of alleles, restricting their sphere of utility. In vitro approaches include stimulating T cells with peptide sequences from the therapeutic protein, using either a panel of normal volunteers or previously treated patients as donors. Again, low-throughput limits the utility of this method for screening potential drug candidates.

Computational predictions, based on the peptide sequences of lead molecules, can be used to broadly and rapidly screen numerous candidates for possible immunogenic effects. Because the TCR is intrinsically variable, computational rules governing the specific interaction between TCR and peptide-MHC class II complexes have not been developed. Thus, in silico identification of potentially immunogenic peptides currently relies on predicting which peptides bind to MHC class II molecules (126–129). Only a subset of the peptide-MHC class II complexes will also bind and activate a TCR. As antigen presentation is necessary, but not sufficient, for T-cell activation, such prediction methods will systematically produce many false positives.

Qualitative prediction methods calculate an immunogenicity score which, combined to a threshold, provide a list of binder and nonbinder peptides (130). The sequence-based approaches often have difficulty in detecting binding motifs because of the highly variable amino acid composition of binding peptides. However, some clear rules have been established, such as the presence of specific anchor residues required for binding to the MHC groove, at well-defined positions in the peptide (131).

A very popular and relatively easy method based on the concept of "virtual matrices" was introduced in 1999 by Sturniolo et al. (132). The interaction between a peptide in an extended conformation and the MHC class II molecule involves several pockets along the groove of MHC, one for each amino acid side chain of the peptide. The method assumes that each pocket binds independently to an amino acid of the peptide. The strength of the method results from treating MHC pockets as independent units that are shared between HLA alleles. By generating pocket profiles for a limited set of pockets, the authors estimated quantitative matrices for a range of HLA alleles, using sequence comparison of the residues that make up each of the variable pockets. Affinity of a peptide for a particular HLA allele was predicted by summing the individual matrix-derived scores of the amino acids at each position along the peptide. The calculated affinity score was shown to be in good agreement with experimentally determined affinity. The virtual matrices have been implemented in the program TEPITOPE developed by the same group (133).

Experimental confirmation of in silico immunogenicity predictions can be labor intensive. A recent study compared serum anti-drug antibody levels with ex vivo responses of patient T cells stimulated with drug peptide sequences and correlated the magnitude of these responses with the patients' MHC allele carriage. Good correlation was found between the predictions made by in silico screening of eight common MHC alleles against overlapping peptide fragments and the in vivo as well as ex vivo antibody responses to drug (134). Another approach to validate an in silico screening method is to screen proteins with well-known immunogenic epitopes and observe the predictive value of the algorithm to identify those peptides. In some cases, additional empirical evidence, such as co-crystallization of peptides with MHC class II molecules, may be used to qualify the algorithm and to benchmark the score thresholds that are used to identify peptides most likely to bind.

FUTURE VALUE OF PHARMACOGENOMICS
Intersection Between Stratified Medicine and Targeted Therapies
Genetic variability may affect the PK and PD of protein therapeutics and may thus have the potential to affect patient responses. To achieve broader utilization of Pgx tools, it is critical that the scientific community understand what such information will really bring to clinical practice. Individualized medicine has received great publicity in recent years, with articles forecasting the use of genomic tests to tailor new drugs and provide patients with targeted treatment. Although this type of statement is appealing, it must be tempered by the fact that Pgx data are expected to contribute to clinical decision-making but should not be expected to supplant other forms of clinical information.

Consider a Pgx test with regard to its predictive capacity at the patient level. In general, evidence-based medicine requires a population analysis of predictive value for any diagnostic test. Individual treatment decisions may be based on the test results and an understanding of the given test, plus a constellation of additional information that is unique to the patient. While a nominal hazard ratio may be associated with a positive test result, this prediction can never be applied to a single individual, since the information is probabilistic. Weiss et al. provides a review of Pgx tests in the context of clinical decision-making (135).

Looking forward, the use of Pgx tests seems more likely in cases where several treatment options exist. Given recent advances in drug development, molecular therapies are expected to dramatically increase the number of potential treatments for many disease indications. However, in some indications, restrictions in insurance coverage may limit the number of protein drugs a patient may try. On the basis of the price of protein therapeutics and the obvious benefit to patients when optimal therapy is selected early in treatment, new therapeutics may be subject to increased scrutiny by regulatory agencies.

Academic researchers have enthusiastically embraced Pgx as a method to aid in treatment decisions, with positive examples of genetic associations in both target- and exposure-mediated response factors. Regulatory agencies have added the results of such scientific research to drug labels (e.g., *CYP2C9* and *VKORC* alleles in warfarin dose finding) and in some cases requested inclusion of Pgx information prior to licensure (e.g., *KRAS* mutation status in panitumumab response).

The question of how best to select patients for clinical development and rapid market entry is still unsolved, as stratification by candidate Pgx markers may require larger treatment populations in early clinical trials. In some cases, a putative predictive marker may be associated with a patient's prognosis and not related to the specific therapeutic (136). The speed of development may be compromised to more fully understand the role of the Pgx marker in the proposed treatment population. On the other hand, for some drugs developed in specific patient subsets (with label restrictions), post-marketing Pgx studies have shown that some patients testing negative for the marker may also benefit from treatment. This has been noted in the cases of trastuzumab and cetuximab (97,137). Refinements to the diagnostic testing paradigm may alter the indication as new information emerges from expanded clinical use, as in the case of trastuzumab (13,14). The rapid pace of drug development may preclude validation of Pgx-based diagnostics until after product launch if clear associations are not established in early trials. In this case, postlaunch strategies might use Pgx markers to refine the indication.

Pharmacogenomic Strategies in Clinical Care

Information on variability in the target and the PK/PD behavior for a therapeutic protein will promote development of drugs with the highest value for patients. Such information may guide the decision of which drug to prescribe as first choice for a given patient. As the literature on Pgx studies of protein therapeutics matures, it is likely that certain genotypes will be associated with responses in several indications, as has been noted for *FCGR3A* genotype and Fc-containing drugs. The ideal situation would allow physicians to use Pgx data from multiple genotypes and multiple potential treatment options within a given indication to aid in decisions about care. This would require well-controlled studies that differentiate between prognostic markers and markers that predict specific treatment responses (136). The information for multiple treatments will not become available all at once. Nonetheless, all stakeholders, including industry, academics, regulatory agencies, payers, and patients, should promote the generation of this data.

Ultimately, one could envision that several markers will define subsets of patients within an indication, and information on the best treatment for a given patient subset will be available to physicians and patients. Patient subsets would be defined by a predominant molecular pathology that indicates certain target pathways for therapeutic intervention. In this scenario, the treatment options targeting multiple pathways could be selected on the basis of results of diagnostic assays. This is similar to the current practice of customized care in oncology settings. A large research effort will be required in the coming decades to define novel therapeutic targets and to characterize responsive patient subsets. Our hope is that Pgx strategies can help to shift the treatment paradigm for most indications, resulting in improved patient responses, earlier treatment with effective medications, and lower incidence of drug-related adverse events.

Pharmacogenomics as a Business Strategy

Pharmacogenomic data may have different impacts on drug development and on the overall business strategy. Often, the potential to reduce cost is invoked as a rationale for Pgx approaches. This may be true if sufficient evidence is available in advance of pivotal clinical trials. Adding Pgx assessments will increase study costs, and currently microarray analysis costs between $200 and $1000 per patient. This expense may be offset by selection of patients most likely to benefit from treatment, which could potentially reduce both the patient numbers estimated to provide a specified hazard ratio and the time required to observe a clinically relevant treatment effect.

A second strategy is to rescue phase II and III trials that fail to reach the primary end point for the entire population but show promise in a subset of patients. In this case, the financial saving may correspond to the total cost of the research and development program, usually estimated in the hundreds of millions of dollars.

An aspect that is less publicized is the opportunity to manage the drug's life cycle following product launch. Improvement of existing products, through Pgx analysis of utility within an indication (supporting a market niche) and through rapid expansion to additional indications, is a viable strategy to expand a portfolio. If a specific patient subset responds well to the drug as second- or third-line treatment, it may be easier to move the therapeutic into a first-line setting for that subset of marker-positive patients. In addition, higher response rates in patient subsets provide tougher standards of efficacy for competing drugs. If the therapeutic is comarketed with a diagnostic test, this too may afford a measure of protection.

The competitive advantage for a drug with a clearly defined patient population may offset the perceived limitations to label. The real value of the drug will be promoted by providing physicians with the best information predicting who will benefit most from treatment. Both payers and regulatory agencies consider Pgx data as a tool to better understand how a drug should be used, to avoid the use of ineffective treatment and reduce adverse events, and, ultimately, to reduce health care costs. For the industry, the question is how to integrate this information as a competitive advantage and to find the opportunities to optimize the use of current and future products by applying Pgx analysis.

CONCLUSION

The potential for Pgx to be widely applied in patient care will take years to be fully realized. Developers of conventional and protein therapeutics have foreseen value in identifying genomic sources of variability that may lead to different patient outcomes. Integrating Pgx approaches to aid decision-making in drug development and clinical practice will require foresight, patience, and dedication of resources.

This chapter has introduced several aspects of protein therapeutics exposure (PK) and response (PD) that are subject to genetic variability and may benefit from further Pgx investigation. As the field matures, it is expected that Pgx analysis will accompany the development programs of more protein therapeutics, as it has for conventional drugs. The grand promise of Pgx is that these efforts will prove quite valuable in the long run by improving safety and efficacy of new therapeutics, providing market opportunity for drugs to treat selected patient subsets, and revealing new mechanisms by which disease pathways can be targeted.

REFERENCES

1. European Medicines Agency. E15 definitions for genomic biomarkers, pharmaco-genomics, pharmacogenetics, genomic data and sample coding categories EMEA/ CHMP/ICH/437986/2006. 2008. http://www.emea.europa.eu/pdfs/human/ich/ 43798606en.pdf
2. Food and Drug Administration. Guidance to industry: pharmacogenomic data sub-missions. 2005. http://www.fda.gov/downloads/RegulatoryInformation/Guidances/ UCM126957.pdf
3. Lacana E, Amur S, Mummanneni P, et al. The emerging role of pharmacogenomics in biologics. Clin Pharmacol Ther 2007; 82(4):466–471.
4. Leader B, Baca QJ, Golan DE. Protein therapeutics: a summary and pharmacological classification. Nat Rev Drug Discov 2008; 7(1):21–39.
5. Baudhuin LM, Langman LJ, O'Kane DJ. Translation of pharmacogenetics into clinically relevant testing modalities. Clin Pharmacol Ther 2007; 82(4):373–376.
6. Baudhuin LM, Highsmith WE, Skierka J, et al. Comparison of three methods for genotyping the UGT1A1 (TA)n repeat polymorphism. Clin Biochem 2007; 40(9–10): 710–717.
7. Jain KK. Applications of AmpliChip CYP450. Mol Diagn 2005; 9(3):119–127.
8. Gage BF, Eby C, Johnson JA, et al. Use of pharmacogenetic and clinical factors to predict the therapeutic dose of warfarin. Clin Pharmacol Ther 2008; 84(3):326–331.
9. Roses AD. Pharmacogenetics in drug discovery and development: a translational perspective. Nat Rev Drug Discov 2008; 7(10):807–817.
10. Dobbin KK, Zhao Y, Simon RM. How large a training set is needed to develop a classifier for microarray data? Clin Cancer Res 2008; 14(1):108–114.
11. Amado RG, Wolf M, Peeters M, et al. Wild-type KRAS is required for panitumumab efficacy in patients with metastatic colorectal cancer. J Clin Oncol 2008; 26(10): 1626–1634.
12. Freeman DJ, Juan T, Reiner M, et al. Association of K-ras mutational status and clinical outcomes in patients with metastatic colorectal cancer receiving pan-itumumab alone. Clin Colorectal Cancer 2008; 7(3):184–190.
13. Tsuda H. HER-2 (c-erbB-2) test update: present status and problems. Breast Cancer 2006; 13(3):236–248.
14. Wolff AC, Hammond ME, Schwartz JN, et al. American Society of Clinical Oncology/ College of American Pathologists Guideline Recommendations for Human Epidermal Growth Factor Receptor 2 Testing in Breast Cancer. Arch Pathol Lab Med 2007; 131(1):18.
15. Krejsa C, Rogge M, Sadee W. Protein therapeutics: new applications for pharma-cogenetics. Nat Rev Drug Discov 2006; 5(6):507–521.
16. Stein LD. Human genome: end of the beginning. Nature 2004; 431(7011):915–916.
17. Conrad DF, Jakobsson M, Coop G, et al. A worldwide survey of haplotype variation and linkage disequilibrium in the human genome. Nat Genet 2006; 38(11):1251–1260.
18. Breunis WB, van Mirre E, Bruin M, et al. Copy number variation of the activating FCGR2C gene predisposes to idiopathic thrombocytopenic purpura. Blood 2008; 111(3):1029–1038.
19. Aitman TJ, Dong R, Vyse TJ, et al. Copy number polymorphism in Fcgr3 predis-poses to glomerulonephritis in rats and humans. Nature 2006; 439(7078):851–855.

20. Fanciulli M, Norsworthy PJ, Petretto E, et al. FCGR3B copy number variation is associated with susceptibility to systemic, but not organ-specific, autoimmunity. Nat Genet 2007; 39(6):721–723.

21. Bayani J, Selvarajah S, Maire G, et al. Genomic mechanisms and measurement of structural and numerical instability in cancer cells. Semin Cancer Biol 2007; 17(1):5–18.

22. Degenhardt YY, Wooster R, McCombie RW, et al. High-content analysis of cancer genome DNA alterations. Curr Opin Genet Dev 2008; 18(1):68–72.

23. Pusztai L, Anderson K, Hess KR. Pharmacogenomic predictor discovery in phase II clinical trials for breast cancer. Clin Cancer Res 2007; 13(20):6080–6086.

24. Trepicchio WL, Essayan D, Hall ST, et al. Designing prospective clinical pharma-cogenomic (PG) trials: meeting report on drug development strategies to enhance therapeutic decision making. Pharmacogenomics J 2006; 6(2):89–94.

25. Simon R. Development and evaluation of therapeutically relevant predictive clas-sifiers using gene expression profiling. J Natl Cancer Inst 2006; 98(17):1169–1171.

26. Dupuy A, Simon RM. Critical review of published microarray studies for cancer outcome and guidelines on statistical analysis and reporting. J Natl Cancer Inst 2007; 99(2):147–157.

27. Jiang W, Freidlin B, Simon R. Biomarker-adaptive threshold design: a procedure for evaluating treatment with possible biomarker-defined subset effect. J Natl Cancer Inst 2007; 99(13):1036–1043.

28. Hemar A, Subtil A, Lieb M, et al. Endocytosis of interleukin 2 receptors in human T lymphocytes: distinct intracellular localization and fate of the receptor alpha, beta, and gamma chains. J Cell Biol 1995; 129(1):55–64.

29. Dubois S, Mariner J, Waldmann TA, et al. IL-15Ralpha recycles and presents IL-15 In trans to neighboring cells. Immunity 2002; 17(5):537–547.

30. Marijanovic Z, Ragimbeau J, Kumar KG, et al. TYK2 activity promotes ligand-induced IFNAR1 proteolysis. Biochem J 2006; 397(1):31–38.

31. Kumar KG, Barriere H, Carbone CJ, et al. Site-specific ubiquitination exposes a linear motif to promote interferon-alpha receptor endocytosis. J Cell Biol 2007; 179(5):935–950.

32. Kumar KG, Krolewski JJ, Fuchs SY. Phosphorylation and specific ubiquitin acceptor sites are required for ubiquitination and degradation of the IFNAR1 subunit of type I interferon receptor. J Biol Chem 2004; 279(45):46614–46620.

33. Babon JJ, Sabo JK, Soetopo A, et al. The SOCS box domain of SOCS3: structure and interaction with the elonginBC-cullin5 ubiquitin ligase. J Mol Biol 2008; 381(4):928–940.

34. Subramaniam PS, Johnson HM. The IFNAR1 subunit of the type I IFN receptor complex contains a functional nuclear localization sequence. FEBS Lett 2004; 578(3):207–210.

35. Imanaka K, Tamura S, Fukui K, et al. Enhanced expression of suppressor of cytokine signalling-1 in the liver of chronic hepatitis C: possible involvement in resistance to interferon therapy. J Viral Hepat 2005; 12(2):130–138.

36. Persico M, Capasso M, Persico E, et al. Suppressor of cytokine signaling 3 (SOCS3) expression and hepatitis C virus-related chronic hepatitis: insulin resistance and response to antiviral therapy. Hepatology 2007; 46(4):1009–1015.

37. Persico M, Capasso M, Russo R, et al. Elevated expression and polymorphisms of SOCS3 influence patient response to antiviral therapy in chronic hepatitis C. Gut 2008; 57(4):507–515.

38. Walsh MJ, Jonsson JR, Richardson MM, et al. Non-response to antiviral therapy is associated with obesity and increased hepatic expression of suppressor of cytokine signaling 3 (SOCS-3) in patients with chronic hepatitis C, viral genotype 1. Gut 2006; 55(4):529–535.

39. Irandoust MI, Aarts LH, Roovers O, et al. Suppressor of cytokine signaling 3 con-trols lysosomal routing of G-CSF receptor. EMBO J 2007; 26(7):1782–1793.

40. Rui L, Yuan M, Frantz D, et al. SOCS-1 and SOCS-3 block insulin signaling by ubiq-uitin-mediated degradation of IRS1 and IRS2. J Biol Chem 2002; 277(44):42394–42398.

41. Flores-Morales A, Greenhalgh CJ, Norstedt G, et al. Negative regulation of growth hormone receptor signaling. Mol Endocrinol 2006; 20(2):241–253.

42. Howard JK, Flier JS. Attenuation of leptin and insulin signaling by SOCS proteins. Trends Endocrinol Metab 2006; 17(9):365–371.
43. Mager DE, Jusko WJ. General pharmacokinetic model for drugs exhibiting target-mediated drug disposition. J Pharmacokinet Pharmacodyn 2001; 28(6):507–532.
44. Ng CM, Joshi A, Dedrick RL, et al. Pharmacokinetic-pharmacodynamic-efficacy analysis of efalizumab in patients with moderate to severe psoriasis. Pharm Res 2005; 22(7):1088–1100.
45. Joshi A, Bauer R, Kuebler P, et al. An overview of the pharmacokinetics and pharmacodynamics of efalizumab: a monoclonal antibody approved for use in psoriasis. J Clin Pharmacol 2006; 46(1):10–20.
46. Sun YN, Lu JF, Joshi A, et al. Population pharmacokinetics of efalizumab (human-ized monoclonal anti-CD11a antibody) following long-term subcutaneous weekly dosing in psoriasis subjects. J Clin Pharmacol 2005; 45(4):468–476.
47. Emens LA, Davidson NE. Trastuzumab in breast cancer. Oncology (Williston Park) 2004; 18(9):1117–1128; discussion 31–32, 37–38.
48. Lammerts van Bueren JJ, Bleeker WK, Bogh HO, et al. Effect of target dynamics on pharmacokinetics of a novel therapeutic antibody against the epidermal growth factor receptor: implications for the mechanisms of action. Cancer Res 2006; 66(15): 7630–7638.
49. Im SA, Kim SB, Lee MH, et al. Docetaxel plus epirubicin as first-line chemotherapy in MBC (KCSG 01-10-05): phase II trial and the predictive values of circulating HER2 extracellular domain and vascular endothelial growth factor. Oncol Rep 2005; 14(2): 481–487.
50. Esteva FJ, Cheli CD, Fritsche H, et al. Clinical utility of serum HER2/neu in mon-itoring and prediction of progression-free survival in metastatic breast cancer patients treated with trastuzumab-based therapies. Breast Cancer Res 2005; 7(4): R436–R443.
51. Ali SM, Carney WP, Esteva FJ, et al. Serum HER-2/neu and relative resistance to trastuzumab-based therapy in patients with metastatic breast cancer. Cancer 2008; 113(6):1294–1301.
52. Gregorc V, Ceresoli GL, Floriani I, et al. Effects of gefitinib on serum epidermal growth factor receptor and HER2 in patients with advanced non-small cell lung cancer. Clin Cancer Res 2004; 10(18 pt 1):6006–6012.
53. Asgeirsson KS, Agrawal A, Allen C, et al. Serum epidermal growth factor receptor and HER2 expression in primary and metastatic breast cancer patients. Breast Cancer Res 2007; 9(6):R75.
54. Baselga J, Pfister D, Cooper MR, et al. Phase I studies of anti-epidermal growth factor receptor chimeric antibody C225 alone and in combination with cisplatin. J Clin Oncol 2000; 18(4):904–914.
55. Baselga J. Phase I and II clinical trials of trastuzumab. Ann Oncol 2001; 12(suppl 1): S49–S55.
56. Glennie MJ, French RR, Cragg MS, et al. Mechanisms of killing by anti-CD20 monoclonal antibodies. Mol Immunol 2007; 44(16):3823–3837.
57. Cragg MS, Walshe CA, Ivanov AO, et al. The biology of CD20 and its potential as a target for mAb therapy. Curr Dir Autoimmun 2005; 8:140–174.
58. Berinstein NL, Grillo-Lopez AJ, White CA, et al. Association of serum Rituximab (IDEC-C2B8) concentration and anti-tumor response in the treatment of recurrent low-grade or follicular non-Hodgkin's lymphoma. Ann Oncol 1998; 9(9):995–1001.
59. Ng CM, Bruno R, Combs D, et al. Population pharmacokinetics of rituximab (anti-CD20 monoclonal antibody) in rheumatoid arthritis patients during a phase II clinical trial. J Clin Pharmacol 2005; 45(7):792–801.
60. Ghetie V, Hubbard JG, Kim JK, et al. Abnormally short serum half-lives of IgG in beta 2-microglobulin-deficient mice. Eur J Immunol 1996; 26(3):690–696.
61. Junghans RP, Anderson CL. The protection receptor for IgG catabolism is the beta2-microglobulin-containing neonatal intestinal transport receptor. Proc Natl Acad Sci U S A 1996; 93(11):5512–5516.

62. Prabhat P, Gan Z, Chao J, et al. Elucidation of intracellular recycling pathways leading to exocytosis of the Fc receptor, FcRn, by using multifocal plane microscopy. Proc Natl Acad Sci U S A 2007; 104(14):5889–5894.
63. Dickinson BL, Claypool SM, D'Angelo JA, et al. Ca2+-dependent Calmodulin Binding to FcRn Affects Immunoglobulin G Transport in the Transcytotic Pathway. Mol Biol Cell 2008; 19(1):414–423.
64. Chaudhury C, Brooks CL, Carter DC, et al. Albumin binding to FcRn: distinct from the FcRn-IgG interaction. Biochemistry 2006; 45(15):4983–4990.
65. Dall'Acqua WF, Kiener PA, Wu H. Properties of human IgG1s engineered for enhanced binding to the neonatal Fc receptor (FcRn). J Biol Chem 2006; 281(33): 23514–23524.
66. Hinton PR, Xiong JM, Johlfs MG, et al. An engineered human IgG1 antibody with longer serum half-life. J Immunol 2006; 176(1):346–356.
67. Bitonti AJ, Dumont JA. Pulmonary administration of therapeutic proteins using an immunoglobulin transport pathway. Adv Drug Deliv Rev 2006; 58(9–10):1106–1118.
68. Roopenian DC, Akilesh S. FcRn: the neonatal Fc receptor comes of age. Nat Rev Immunol 2007; 7(9):715–725.
69. Gunraj CA, Fernandes BJ, Denomme GA. Synonymous nucleotide substitutions in the neonatal Fc receptor. Immunogenetics 2002; 54(2):139–140.
70. Sachs UJ, Socher I, Braeunlich CG, et al. A variable number of tandem repeats polymorphism influences the transcriptional activity of the neonatal Fc receptor alpha-chain promoter. Immunology 2006; 119(1):83–89.
71. Nakamura A, Kojo T, Arahata K, et al. Reduction of serum IgG level and peripheral T-cell counts are correlated with CTG repeat lengths in myotonic dystrophy patients. Neuromuscul Disord 1996; 6(3):203–210.
72. Pan-Hammarstrom Q, Wen S, Ghanaat-Pour H, et al. Lack of correlation between the reduction of serum immunoglobulin concentration and the CTG repeat expansion in patients with type 1 dystrophia [correction of Dystrofia] myotonica. J Neuroimmunol 2003; 144(1–2):100–104.
73. Laegreid WW, Heaton MP, Keen JE, et al. Association of bovine neonatal Fc receptor alpha-chain gene (FCGRT) haplotypes with serum IgG concentration in newborn calves. Mamm Genome 2002; 13(12):704–710.
74. Managit C, Kawakami S, Yamashita F, et al. Effect of galactose density on asialoglycoprotein receptor-mediated uptake of galactosylated liposomes. J Pharm Sci 2005; 94(10):2266–2275.
75. Patel S, Stein R, Ong GL, et al. Enhancement of tumor-to-nontumor localization ratios by hepatocyte-directed blood clearance of antibodies labeled with certain residualizing radiolabels. J Nucl Med 1999; 40(8):1392–1401.
76. Keck R, Nayak N, Lerner L, et al. Characterization of a complex glycoprotein whose variable metabolic clearance in humans is dependent on terminal N-acetylglucosamine content. Biologicals 2008; 36(1):49–60.
77. Komoriya K, Kato Y, Hayashi Y, et al. Characterization of the hepatic disposition of lanoteplase, a rationally designed variant of tissue plasminogen activator in rodents. Drug Metab Dispos 2007; 35(3):469–475.
78. Webster R, Edgington A, Phipps J, et al. Pharmacokinetics and clearance processes of UK-279,276 (rNIF) in rat and dog: comparison with human data. Xenobiotica 2006; 36(4):341–349.
79. Minagar A, Adamashvili I, Jaffe SL, et al. Soluble HLA class I and class II molecules in relapsing-remitting multiple sclerosis: acute response to interferon-beta1a treatment and their use as markers of disease activity. Ann N Y Acad Sci 2005; 1051:111–120.
80. Fusco C, Andreone V, Coppola G, et al. HLA-DRB1*1501 and response to copolymer-1 therapy in relapsing-remitting multiple sclerosis. Neurology 2001; 57(11):1976–1979.
81. Cunningham S, Graham C, Hutchinson M, et al. Pharmacogenomics of responsiveness to interferon IFN-beta treatment in multiple sclerosis: a genetic screen of 100 type I interferon-inducible genes. Clin Pharmacol Ther 2005; 78(6):635–646.
82. Sellebjerg F, Datta P, Larsen J, et al. Gene expression analysis of interferon-beta treatment in multiple sclerosis. Mult Scler 2008; 14(5):615–621.

83. Bertolotto A, Gilli F, Sala A, et al. Persistent neutralizing antibodies abolish the interferon beta bioavailability in MS patients. Neurology 2003; 60(4):634–639.

84. Carrascosa A, Audi L, Fernandez-Cancio M, et al. The exon 3-deleted/full-length growth hormone receptor polymorphism did not influence growth response to growth hormone therapy over two years in prepubertal short children born at term with adequate weight and length for gestational age. J Clin Endocrinol Metab 2008; 93(3):764–770.

85. Pantel J, Grulich-Henn J, Bettendorf M, et al. Heterozygous nonsense mutation in exon 3 of the growth hormone receptor (GHR) in severe GH insensitivity (Laron syndrome) and the issue of the origin and function of the GHRd3 isoform. J Clin Endocrinol Metab 2003; 88(4):1705–1710.

86. Pantel J, Machinis K, Sobrier ML, et al. Species-specific alternative splice mimicry at the growth hormone receptor locus revealed by the lineage of retroelements during primate evolution. J Biol Chem 2000; 275(25):18664–18669.

87. Pilotta A, Mella P, Filisetti M, et al. Common polymorphisms of the growth hormone (GH) receptor do not correlate with the growth response to exogenous recombinant human GH in GH-deficient children. J Clin Endocrinol Metab 2006; 91(3):1178–1180.

88. Carrascosa A, Esteban C, Espadero R, et al. The d3/fl-growth hormone (GH) receptor polymorphism does not influence the effect of GH treatment (66 microg/kg per day) or the spontaneous growth in short non-GH-deficient small-for-gestational-age children: results from a two-year controlled prospective study in 170 Spanish patients. J Clin Endocrinol Metab 2006; 91(9):3281–3286.

89. Criswell LA, Lum RF, Turner KN, et al. The influence of genetic variation in the HLA-DRB1 and LTA-TNF regions on the response to treatment of early rheumatoid arthritis with methotrexate or etanercept. Arthritis Rheum 2004; 50(9):2750–2756.

90. Fabris M, Di PE, D'Elia A, et al. Tumor necrosis factor-alpha gene polymorphism in severe and mild-moderate rheumatoid arthritis. J Rheumatol 2002; 29(1):29–33.

91. Fabris M, Tolusso B, Di Poi E, et al. Tumor necrosis factor-alpha receptor II polymorphism in patients from southern Europe with mild-moderate and severe rheumatoid arthritis. J Rheumatol 2002; 29(9):1847–1850.

92. Louis E, Franchimont D, Piron A, et al. Tumour necrosis factor (TNF) gene polymorphism influences TNF-alpha production in lipopolysaccharide (LPS)-stimulated whole blood cell culture in healthy humans. Clin Exp Immunol 1998; 113(3):401–406.

93. Leyva L, Fernandez O, Fedetz M, et al. IFNAR1 and IFNAR2 polymorphisms confer susceptibility to multiple sclerosis but not to interferon-beta treatment response. J Neuroimmunol 2005; 163(1–2):165–171.

94. Benvenuti S, Sartore-Bianchi A, Di Nicolantonio F, et al. Oncogenic activation of the RAS/RAF signaling pathway impairs the response of metastatic colorectal cancers to anti-epidermal growth factor receptor antibody therapies. Cancer Res 2007; 67(6):2643–2648.

95. Cappuzzo F, Varella-Garcia M, Finocchiaro G, et al. Primary resistance to cetuximab therapy in EGFR FISH-positive colorectal cancer patients. Br J Cancer 2008; 99(1):83–89.

96. Italiano A, Follana P, Caroli FX, et al. Cetuximab shows activity in colorectal cancer patients with tumors for which FISH analysis does not detect an increase in EGFR gene copy number. Ann Surg Oncol 2008; 15(2):649–654.

97. Chung KY, Shia J, Kemeny NE, et al. Cetuximab shows activity in colorectal cancer patients with tumors that do not express the epidermal growth factor receptor by immunohistochemistry. J Clin Oncol 2005; 23(9):1803–1810.

98. Li X, Luwor R, Lu Y, et al. Enhancement of antitumor activity of the anti-EGF receptor monoclonal antibody cetuximab/C225 by perifosine in PTEN-deficient cancer cells. Oncogene 2006; 25(4):525–535.

99. Valabrega G, Montemurro F, Aglietta M. Trastuzumab: mechanism of action, resistance and future perspectives in HER2-overexpressing breast cancer. Ann Oncol 2007; 18(6):977–984.

100. Jhawer M, Goel S, Wilson AJ, et al. PIK3CA mutation/PTEN expression status predicts response of colon cancer cells to the epidermal growth factor receptor inhibitor cetuximab. Cancer Res 2008; 68(6):1953–1961.
101. Treon SP, Hansen M, Branagan AR, et al. Polymorphisms in FcgammaRIIIa (CD16) receptor expression are associated with clinical response to rituximab in Waldenstrom's macroglobulinemia. J Clin Oncol 2005; 23(3):474–481.
102. Weng WK, Levy R. Two immunoglobulin G fragment C receptor polymorphisms independently predict response to rituximab in patients with follicular lymphoma. J Clin Oncol 2003; 21(21):3940–3947.
103. Hatjiharissi E, Hansen M, Santos DD, et al. Genetic linkage of Fc gamma RIIa and Fc gamma RIIIa and implications for their use in predicting clinical responses to CD20-directed monoclonal antibody therapy. Clin Lymphoma Myeloma 2007; 7(4):286–290.
104. Anolik JH, Campbell D, Felgar RE, et al. The relationship of FcgammaRIIIa genotype to degree of B cell depletion by rituximab in the treatment of systemic lupus erythematosus. Arthritis Rheum 2003; 48(2):455–459.
105. Cartron G, Dacheux L, Salles G, et al. Therapeutic activity of humanized anti-CD20 monoclonal antibody and polymorphism in IgG Fc receptor FcgammaRIIIa gene. Blood 2002; 99(3):754–758.
106. Dall'Ozzo S, Tartas S, Paintaud G, et al. Rituximab-dependent cytotoxicity by natural killer cells: influence of FCGR3A polymorphism on the concentration-effect relationship. Cancer Res 2004; 64(13):4664–4669.
107. Hatjiharissi E, Xu L, Santos DD, et al. Increased natural killer cell expression of CD16, augmented binding and ADCC activity to rituximab among individuals expressing the Fc{gamma}RIIIa-158 V/V and V/F polymorphism. Blood 2007; 110(7):2561–2564.
108. Scallon BJ, Moore MA, Trinh H, et al. Chimeric anti-TNF-alpha monoclonal antibody cA2 binds recombinant transmembrane TNF-alpha and activates immune effector functions. Cytokine 1995; 7(3):251–259.
109. Louis EJ, Watier HE, Schreiber S, et al. Polymorphism in IgG Fc receptor gene FCGR3A and response to infliximab in Crohn's disease: a subanalysis of the ACCENT I study. Pharmacogenet Genomics 2006; 16(12):911–914.
110. Louis E, El Ghoul Z, Vermeire S, et al. Association between polymorphism in IgG Fc receptor IIIa coding gene and biological response to infliximab in Crohn's disease. Aliment Pharmacol Ther 2004; 19(5):511–519.
111. Willot S, Vermeire S, Ohresser M, et al. No association between C-reactive protein gene polymorphisms and decrease of C-reactive protein serum concentration after infliximab treatment in Crohn's disease. Pharmacogenet Genomics 2006; 16(1):37–42.
112. Beano A, Signorino E, Evangelista A, et al. Correlation between NK function and response to trastuzumab in metastatic breast cancer patients. J Transl Med 2008; 6:25.
113. Zhang W, Gordon M, Schultheis AM, et al. FCGR2A and FCGR3A polymorphisms associated with clinical outcome of epidermal growth factor receptor expressing metastatic colorectal cancer patients treated with single-agent cetuximab. J Clin Oncol 2007; 25(24):3712–3718.
114. Varchetta S, Gibelli N, Oliviero B, et al. Elements related to heterogeneity of antibody-dependent cell cytotoxicity in patients under trastuzumab therapy for primary operable breast cancer overexpressing Her2. Cancer Res 2007; 67(24):11991–11999.
115. Musolino A, Naldi N, Bortesi B, et al. Immunoglobulin G fragment C receptor polymorphisms and clinical efficacy of trastuzumab-based therapy in patients with HER-2/neu-positive metastatic breast cancer. J Clin Oncol 2008; 26(11):1789–1796.
116. Suzuki E, Niwa R, Saji S, et al. A nonfucosylated anti-HER2 antibody augments antibody-dependent cellular cytotoxicity in breast cancer patients. Clin Cancer Res 2007; 13(6):1875–1882.
117. Peipp M, Lammerts van Bueren JJ, Schneider-Merck T, et al. Antibody fucosylation differentially impacts cytotoxicity mediated by NK and PMN effector cells. Blood 2008; 112(6):2390–2399.
118. Schellekens H. Immunogenicity of therapeutic proteins: clinical implications and future prospects. Clin Ther 2002; 24(11):1720–1740; discussion 19.

119. Mire-Sluis AR, Barrett YC, Devanarayan V, et al. Recommendations for the design and optimization of immunoassays used in the detection of host antibodies against biotechnology products. J Immunol Methods 2004; 289(1–2):1–16.
120. Koren E, Smith HW, Shores E, et al. Recommendations on risk-based strategies for detection and characterization of antibodies against biotechnology products. J Immunol Methods 2008; 333(1–2):1–9.
121. Koren E, Mytych D, Koscec M, et al. Strategies for the preclinical and clinical characterization of immunogenicity. Dev Biol (Basel) 2005; 122:195–200.
122. De Groot AS, Scott DW. Immunogenicity of protein therapeutics. Trends Immunol 2007; 28(11):482–490.
123. Complete sequence and gene map of a human major histocompatibility complex. The MHC sequencing consortium. Nature 1999; 401(6756):921–923.
124. Barbosa MD, Vielmetter J, Chu S, et al. Clinical link between MHC class II haplotype and interferon-beta (IFN-beta) immunogenicity. Clin Immunol 2006; 118(1):42–50.
125. Lucas A, Nolan D, Mallal S. HLA-B*5701 screening for susceptibility to abacavir hypersensitivity. J Antimicrob Chemother 2007; 59(4):591–593.
126. Salomon J, Flower DR. Predicting Class II MHC-Peptide binding: a kernel based approach using similarity scores. BMC Bioinformatics 2006; 7:501.
127. Flower DR. Towards in silico prediction of immunogenic epitopes. Trends Immunol 2003; 24(12):667–674.
128. Nielsen M, Lundegaard C, Lund O. Prediction of MHC class II binding affinity using SMM-align, a novel stabilization matrix alignment method. BMC Bioinformatics 2007; 8:238.
129. Rajapakse M, Schmidt B, Feng L, et al. Predicting peptides binding to MHC class II molecules using multi-objective evolutionary algorithms. BMC Bioinformatics 2007; 8:459.
130. Wan J, Liu W, Xu Q, et al. SVRMHC prediction server for MHC-binding peptides. BMC Bioinformatics 2006; 7:463.
131. Sinigaglia F, Hammer J. Defining rules for the peptide-MHC class II interaction. Curr Opin Immunol 1994; 6(1):52–56.
132. Sturniolo T, Bono E, Ding J, et al. Generation of tissue-specific and promiscuous HLA ligand databases using DNA microarrays and virtual HLA class II matrices. Nat Biotechnol 1999; 17(6):555–561.
133. Bian H, Reidhaar-Olson JF, Hammer J. The use of bioinformatics for identifying class II-restricted T-cell epitopes. Methods 2003; 29(3):299–309.
134. Koren E, De Groot AS, Jawa V, et al. Clinical validation of the "in silico" prediction of immunogenicity of a human recombinant therapeutic protein. Clin Immunol 2007; 124(1):26–32.
135. Weiss ST, McLeod HL, Flockhart DA, et al. Creating and evaluating genetic tests predictive of drug response. Nat Rev Drug Discov 2008; 7(7):568–574.
136. Clark GM. Interpreting and integrating risk factors for patients with primary breast cancer. J Natl Cancer Inst Monogr 2001; (30):17–21.
137. Mass RD, Press MF, Anderson S, et al. Evaluation of clinical outcomes according to HER2 detection by fluorescence in situ hybridization in women with metastatic breast cancer treated with trastuzumab. Clin Breast Cancer 2005; 6(3):240–246.

Index